THE LIBRARY
ST. MARY'S COLLEGE OF MARYLAND
ST. MARY'S CITY, MARYLAND 20686

THE HEMIPTERA

THE HEMIPTERA

by
W.R. Dolling

Illustrated by J.H. Martin

Natural History Museum Publications

OXFORD UNIVERSITY PRESS
1991

Oxford University Press, Walton Street, Oxford OX2 6DP
Oxford New York Toronto
Delhi Bombay Calcutta Madras Karachi
Petaling Jaya Singapore Hong Kong Tokyo
Nairobi Dar Es Salaam Cape Town
Melbourne Auckland

and associated companies in
Berlin & Ibadan

Oxford is a trade mark of Oxford University Press

Published in the United States
by Oxford University Press, New York

All rights reserved. No part of this publication may be reproduced,
stored in a retrieval system, or transmitted, in any form or by any means,
electronic, mechanical, photocopying, recording, or otherwise, without
the prior permission of the publisher

© British Museum (Natural History) 1991

A catalogue record for this book is available from the British Library

Library of Congress Cataloging-in-Publication Data
Dolling, W. R.
 The hemiptera / by W. R. Dolling.
 p. cm.
 Includes bibliographical references and index.
 ISBN 0-19-854016-7 (hbk) : $60.00
 1. Hemiptera. 2. Hemiptera—Great Britain. I. Title.
QL521.D65 1991   91-12973
595.7'54—dc20   CIP

ISBN 0-19-854016-7

Typeset in palatino by Cambridge Photosetting Services
Printed by St Edmundsbury Press, Bury St Edmunds, Suffolk

CONTENTS

Preface	vii
Foreword	viii
Acknowledgements	ix
1. Introduction	1
2. Food	2
3. Symbiotic relationships	12
4. Diseases of bugs	16
5. Natural enemies	20
6. Defence	47
7. Biorhythms	50
8. Dispersal	53
9. Distribution	55
10. The British Hemiptera as a sample of the world fauna	60
11. Morphology	63
12. Classification	80
13. Key to suborders of British Hemiptera	84
14. Heteroptera	85
15. Auchenorrhyncha	143
16. Sternorrhyncha	163
17. Techniques for collecting and preserving Hemiptera	225
18. Glossary	235
19. Information sources	248
References	250
Index	267

PREFACE

Eighty thousand species of Hemiptera (true bugs) have been described and 1700 of these are known to occur in the British Isles. Most species feed on plant sap, a habit that gives this order of insects considerable importance in agriculture, both by direct damage to plants and, more insidiously, by the transmission of plant diseases. Some predaceous Hemiptera, on the other hand, are beneficial to agriculture by feeding on pests. A few species suck the blood of vertebrates, bringing the order to the notice of medical and veterinary scientists as well as the immediate victims of their depredations.

The variety of lifestyles and habitats exploited by Hemiptera and the diversity of their life-histories has attracted many people to study them. This same variety has often caused hemipterists to concentrate their studies on only part of the order and it is one of the aims of this book to give a balanced view of the Hemiptera as a whole. The chapters on the relationships of Hemiptera with other organisms are intended as a guide to the ecological relations of the group wherever it occurs, though examples have been drawn from the British fauna when possible. Only British families of Hemiptera have been included in the keys and the systematic accounts in the latter part of the book but it should still be possible to identify most Hemiptera from anywhere in the world to family level, particularly if the keys are used in conjunction with the chapter setting the British Hemiptera in a World context. The accounts of the biology of the various families represented in Britain are, in general, representative of those families everywhere, though details vary from place to place. An extensive bibliography is given to enable the user of this book to gain an entry into the recent literature on the topics covered. The interested reader's attention is drawn to the chapter on information sources that precedes the references.

FOREWORD

The biological sciences are served by taxonomy (the main scientific activity of the Natural History Museum) at two levels. First, through providing names and identifications, taxonomy enables biologists to communicate their observations about organisms to one another. Second, by organising biological knowledge into an evolutionary framework or classification, it enables us to comprehend and investigate the astounding diversity of the living world. Ten million species of insects, one current estimate of the world fauna, would represent informational chaos for all biologists, without a good synthetic and systematic classification. The synthetic approach also finds expression in biological overviews of particular groups, as in the volume presented here. The Natural History Museum is an ideal place to produce such synoptic works, with our collections and libraries rich in data, our staff experienced in fieldwork throughout the world, and our constant stream of scientific visitors involved in both theoretical and practical biological issues. *The Hemiptera* is one of a series of books we are producing with the intention of introducing to the serious student current ideas on the biology and systematics of insects. Drawing examples mainly from the well-studied British fauna, this book will be useful worldwide to anyone who wishes to ask a question about the biology and identity of bugs. This enormous group, probably numbering 100,000 or more species, has a wide range of biologies, from fungus feeding to blood sucking, and is found in every sort of habitat from mosses to bat skins and from sea water to deserts. The author conceals his scholarship behind a text that is a delight to read, to browse in, or to use as an information source. For many a good reason, this is a book that will be turned to time and time again. I warmly recommend *The Hemiptera* to anyone with an interest in natural history.

Laurence Mound
Keeper of Entomology

ACKNOWLEDGEMENTS

It is a pleasure to thank the many colleagues whose specialised knowledge and helpful suggestions have contributed to the process of converting early drafts of this book into the finished text. Even so, their advice has not always been followed and such inaccuracies and deficiencies as still remain should not be blamed on our colleagues R. L. Blackman, J. M. Cox, V. F. Eastop, D. Hollis, W. J. Knight, L. A. Mound and M. D. Webb at the Natural History Museum, nor on D. J. Williams and M. R. Wilson of the CAB International Institute of Entomology. C. P. Malumphy kindly made the unpublished results of his research on *Pulvinaria* available. Sincere thanks are due to all of these people and to many others too numerous or too modest to mention.

The great majority of the text-figures are the work of J. H. Martin. Most of the aphid figures have been published before, in *Tropical Pest Management* Vol. 29 (1983) and a paper by V. F. Eastop in *Australian Journal of Zoology* Vol. 14 (1966). We have also used some aphid figures by C. A. Gosney from the latter, where Figs 101, 102, 105 and 106, by A. Smith, were also published. Fig. 127, by J. H. Martin, was commissioned for the cover of Stroyan (1984). Figs 90 and 92, by J. M. Palmer, were commissioned for the covers of Hodkinson & White (1979) and White & Hodkinson (1982). Figs 103, 104, 107 and 108, by C. I. Carter, appeared in Carter (1971). Figs 153, 161 and 163 were published in Boratynski (1952, 1961 and 1957, respectively), Fig. 154 in J. M. Cox, *Bulletin of the British Museum (Natural History)* (Entomology) Vol. 58 (1989) and Fig. 155 in Williams (1985). We thank the copyright holders for permission to reproduce them here. Plate 1, fig. 5 is from a painting by A. J. E. Terzi.

Photographic credits:

Natural History Museum Photographic unit (mainly by Harry Taylor): Plates 1.3, 1.5, 2.1–4, 3.1–4, 4.3, 5.3, 5.6, 7.3, 8.1, 8.2, 8.4

K. G. Preston-Mafham/Premaphotos Wildlife: Plates 1.1, 1.2, 1.4, 4.1, 4.2, 4.4–6, 5.1, 5.2, 5.4, 5.5, 6.1–4, 7.1, 7.4–6

J. H. Martin: Plate 7.2

Royal Horticultural Society: Plate 8.3

1

INTRODUCTION

The Hemiptera occupy an enormous range of habitats and display as wide a variety of lifestyles as any other insect order. Forests, meadows, moors and marshes all have characteristic hemipteran faunas and so do ponds, streams, bare soil and rockpools. In the exploitation of plant sap as a food source, and of the surface film of water as a habitat, they are unequalled by any other group. The plant-feeding species include some of the more destructive pests while the predaceous ones may be beneficial to agriculture, horticulture and forestry.

Bugs are traditionally regarded as hemimetabolous insects, lacking the complete metamorphosis characteristic of butterflies, beetles and the other holometabolous groups. Most bugs do, in fact, exhibit a progressive development towards the adult form with each moult. Whiteflies (Aleyrodidae) and male scale insects (Coccoidea) are exceptions: their early stages differ little except in size, with no indication of developing wing-pads, and a drastic reorganization of tissues into the adult form takes place in one or two non-feeding pupal instars. Despite the absence, in most bugs, of a clear morphological differentiation between the growing, immature stages and the reproductive adult, the order is a successful one: about one insect species in twelve is a bug.

There are probably two main reasons for this being the largest and most diverse of the hemimetabolous orders. First, the body form is compact and the central nervous system is greatly concentrated, as in the adults of the most successful holometabolous insects, allowing a similar rapidity of response and a similar degree of behavioural complexity. Secondly, the mouthparts are of a unique and highly specialized type adapted for piercing and sucking. Most bugs feed on plants and this phytophagous habit is probably ancestral, since the related booklice (Psocoptera) and thunderflies (Thysanoptera) are also plant feeders. Nevertheless, predaceous bugs occupy a wider range of habitats than do the plant-feeding members of the order.

The most recent checklist of the British Hemiptera is that of Kloet & Hincks (1964). Identification guides that cover the whole order are almost non-existent, the only reasonably comprehensive one that covers most of the European species being that of Bei-Bienko (1964; English translation 1967). Weber (1930) gave a detailed introduction to many aspects of the biology of the Hemiptera, but in the 60 years since the publication of his book there has been much progress in this field. Villiers (1945, 1947) provided an introduction to the French fauna, heavily biased towards the larger and more conspicuous forms, but not completely neglecting the Sternorrhyncha. Waterhouse (1970), although Australian in emphasis, is a good general introductory work. Weber (1929, 1931, 1935) dealt with the biology of the German hemipterous fauna and Poisson & Pesson (1951) covered many aspects of the order on a worldwide basis.

The reader should bear in mind that generalizations made in the present work often refer only to the British fauna and may not necessarily be true in a wider context. This is particularly true of the keys to, and accounts of, the various families, the British representatives of which may show only a small part of the diversity of structure and habits that are known in other parts of the world. The British bug fauna, however, is a well balanced one and a thorough study of its members should provide a good introduction to the structure, biology and ecology of Hemiptera everywhere.

2

FOOD

The great majority of Hemiptera feed on vascular plants. Very few attack non-vascular plants and none is obligately saprophagous. Predation is confined to the suborder Heteroptera, most families of which consist entirely of predators. Even so, the total number of plant-feeding species in this suborder is greater than the number of predaceous ones. A few Heteroptera have the habit of sucking the blood of vertebrates. Although this is a very unusual mode of existence in the order, the intrusion of the bedbug into human affairs has led to the adoption of the name 'bug' for all members of the order Hemiptera.

Algae, Fungi and Lichens

The only Hemiptera that have been reported to feed on algae are the aquatic Corixidae. Remains of both unicellular and filamentous kinds have been found in their intestines.

Fungi are the normal food of most Aradidae; the sap-sucking *Aradus cinnamomeus* is an exception. Aradids have sometimes been found with their stylets inserted into the fruiting bodies of the larger fungi but they usually feed among the mycelium in rotten wood. Some of their food may consist of the products of fungal external digestion of the substrate rather than the hyphae themselves. Several scale-insects of the family Ortheziidae have been associated with fungi and one has been seen with its mouthparts inserted into the stipe of a basidiomycete fruiting body. Another was observed feeding on mycorrhizal roots. Among British aphids, the only known feeding association involving a fungus is that of *Mimeuria ulmiphila* (Pemphigidae), whose secondary host is elm, on which it lives in mycorrhizal cysts on the roots. The sexual stages of this aphid feed on the leaves of maples.

No Hemiptera actually feed on lichens but certain predaceous Heteroptera, principally Microphysidae and the genus *Temnostethus* (Anthocoridae), are associated with lichens on trees and rocks, preying on the small arthropods that find shelter among them. *Physatocheila* species are associated more often with lichen-covered host trees than with those free of epiphytes. Perhaps in this case the lichens afford shelter from natural enemies.

Mosses and liverworts

Among Heteroptera, only some Tingidae appear to be dependent on mosses for food as well as shelter. Probably all species of *Acalypta* are restricted to these host plants. They can be reared on mosses alone and have been observed feeding on the capsules. It has been suggested that mosses are the hosts of some other lacebugs whose feeding habits are not known for certain. Ortheziidae are often found among mosses but there is no hard evidence that they actually feed on them. Surprisingly for sap-sucking insects, several genera of aphids are known to feed on mosses, which lack true vascular tissues. Müller (1973) details six European species of Aphididae feeding on mosses. They belong to five genera: *Decorosiphon*, *Pseudacaudella*, *Myzodium*, *Muscaphis* and *Aspidaphium*; all six species are British. *Aspidaphium cuspidati* is associated with the semi-aquatic moss *Acrocladium cuspidatum*, on which it usually feeds below the surface of the water. Müller also produced evidence that *Jacksonia papillata*, polyphagous on low plants, may sometimes feed on mosses. In North America another aphid, *Melaphis rhois* (Pemphigidae), migrates between *Rhus* species, its primary hosts, and mosses. It has been accidentally introduced into Europe, where it apparently lives on mosses throughout the year, like the other European moss-feeding aphids. No Hemiptera seem to be associated with liverworts.

Ferns and horsetails

Some Miridae feed on ferns, especially the developing sori. *Bryocoris pteridis* lives mainly on species of *Dryopteris* and *Athyrium*, while *Monalocoris filicis* prefers bracken, though either species may sometimes be found on the other's preferred host plants. The delphacid *Ditropis pteridis* is monophagous on bracken while the typhlocybine cicadellid *Eupteryx filicum* is widespread on various ferns, particularly on *Polypodium*. Three aphids feed on ferns in Britain: *Amphorophora ampullata* is found on species of *Dryopteris* and some other genera; the recently introduced *Sitobion ptericolens* occurs sporadically on bracken; and *Idiopterus nephrolepidis* has long been known as a nuisance on ferns grown indoors. Three whiteflies specific to ferns have also been encountered under glass, namely *Aleuropteridis filicicola*, *Filicaleyrodes williamsi* and *Aleurotulus nephrolepidis*, and so have some fern-feeding coccoids including the mealybug *Spilococcus filicicola* on *Trichomanes spicatum*.

Several common, polyphagous Sternorrhyncha that are pests in glasshouses and conservatories include ferns among their hosts. Chief among them are *Coccus hesperidum* (Coccidae) and *Planococcus citri* (Pseudococcidae); *Chrysomphalus aonidum* (Diaspididae); the glasshouse whitefly, *Trialeurodes vaporariorum*; and the aphid *Aulacorthum circumflexum*. Another aphid, *Rhopalosiphum nymphaeae*, lives out of doors on a wide range of aquatic plants and is known to include the water-fern *Azolla* among its host plants. Another polyphagous aphid, *Aphis fabae*, and the equally polyphagous froghopper *Philaenus spumarius* are both found occasionally on ferns, including bracken.

Some authors have suggested that a cicadellid, *Macrosteles frontalis*, feeds on horsetails, *Equisetum* species. It is the only British hemipteran to have been associated with Equisetaceae.

Conifers

The evergreen foliage of many conifers provides shelter during the winter for a number of insects that do not breed there. In particular, many Psylloidea migrate to conifers in winter, though no British species breeds on these hosts. Similarly, Typhlocybine leafhoppers of the genera *Zygina* and *Empoasca*, and also *Linnavuoriana decempunctata*, can be beaten from conifers in the winter months. It seems likely, considering the length of time that these insects spend on conifers, that they take at least a little food from them.

A second category of insects associated with conifers but not directly dependent on them for food is that of the conifer-dwelling predators. Anthocoridae of the genera *Elatophilus*, *Tetraphleps* and *Acompocoris* are specific to conifers, as are some Miridae. The precise food requirements of many of the mirids are unknown but *Alloeotomus gothicus*,

Pilophorus cinnamopterus and *Phytocoris pini*, all confined to Scots Pine, are certainly almost wholly predaceous.

Heteroptera feeding directly on conifers include the remarkable pine flatbug, *Aradus cinnamomeus*, which sucks the phloem sap of young stems and branches of Scots Pine; the juniper shieldbug, *Elasmostethus tristriatus*, living on wild and cultivated junipers and cypresses; and two Lygaeidae, *Gastrodes grossipes* and *G. abietum*, living in the cones of pine and spruce respectively. A recent arrival, the lygaeid *Orsillus depressus*, has been found breeding on cypress. Some other rare lygaeids of the genus *Eremocoris* and the shieldbug *Pitedia juniperina*, are also associated with conifers. Among the conifer-inhabiting Miridae, those that are known to feed directly on their host plants include *Camptozygum pinastri* and two species of *Orthops*.

Only two British species of Auchenorrhyncha feed on conifers. Both are cicadellids and they both feed exclusively on pine. They are *Grypotes puncticollis*, a deltocephaline, and the typhlocybine *Wagneripteryx* (or *Aguriahana*) *germari*.

There are no conifer-feeding whiteflies apart from a record of the glasshouse whitefly, *Trialeurodes vaporariorum*, from a cycad.

Several scale insects feed on conifers. Two *Carulaspis* species (Diaspididae) may build up sufficiently large populations on cypresses and other Cupressaceae to cause noticeable damage. *Phenacaspis pinifoliae* and *Dynaspidiotus abietis* (also Diaspididae) occur on various genera of Pinaceae and *D. britannicus*, which usually occurs on broadleaved evergreens like laurel and box, has occasionally been found on pine and yew. *Parthenolecanium pomeranicum* (=*P. cornicrudum*), in the family Coccidae, is the only hemipteran living habitually on yew. *Physokermes piceae*, in the same family, lives on spruce. *Matsucoccus pini* (Margarodidae), on pine, is the only other British species of Coccoidea associated with conifers.

All Adelgidae feed only on conifers. Their peculiar kind of host-alternation, with the sexual and gall-forming generations on spruce and some asexual, non-galling generations on other coniferous genera, is outlined in the section of this book dealing with the family. The related Phylloxeridae never feed on conifers.

In Aphidoidea, all Cinarini (Lachnidae) feed on conifers and so do *Mindarus* species (Mindaridae). The latter produce copious wax. *M. obliquus* is confined to *Picea* and *M. abietinus* distorts the needles of *Abies*. *Asiphum*, *Gootiella* and *Prociphilus* species (Pemphigidae) alternate from their primary, broadleaved hosts to the roots (or, in the case of *P. bumeliae*, the branches) of various conifers. *Elatobium abietinum* (Aphididae) is holocyclic on spruce (*Picea*), which may shed its needles in response to heavy infestations. The closely related *Illinoia morrisoni*, which is native to North America but has recently appeared in Britain, feeds on *Sequoia* and other Cupressaceae. In addition to these insects specific to conifers, the highly polyphagous *Aphis fabae* and *Aulacorthum circumflexum* (Aphididae) have been found feeding on the new shoots of conifers of many genera, especially on lush growth forced under glass.

Grasses, sedges and rushes

Grasslands are a major feature of the British landscape. They owe their present-day predominance to human activities in managing the countryside for grazing pastures and hay meadows and to the fact that our major food crops are cereals. Parks, lawns, sports fields, recreation grounds, airfields and the cutting and mowing of railway and canal banks and roadside verges all contribute to the maintenance of types of vegetation in which grasses predominate. Before Man cleared the forests, grasses must have covered a very small fraction of the area now dominated by them. Windswept moors, marshes and the early stages of succession on the sea shore would have been the major grassland habitats at the time of the forest maximum. In such areas, sedges and rushes still hold their own in competition with grasses and they are frequent, if unwelcome, constituents of the wetter pasture lands. No sedge or rush is deliberately grown as a crop.

Most of the major plant-feeding groups of Hemiptera, apart from the entirely tree-dwelling Phylloxeridae and Adelgidae, have members that feed on grasses. No native British Aleyrodidae or Psylloidea feed on true grasses but one of the two species of Liviidae feeds on rushes and the other on sedges.

Among the aphidoids, all species of the subfamily Siphinae (Chaitophoridae) feed on grasses except for the sedge-feeding *Caricosipha paniculata*. Saltusaphidinae (Callaphididae) are mainly sedge-feeders, with the exception of *Juncobia leegei* on rushes and *Takecallis* species on bamboos (which are primitive grasses). More than 30 British species of Aphididae feed on grasses. This family contains all of the serious hemipteran pests of cereal crops in the British Isles. Some are holocyclic on grasses while others alternate from woody primary hosts. Four British species of Aphididae live on sedges and one lives on woodrushes (*Luzula* species). The Anoeciidae live in summer on the roots of grasses or (*Paranoecia pskovica*) those of sedges; some of them also have their sexual stages there while others migrate to *Cornus* (dogwood). Most species of the subfamily Fordinae (Pemphigidae) are anholocyclic on grass roots and two species of the related subfamily Eriosomatinae alternate between elm and the roots of sedges and grasses.

Several different groups of scale insects feed on grasses. Most species of the subfamily Filippiinae (Coccidae) are found on the lower parts of the leaves, leaf sheaths or stems of grasses or sedges. About half of the dozen or so Eriococcidae that are found in Britain and most of the Pseudococcidae are grass-feeders, some living on the roots or underground rhizomes.

Auchenorrhyncha form a major constituent of the grassland fauna in the British Isles. Many are known to have quite precisely defined habitat requirements that may well reflect a restricted range of host plants. Almost all Delphacidae (with only three known exceptions) feed exclusively on grasses, sedges and rushes. The food of the subterranean nymphs of Cixiidae is unknown in most instances but those of one species of *Cixius* have been found on the underground parts of sedges and *Oliarus leporinus* is strongly associated with reeds, so there is a suspicion that these are its host plants. Both adults and nymphs of the common froghopper, *Philaenus spumarius*, sometimes feed on grasses although dicotyledonous plants are more commonly used in Britain. In North America and New Zealand, to which *P. spumarius* has been introduced, it is a pest of pasture grasses. The related *Cercopis vulneraria* may also feed on grasses, at least as an adult. Froghoppers of the genus *Neophilaenus* are all grass-feeders but, since they favour different habitats, each probably specializes in a different range of grass species. The leafhopper family, Cicadellidae, contains more grass and sedge-feeders than any other hemipteran family. The majority of deltocephaline leafhoppers feed on grasses and a few of them on sedges. These plant families are also hosts to some Aphrodinae, a few Typhlocybinae and the genera *Cicadella* (Cicadellinae) and *Eupelix* (Dorycephalinae).

Some shieldbugs are major pests of cereals in Europe and the Middle East. Prominent among these are the genera *Eurygaster* (Scutelleridae) and *Aelia* (Pentatomidae). In regions with a cooler climate, these bugs feed mostly on the vegetative parts and grains of wild and pasture grasses rather than those of cereal crops. *Neottiglossa* and *Podops* (both Pentatomidae) are probably also grass-feeders. The grass-feeding tribe Chorosomini (Rhopalidae), is represented by *Myrmus miriformis* and *Chorosoma schillingi*, the former species feeding on smaller, softer and more lush grasses than the robust and usually maritime ones that support the latter. The subfamily Blissinae (Lygaeidae) has many species attached to grasses in most parts of the world, but only two of them, both in the genus *Ischnodemus*, have reached Britain. *Ischnodemus sabuleti* is much the commoner of the two. It often swarms in huge numbers on reeds and other grasses of wetlands but it has also been found on rushes and bur-reed (*Sparganium*) in swampy places and it even breeds on cocksfoot grass in damp, acid grassland. Most grass-feeding Heteroptera belong to the family Miridae. These include all 15 British members of the specialized tribe Stenodemini in the subfamily Mirinae (genera *Acetropis, Stenodema, Notostira, Megaloceraea, Trigonotylus, Teratocoris* and *Leptopterna*). A few of these, especially *Teratocoris* species, prefer sedges or rushes to the true grasses. Several other mirids feed on grasses. The best known of them are *Capsus* and *Stenotus* in Mirinae and some species of *Conostethus* and *Amblytylus* in Phylinae.

Monocotyledones other than grasses, sedges and rushes

A few bryocorine Miridae of the genus *Tenthecoris* have been found on hothouse orchids, having been introduced with their host plants from South America. As mentioned above, the mainly grass-feeding lygaeid *Ischnodemus sabuleti* sometimes lives on bur-reed but the only British heteropteran that habitually feeds on non-grasslike monocotyledones is another lygaeid, *Chilacis typhae*, on reedmace (*Typha*). It has recently turned up in North America on the same host.

Among the Auchenorrhyncha, the delphacid *Megalomelodes quadrimaculatus*, which feeds on a wide range of aquatic emergents and marsh plants, probably feeds sometimes on non-grasslike Monocotyledones. The same may be true of the marsh-inhabiting cicadellid *Paramesus nervosus*. *Mocuellus metrius*, another cicadellid, is reported from Alismataceae and the related *Erotettix cyane* lives on the floating leaves of *Potamogeton*, though it is also sometimes encountered on water-lilies.

A number of Aphididae feed on petaloid Monocotyledones. Examples are *Aulacorthum speyeri*, monophagous on lily-of-the-valley; *Aphis newtoni*, holocyclic on *Iris*; and *Aphis epipactis* on marsh helleborine (an orchid). *Rhopalosiphum nymphaeae* alternates between *Prunus* and a variety of aquatic herbs including some monocotyledonous ones such as *Alisma, Potamogeton* and even *Lemna* (duckweed). Bulbous and cormous Monocotyledones of many sorts are attacked, especially under glass, by some of the more polyphagous Aphididae, especially *Myzus persicae* and *Aulacorthum circumflexum*. *Dysaphis tulipae* survives underground or in store on bulbs and corms and attacks the growing parts in the spring. *Myzus ascalonicus* is a pest of shallots, leeks and chives but has many other hosts including Dicotyledones. *Rhopalosiphoninus latysiphon* and *R. staphyleae* are often found on tulip bulbs and other bulbs and corms. They are also common on stored potatoes and mangolds respectively. No British Sternorrhyncha other than Aphididae live on non-grasslike monocotyledons but introduced Aphidoidea of other families (notably *Cerataphis* in Hormaphididae), some Aleyrodoidea and a great variety of Coccoidea attack orchids, palms, aroids, bromeliads and related plants in hothouses.

Dicotyledones

Dicotyledonous plants are the dominant form of vegetation in most natural habitats in the British Isles. They greatly outnumber the other vascular plants in terms of species. From the ecological point of view they may conveniently be divided into two groups depending on whether they are woody (trees and shrubs) or non-woody (herbs). Many non-woody Dicotyledones die down in the winter, in contrast to their woody relatives, thus depriving their dependent insect fauna of food and shelter for a large part of the year. Herbs are also structurally simpler than trees and shrubs, providing fewer ecological niches. For these reasons, herbs generally have fewer species of insects living on them than have woody plants but there are many more species of herbs so that roughly equal numbers of insects are associated with the two structural types.

Aphid biology is closely tied up with the physiology of woody plants. Typical aphid life cycles are either confined to trees and shrubs or involve alternation between them and herbaceous plants. Many aphids live permanently on herbaceous plants, having abandoned the sexual generations, and only a few have successfully transferred the whole cycle to herbs. All Phylloxeridae live on broadleaved trees or vines but no Adelgidae live on dicotyledonous plants of any kind.

In the superfamily Psylloidea, most British Aphalaridae, some Triozidae, the single species of Calophyidae, Homotomidae and Spondyliaspididae (all three presumably introduced with their non-native host plants) and all Psyllidae reproduce on woody Dicotyledones. The remaining aphalarids and triozids live on herbaceous Dicotyledones.

Aleyrodidae live mainly on woody Dicotyledones, but *Aleyrodes proletella* and *A. lonicerae* live on herbaceous ones, the former all the year round and the latter in the summer, migrating to woody plants for the winter.

Most Coccoidea are associated with dicotyledonous hosts. Diaspididae are confined to trees and shrubs. *Asterodiaspis* (Asterolecaniidae) is confined to oak but the related *Planchonia* lives on *Thesium*, a low-growing plant with perennial stems. Kermesidae and those Coccidae that do not live on grasses or conifers live on broadleaved trees and shrubs but not on herbs. A few Eriococcidae and Pseudococcidae feed on both woody and non-woody dicotyledonous plants. *Steingelia gorodetskia* (Margarodidae) lives on birch and *Orthezia urticae* is known to feed on various dicotyledonous herbs. The host-range of most Ortheziidae is unknown.

Among the British Auchenorrhyncha, the food of the underground nymphs of *Cicadetta* (Cicadidae) and *Cercopis* (Cercopidae) is not known with any certainty though it has been suggested that the former prefer the roots of trees. Adult *Cicadetta* certainly feed on the sap of tree branches, though the females descend to lay their eggs near the ground. *Philaenus spumarius* (Cercopidae) is very polyphagous, feeding on ferns and grasses as well as broadleaved herbs and shrubs. The four *Aphrophora* species (Cercopidae) are restricted to dicotyledonous plants. The commonest of them, *A. alni*, feeds as a nymph on herbs, very close to the ground, but the adult may feed on woody plants. The other three species are confined to trees and shrubs. *Centrotus cornutus* and *Gargara genistae* (Membracidae) also feed on dicotyledonous plants. The large family Cicadellidae is well represented in the fauna of Dicotyledones, with several subfamilies mainly or wholly on woody ones, but with relatively few species on herbs. Delphacidae, the second largest family of Auchenorrhyncha in the British Isles, have only a couple of species feeding on dicotyledonous herbs and none at all on trees or shrubs. *Issus coleoptratus* lives on broadleaved trees and ivy but the food preferences of *I. muscaeformis* are unknown, as are those of *Tettigometra impressopunctata* (Tettigometridae) and most cixiids.

Many species of the largest heteropteran family, Miridae, feed on dicotyledonous plants, both woody and herbaceous. Stenocephalidae live on Euphorbiaceae and *Pyrrhocoris apterus*, like most Pyrrhocoridae, on Malvaceae. Berytidae and Piesmidae feed on non-woody Dicotyledones and so do all but the two grass-feeding Rhopalidae and the alydid *Alydus calcaratus*, though the last may feed also on shrubby Leguminosae. Most Coreidae live on dicotyledonous herbs but *Gonocerus acuteangulatus* feeds principally on box. About half of the species of Pentatomidae feed on Dicotyledones, most of them on herbs. Probably most or all Cydnidae and *Odontoscelis* species (Scutelleridae) feed on dicotyledonous herbs, whereas most Acanthosomatidae live on dicotyledonous trees or rarely shrubs (*Elasmostethus tristriatus*, however, feeds on Cupressaceae). Most Tingidae feed on Dicotyledones, predominantly the herbaceous kinds.

Site and tissue used by plant-feeders

Vascular plants, especially angiosperms (flowering plants) present a great diversity of structures and tissues, most of which are exploited in some way by specialist Hemiptera. Roots, stems, leaves, buds and reproductive organs are all attacked. The main constraint on feeding by Hemiptera is the nature of their mouthparts, which allow only liquids and fine particles in suspension to be swallowed. The lack of biting mandibles rules our stem-boring and leaf-mining as ways of life, but the internal tissues of stems and leaves can be reached by inserting the stylets from the outside. The inability to gnaw also means that escape from closed galls is impossible but many Sternorrhyncha, principally species of Aphidoidea and Psylloidea, cause the formation of open gall-like structures or induce characteristic crumpling or rolling of leaves.

The phloem vessels of roots are tapped by many Aphidoidea, a few Coccoidea and Psylloidea, some stages of the Vine Phylloxera, some Auchenorrhyncha (nymphs of Cixiidae) and some Heteroptera (Cydnidae), while root xylem provides the nymphal diet of *Cicadetta* (Cicadidae) and *Cercopis* (Cercopidae). The great majority of Sternorrhyncha, Auchenorrhyncha and plant-feeding Heteroptera suck phloem sap from stems and leaves. Exceptions are the xylem-feeding Cicadidae (adults), Cercopidae and the cicadellid subfamilies Cicadellinae and Evacanthinae, with occasional xylem-feeding by other cicadellids. Xylem sap probably contains fewer secondary plant substances, such as alkaloids, than phloem sap does and this difference in composition may account for the generally wider host-plant range of xylem-feeders as compared with phloem-feeders. Xylem sap is also much more dilute, so that the xylem-feeders need to consume greater amounts of fluid than their phloem-feeding relatives, and produce much larger quantities of much more dilute excreta. The continuous shower of watery drops from the branches of trees supporting large populations of Cercopidae nymphs is the foundation of the legend of the 'rain tree' that waters the ground in its shade. The flow is said to stop at dusk and to resume at dawn, presumably in response to variations in the tree's transpiration rate. The contents of mesophyll cells are the usual food of most Cicadellidae: Typhlocybinae, Tingidae and the small tropical family Malcidae. Some Miridae: Stenodemini feed on the mesophyll of grass blades. Several Pentatomoidea, Coreoidea and Miridae preferentially attack the concentrations of nutrients that occur in buds, fruits, shoot-tips and ripening seeds. Fully ripe seeds are unavailable as a food source to most bugs with the exception of Lygaeidae, most of which specialize in this food source. Pollen and spores are difficult foods to manage but maturing anthers and sporangia are exploited by some anthocorids and by Miridae: Bryocorinae respectively.

Animal food

The use of food of animal origin by members of the Hemiptera is confined to Heteroptera. The predaceous habit is almost universal in the aquatic groups, except for some Corixidae, and in Leptopodomorpha and Dipsocoromorpha. With the major exceptions of Tingidae and some Miridae, Cimicomorpha are also mainly predaceous but the only predators in Pentatomomorpha are the asopine Pentatomidae and some non-British Lygaeidae and Pyrrhocoridae. Prey consists mainly of small arthropods.

Cimicidae feed only on the blood of vertebrates, usually that of birds, bats or man. This family includes the Bedbug. The bite is painless when the bug is feeding, but the effect of the residual anticoagulants from the bug's saliva is usually to raise a small, itching spot at the feeding site. Some Anthocoridae, notably *Lyctocoris campestris*, are facultative bloodsuckers and a few Miridae will gorge on vertebrate blood if allowed to do so, but most bites by bugs other than Cimicidae are exploratory only and do not lead to prolonged feeding. Other blood-sucking bugs are the small family Polyctenidae, which all feed on bats; some Lygaeidae of the tribe Cleradini, which suck the blood of nestling birds; and the 'kissing bugs' that constitute the subfamily Triatominae of the family Reduviidae and feed on the blood of Man and various hole-nesting or burrow-inhabiting mammals, birds and reptiles. These groups are mainly found in tropical countries.

Saprophagy is uncommon in Hemiptera. Some phloem-feeding shieldbugs, coreids and alydids are known to be attracted occasionally to carrion or the excrement of birds or mammals. It may be that they acquire from these unusual foods certain essential substances that are lacking in plant sap. Apart from this phenomenon, saprophagy is at least occasionally practised by a number of Gerromorpha and Corixidae that feed on dead or moribund arthropods and possibly by some mirids and anthocorids.

Choice of habitat, feeding site and food

The environment in which a newly-hatched bug finds itself depends on the oviposition site selected by its mother or, in the case of viviparous aphids and those coccids that have portable egg-sacs, the whereabouts of the mother at the time of birth or hatching. In some instances the food source is perennial and relatively permanent, as in the case of tree-dwelling coccids or whiteflies on evergreen shrubs, and there is little reason to leave the hostplant on which the previous generation grew up. Colonies in such situations may persist for many years. Flightless insects, of course, have little choice in the matter of selecting a suitable environment for their offspring and species with wingless females can disperse only slowly on foot.

The importance of maternal choice in determining nymphal host plants was graphically demonstrated in experiments on *Ribautiana ulmi* by Claridge & Wilson (1978a). This leafhopper is normally monophagous on elm, which is the only plant chosen for oviposition. Nevertheless, nymphs showed good survival on a variety of other trees and even showed a preference for some of them over their native elm in choice experiments.

Often, suitable food is available only seasonally and overwintering sites remote from the food source are selected. The adults are faced with the problem of re-finding a suitable feeding place in the spring. Flight plays a major part in this process. After flying for a while, the insect will settle apparently almost at hazard, selecting only a very ill-defined type of landing site such as tree canopy, bushes or the ground. Aquatic insects are attracted to shiny surfaces that may be water but could equally well be a wet road or the roof of a greenhouse or car. It is obvious from the poor degree of discrimination displayed at this stage that vision is able supply only limited information about colour, light intensity, reflectivity and vegetation profile. Once the insect has settled, the chemical and mechanical senses come into play. If the bug is a predator it may begin to search for prey, whereas a plant-feeder will usually sample the substrate by probing with the rostrum and sucking up a little sap or cell contents. If suitable prey is available or the plant is acceptable as food the insect may settle for a while, though a general restlessness may drive it to move on even from apparently satisfactiory sites until the urge to wander has faded.

The site selected for egg-laying may not be close to the food normally used by the adult. Obvious examples are the groups of Auchenorrhyncha whose nymphs live below ground but whose adults live on the aerial parts of plants (*Cicadetta, Cercopis* and some Cixiidae). There are many other instances of well marked differences in feeding sites between adults and nymphs that require females to abandon their normal haunts when the eggs are due to be laid. Adults of the froghopper *Aphrophora alni*, for example, often feed on trees and bushes but the eggs are always laid very close to ground level, where nymphal development takes place. The predatory anthocorid bug *Anthocoris gallarumulmi* always lays its eggs on leaf-rolls or galls induced by aphids on elm and the nymphs develop there, feeding on the occupants, but the galls are not available during most of the bug's adult life, when it roams widely in search of many different arthropods on various trees.

A female hemipteran may regularly select for oviposition a different species of plant from that on which she fed during her own development. Examples of such behaviour are found in three typhlocybine leafhoppers, *Lindbergiana aurovittata, Edwardsiana rosae* and *Eupteryx aurata*, which lay overwintering eggs in the autumn on bramble or holm oak, on rose and on nettle, respectively, so the first generation of each species in the spring is obliged to feed only on the plant so selected. Adults of the new generation of each of these species, however, do not remain to breed on the plants on which they were reared. They disperse to a wide range of host plants on which the summer generations feed and breed. Their descendants return to the specialized hosts in the autumn.

Alternation of host plants in different generations reaches its greatest degree of development in Adelgidae, Aphididae, Pemphigidae and Anoeciidae. Different generations of species in these families may have precisely defined and unrelated host plants and members of one generation are unable to develop satisfactorily on another's host. A fuller treatment of these elaborate life-cycles is given in the accounts of the various families in a later chapter.

One puzzling phenomenon is the occurrence of differing degrees of host-specificity shown by related plant-feeding bugs. One member of a genus may live only on a single plant species while a close relative may accept a wide variety of plants even of different families. The genus *Aphis* contains a number of strictly monophagous species but also one of the most widely polyphagous of all phloem-feeding insects, *Aphis fabae*. In this connection, it is probably relevant to note the circumstances in which some usually monophagous aphids are sometimes found on plants other than their normal hosts: the plants involved are either etiolated in deep shade under logs and stones or forced into rapid growth under glass. In such situations, secondary plant substances that may inhibit feeding are present in the sap in much weaker concentrations than usual. Perhaps widely polyphagous insects like *A. fabae* are insensitive to such substances even in normal concentrations.

The physiological state of the hostplant at the time of host-selection may be important in determining preferences. Woodroffe (1961) reports working the same willow trees for years and finding only small numbers of four willow-feeding species of Miridae on them. After ten years the same trees sudenly yielded a great abundance of the same Miridae and, in addition, moderate numbers of two species normally living on poplars, one usually associated with birch and one that generally breeds on elm and on nettle and other low plants. A predaceous mirid was also abundant that year on the trees although it had not been taken there previously. Presumably, some change had occurred in the plants in response to unusual climatic conditions that year, making them acceptable to insects that usually would have shunned them.

Plants may have non-chemical defences in addition to the chemical ones. Such defences include waxy leaves that are difficult for most insects to cling to and bristly, woolly or sticky hairs that make it difficult for insects to walk about. Host plant specificity involving such plants reflects the varying degrees of success that insects have in overcoming or circumventing the defences.

Predatory bugs can be broadly classified according to habitat: under water (Nepomorpha), the water surface (Gerromorpha), bare ground (Leptopodomorpha), interstitial, litter and moss (Dipsocoromorpha and a few Anthocoridae and Miridae) and above the ground surface on vascular plants (most predaceous Cimicomorpha and the predatory shieldbugs). Few predaceous bugs occupy the full range of habitats available in the broad categories outlined above. The aquatic species, for example, are affected by such factors as salinity, acidity, organic content, oxygenation, rate of flow, size of water body, degree of shading, plant density, type of substrate (silt, sand, shingle, peat etc) and depth. Predaceous bugs hunting on plants may be confined to a particular vegetation layer or type or even, in the case of some species of Anthocoridae, a single species of plant. All of these factors reduce the range of prey available, but make it possible for the predator to improve the efficiency of its hunting technique by specialization. The type of prey taken depends on its size, the vigour with which it defends itself and the boldness of the predator: raptorial predators such as the water bugs, Reduviidae and Nabidae seize their prey and attempt to subdue it by force, whereas 'timid' predators like Anthocoridae and the pentatomid subfamily Asopinae attack only defenceless prey such as slow-moving insect larvae, pupae and eggs. In either case, the body contents of the prey are broken down by a combination of the mechanical action of the stylets and the digestive action of the bug's saliva, much as in the plant-feeding groups.

Prey is first detected by sight in the subaquatic and some water-surface bugs and in Saldidae. These groups have large, smoothly rounded eyes with many small facets and their visual acuity is much greater than that of the asopine pentatomids, anthocorids, nabids, reduviids, mirids and other hemipteran predators that usually do not recognize prey until they have touched it with their antennae. The numerous hairlike sense organs on the antennae furnish information about the chemical nature of the objects that they touch as well as shape, size and movements. Tremors in the surface film, made by struggling insects that have fallen in the water, attract pondskaters from several centimetres away, further than the distance over which sight operates. The blood-sucking Cimicidae are attracted towards the body warmth of their hosts, but only over very short distances.

3

SYMBIOTIC RELATIONSHIPS

In its broadest sense, the term symbiosis embraces all those relationships in which members of one species exploit those of another in ways other than simply feeding on them or at their expense. It includes such diverse phenomena as pollination, which may benefit both partners in the relationship or only the plant, and phoresy, in which a small animal clings to a larger and more mobile one in the interests of its own dispersal but gives nothing in return and may even be an encumbrance. No Hemipteran is an essential or habitual pollinator although casual transfer of pollen, by Miridae living on the flower-heads of Umbelliferae and Compositae, and by Anthocoridae on a variety of plants, probably occurs. The pondweed bug, *Mesovelia furcata*, is said to play an accidental role in pollinating water-lilies. Phoresy does not figure at all prominently in the lives of bugs but juvenile freshwater bivalve molluscs of the family Sphaeriidae have been found clinging to corixids. The most important symbiotic relationships of Hemiptera with other organisms are those with internal micro-organisms that contribute to their nutrition, those with plant disease organisms that are transmitted by the bugs' feeding activities and those with ants.

Internal symbiotes

Plant-feeding and blood-sucking Hemiptera, but not predaceous ones, frequently harbour symbiotic fungi and bacteria whose primary function, so far as the bug is concerned, is the synthesis of sterols and amino acids. Major reviews of the subject are those of Buchner (1965) and Houk & Griffiths (1980).

There are two main types of symbiosis in Hemiptera. One, confined to Heteroptera, involves the presence of bacteria in specialized crypts or caeca of the midgut. There may be a few, elongate caeca, as in some Lygaeidae, or numerous, short ones arranged in rows along the gut, as in most Pentatomidae. Caeca containing symbiotes are universal in Pentatomoidea, except for the predaceous Asopinae, frequent in Lygaeidae and occasional in Coreidae. Symbiotes are lacking in Miridae, Tingidae, Piesmidae, Corixidae and the wholly predaceous families. The symbiotes are transmitted from one generation to the next through the ingestion of adult faeces by the newly hatched young. Pentatomoid females smear their eggs with faeces at the time of oviposition. In Acanthosomatidae, where the crypts no longer communicate with the gut in the adult, transmission of symbiotes to the surface of the eggs takes place via a special 'lubricating organ' associated with the ovipositor. Gut symbiotes are never included within the egg in any Hemiptera.

The second type of symbiosis is characterized by the inclusion of bacteria or yeast-like fungi within cells of the fatbody. In some bugs the micro-organisms are housed in scarcely modified fat cells and may also be present in the haemolymph, as in some Coccoidea, but they are usually contained in specialized cells called mycetocytes. The mycetocytes of some bugs are scattered among normal cells of the fatbody but in the great majority of cases they are concentrated into discrete organs, the mycetomes. Intracellular symbiotes are present in bedbugs, some Lygaeidae (*Ischnodemus*, *Kleidocerys*, *Nysius*), all

Sternorrhyncha and most Auchenorrhyncha. The main exception in Auchenorrhyncha is the cicadellid subfamily Typhlocybinae in which the diet (mesophyll cells) presumably supplies all necessary nutrients, some of which are lacking from the plant sap consumed by their relatives. Some parenchyma-feeding Coccoidea (principally Diaspididae) also lack symbiotic micro-organisms. The micro-organisms concerned in the intracellular type of symbiosis are transmitted transovarially, that is, they are included within the egg before it is laid.

Most Coccoidea have yeast-like symbiotes (rod-like bacteria in Eriococcidae) in mycetocytes scattered throughout the fatty tissues. In Pseudococcidae, however, the yeast-containing mycetocytes are grouped into a single, brightly coloured mycetome, which may become bilobed or divided in the adult as the gonads develop in the same area of the body. The small, paired, orange mycetomes of whiteflies are closely associated with the genital tract in both sexes. In aphids symbiotic bacteria, often of two or three different kinds, are typically housed in a pair of elongate mycetomes running through several abdominal segments. Auchenorrhyncha usually have two or three different kinds of symbiotes. Buchner's (1965: 752) table shows two kinds present in 55 per cent of species examined, three kinds in 30.5 per cent and a few species with one kind, four to six kinds or (Typhlocybinae only) none at all. The majority of symbiotes in Auchenorrhyncha are bacteria but yeasts are frequently also present. Some general types are common to both Fulgoromorpha and Cicadomorpha but each infraorder typically has in addition a type of micro-organism not found in the other. A further difference between these two infra-orders is that in Fulgoromorpha symbiotes of different types are segregated into separate mycetomes while in Cicadomorpha they occur together.

One group of basidiomycete fungi, not represented in Britain, lives in a symbiotic relationship with diaspidid scale insects. This is the family Septobasidiaceae, which is related to smuts, rusts and jelly fungi. *Septobasidium* encrusts whole colonies of scales, which live under the perennial mats of hyphae. The fungus invades the bodies of the insects and eventually kills them, but not usually until some young crawlers have been produced to keep the colony going. Emigrating crawlers are already infected when they leave the parent colony. The fungus clearly derives nourishment from this association and it is believed that the scale insects benefit from the protection afforded by the fungal crust. Two dozen species of the fungal genus, which was monographed by Couch (1938), are known to have this kind of relationship with almost as many species of scale insects.

Transmission of diseases

Fortunately, the few bloodsucking Hemiptera found in Britain are not known to transmit any disease of humans or of domestic animals. Some Cimicidae have been found to transmit blood-borne diseases of birds in North America, and triatomine Reduviidae in Central and South America carry blood-borne diseases of many vertebrates, including man and are a serious public health problem.

Numerous sap-sucking bugs have been implicated in the transmission of plant diseases. At present, almost nothing is known of the diseases of wild plants (other than those caused by microfungi) or of the ways in which they are transmitted but a large and rapidly growing literature exists dealing with crop diseases and their vectors. In various parts of the world, Hemiptera have been shown to transmit or to facilitate infection by plant diseases caused by Protozoa, fungi, bacteria, spiroplasmas, mycoplasmas, mycoplasma-like organisms, rickettsia-like organisms (now known, oddly, as 'fastidious vascular bacteria') and viruses. In Britain, viruses are the main plant pathogens transmitted by hemipterans.

A primitive sort of dependence on a hemipterous insect is shown by the *Nectria* fungus that causes beech bark disease. The spores are spread mainly by wind and rain but the fungus enters the tree through feeding wounds made by the coccid *Cryptococcus fagi*. A second fungus disease associated with Hemiptera is a species of *Pycnostyanus* (or *Sporocybe*) that causes bud blast of rhododendron, entering via the oviposition punctures made by the leafhopper *Graphocephala fennahi*. Apart from these two instances there is little firm evidence of fungal pathogens being dependent on Hemiptera for their transmission.

Bacterial diseases of plants seem mostly not to rely on specific vectors but many of them enter their hosts through wounds, including those made by insects. The tiny feeding punctures made by Hemiptera are of less importance in this regard than the more extensive damage done by insects with biting mandibles. Nevertheless, a number of mirids and aphids have been shown to be at least occasionally concerned in the spread of some bacterial infections, including fireblight of fruit trees (more usually spread by pollinators) and some rots, cankers and galls. Adelgids, including *Adelges cooleyi*, have been found to be associated with bacterial galls on conifers in Europe and North America.

Hemiptera, particularly aphids, leafhoppers and planthoppers, are the most important vectors of plant pathogenic viruses. A recent, worldwide survey found that bugs of these three families were known to transmit, respectively, 164, 77 and 21 kinds of viruses. There are two major types of virus transmission. The commonest and least specialized kind is the 'noncirculative' type. These viruses contaminate the mouthparts of the insect vectors as they feed on or simply probe infected plant tissues and are lost each time the insect moults. They do not usually persist for long in the insect's body and they are immediately infective if the insect should move on to probe another plant. Alate aphids and other fully winged Hemiptera dispersing in search of new host plants may probe a number of plants of different species in quick succession and so transmit noncirculative viruses from and to plants on which they do not actually feed. The 'circulative' group of viruses show a more highly specialized relationship with their vectors. Unlike the noncirculative viruses, they enter the haemolymph and eventually reach the salivary glands. They are acquired by the insects only by feeding, not by probing; they persist longer in the insect's body, surviving its moults, and undergo an incubation period before becoming infective to plants again. They are injected into the new host plant with the vector's saliva as it begins to feed. Many circulative viruses are 'propagative', multiplying in the insect vectors's body and a few are even passed on transovarially to the next generation. Some circulative viruses produce symptoms of disease in their insect vectors as well as in their plant hosts. Virus transmission has been demonstrated not only in the three major vector groups mentioned above but also in several other families of Hemiptera. In Britain, the glasshouse whitefly is the only member of any of these minor families likely to be of any importance in this regard.

Many plant diseases once thought to be caused by viruses are now known to be due to mycoplasmas or mycoplasma-like organisms. These micro-organisms resemble bacteria without cell walls. Their most frequent vectors are leafhoppers and they are believed to have transmission cycles mainly of the circulative type, like the majority of viruses with auchenorrhynchan vectors. A few psyllids and a planthopper are also known to be vectors in some parts of the world but there is only slender evidence of aphids having this role.

Modern works summarizing current knowledge of the whole topic of hemipteran vectors of plant diseases are those of Harris & Maramorosch (1977) on aphids, Maramorosch & Harris (1979) on leafhoppers and planthoppers and Harris & Maramorosch (1980) on other groups. Nault & Ammar (1989) reviewed leafhopper and planthopper transmission of plant viruses. Chapters 12-15 of Nault & Rodriguez (1985) mention recent advances concerning leafhoppers and planthoppers in this rapidly developing field.

Relationships with ants

Before its origins were understood, the name honeydew was applied to a sugary liquid that mysteriously appeared on the leaves of plants. This liquid is now known to consist of the excreta of phloem-feeding Hemiptera. It is produced in surprisingly large quantities by aphids and some other Sternorrhyncha and is rich in sugars and nitrogenous nutrients. Discarded honeydew on leaves provides food for some microfungi and many flying insects, particularly Diptera and Hymenoptera, for which it is an important source of energy. Honeydew differs chemically from sap, nectar and the products of extrafloral nectaries, most significantly in the presence of the trisaccharides melezitose and glucosucrose. These sugars and some nitrogenous substances in honeydew are attractive to many insects, among them several kinds of natural enemies of Sternorrhyncha.

One group of insects, the ants, have developed a mutually beneficial relationship with certain phloem-feeding Hemiptera. Aphids are the major group of bugs that have entered into this partnership. They provide the ants with energy-rich food and, in return, receive protection from predators and parasites. Four adaptations are characteristic of myrmecophilous aphids: (1) they cluster together into dense flocks that are easy for the ants to tend; (2) instead of discarding their honeydew they release it slowly in response to being stroked by the ants' antennae, retaining it as a drop suspended from the anus, from where it is imbibed by the ants; (3) they release alarm pheromones more readily than species that are not habitually ant-attended; and (4) they do not disperse or fall from the plant when neighbouring aphids release the alarm pheromone. Specialized behaviour on the part of the ants involves soliciting honeydew from the aphids by palpating them with the antennae and responding to the aphids' alarm pheromone by an increased alertness and readiness to attack any intruding natural enemies. Additional protection may be provided by the ants building shelters of soil particles over aphid colonies near the ground. The ants may carry aphids to new feeding sites and store their winter eggs in the nest until they hatch. They seem able to suppress the production of alatae to some extent, perhaps by means of a pheromone, and apparently in some instances by manipulating the population density of their aphid colonies.

There is a darker side to the ant-aphid relationship. Pontin (1978) found that an average nest of the yellow ant *Lasius flavus* consumed, in addition to the honeydew produced by its subterranean aphid flocks, about 3000 first-instar nymphs daily, plus some older aphids. There seemed to be a balance maintained between the quantities of honeydew and aphids consumed, as feeding with artificial honeydew resulted in a higher proportion of the aphids being eaten. The same author (Pontin, 1959) found that the larvae of some kinds of ladybirds and hoverflies could prey on aphids that were attended by ants without themselves being attacked. Larvae of the hoverfly *Pipizella* even lived inside the earth shelters constructed by the ants to protect their charges. The small larvae of cecidomyiid and chamaemyiid flies could escape the attentions of the ants by hiding in crevices and some aphidiid parasites (Hymenoptera) could attack ant-attended aphids with apparent impunity. Adults of one aphidiid species, *Myrmecobosca mandibularis*, live underground in nests of the black ant *Lasius niger*, being fed by them as if they were members of the colony and soliciting honeydew from subterranean aphids.

By no means are all aphid species attended by ants and an aphid that is attended by one species of ant may be ignored or eaten by another. Close associations between ants and other honeydew-producing Hemiptera are also known. In the British fauna, several scale insects and mealybugs are involved (but not diaspid scales, which do not produce honeydew) and at least one species each of Psyllidae and Membracidae.

Buckley (1987) reviewed the tripartite relations between ants, plants and Hemiptera with the exception of the few known associations of tropical Heteroptera with ants.

4

DISEASES OF BUGS

Micro-organisms are encountered within or on the surface of the body of insects in a variety of roles; their association with dead or moribund specimens may reflect their presence as items of food, symbiotes or harmless commensals of healthy insects or they may be saprophytes. These alternative possibilities should be borne in mind during microscopic examination for pathogens. Major textbooks on insect pathology are those of Steinhaus (1946, 1949, 1963) and Cantwell (1974). The identification guides of Weiser (1969) and Poinar & Thomas (1984) give a good introduction to the practical aspects of the subject.

Almost all Hemiptera feed by piercing a membrane (plant or animal epidermis) and sucking the fluids contained by it. It is, therefore, almost impossible for infection to occur orally in plant-feeders while predators could only be infected via the oral route if their prey were itself infected. For this reason almost all recorded epidemic diseases of Hemiptera are fungal: germinating fungal spores secrete enzymes that enable the developing haustoria to penetrate the host's cuticle. Virus diseases are unknown in Hemiptera and bacterioses are very infrequent.

Bacteria

Soft-bodied insects killed by bacteria become limp and darkened; their body contents liquefy and become foul-smelling. Often there is a discharge of foul fluid from the mouth or anus of the dying insects. Liquefaction and colour change are less apparent in hard-bodied hosts.

Bacterial infections are very uncommon in natural populations of Hemiptera but in the crowded conditions of laboratory cultures healthy insects probing the bodies of dead and dying ones may establish a chain of infection. Bacteria are part of the normal gut flora of many hemipterans, as symbiotes or faecal saprophytes or both.

Fungi

Insects killed by fungi normally become hard, dry and brittle; their bodies are packed with hyphae, which usually erupt through the body wall and produce either conidia or true fruiting bodies externally.

A list of fungal pathogens of British insects was published by Leatherdale (1970). His records from Hemiptera, plus a few others, are summarized in Table 1. Madelin (1966) gave a comprehensive review of the subject on a world-wide basis and much useful practical information is available in the treatise of Müller-Kögler (1965). Perhaps the best recent account, with a very extensive bibliography, is that of Evlakhova (1974). The coloured illustrations and the generic keys of Samson et al. (1988) are a valuable aid to identification, though their Plate 12, purporting to show the fungus *Conidiobolus niger* on a dipteran, in fact portrays it on a cicadellid and a delphacid.

Table 1 Fungal pathogens of British Hemiptera. (Based on Leatherdale, 1970)

Fungi	Hosts
Phycomycetes	
Chytridiales	
Myiophagus	*Coccoidea (same species on Coleoptera, Diptera)
Blastocladiales	
Coelomomyces	*Notonectidae
Saprolegniales	
Saprolegnia	††eggs of Corixidae
Entomophthorales	
Entomophthora (including *Tarichium*)	Lygaeidae, Miridae, Cercopidae, Cicadellidae, Psylloidea, Aphidoidea
Ascomycetes	
Hypocreales	
Cordyceps (including *Ophiocordyceps*)	Coccoidea
Cordyceps (conidial forms)	Nabidae, Delphacidae, many undetermined Homoptera, Aphidoidea, Aleyrodoidea, Coccoidea
Sphaerostilbe (as conidial form *Microcera*)	Coccoidea: *Chionaspis*
Myriangiales	
Myriangium	Coccoidea: *Chionaspis*
Laboulbeniales	
Coreomyces	**Corixidae
Autophagomyces	**Mesoveliidae, **Veliidae
Deuteromycetes	
Moniliales	
Beauveria	Lygaeidae, Coccoidea
Paecilomyces	Lygaeidae, Aphidoidea, †eggs of Delphacidae
Aspergillus	Cimicidae, Aphidoidea
Tritirachium	†eggs of Cicadellidae

*Non-British records from Evlakhova (1974)
**Non-British records from Poisson (1957)
†British records from Waloff (1980)
††British record from Crisp (1961)

The majority of records of fungal parasites of Hemiptera refer to the genus *Entomophthora*. This genus has a very wide host range, but many of its species are at least order-specific. Bugs killed by this fungus have a very characteristic appearance as the conidia of the parasite erupt through the arthrodial membranes of the host, appearing especially as conspicuous, granular rings between the abdominal segments. Non-parasitic Entomophthorales are frequent saprophytes in the bodies of bugs that have died from other causes, including *Entomophthora* infections.

Coelomomyces species all attack the aquatic larvae of Diptera with the sole exception of the Russian *C. notonectae* in the water boatman *Notonecta*. Infections by this genus are virtually symptomless externally.

The best known genus of insect-parasitic fungi is *Cordyceps*. Asci of this genus are carried on a large and characteristically club- or drumstick-shaped fruiting body that is frequently longer than the body of the insect on which it grows. Many form-genera of Deuteromycetes that are believed to be conidial forms of *Cordyceps* are reported from British Hemiptera; Leatherdale (1970) recognized *Acremonium* (=*Sporotrichum*),

Cephalosporium, Cladosporium, Gibellula, Hirsutella, Hymenostilbe and *Torrubiella*. The conidia of these forms are carried either on synnemata, bundles of hyphae that radiate from the host's body as a series of thick threads, or on a fuzz of mycelium on the body surface.

Dense and often coalescent blobs of mycelium, frequently drawn out into irregularly shaped, finger-like processes (coremia), on the surface of dead insects are indicative of infection by various Deuteromycetes. The coremia of *Beauveria* are usually white, those of *Paecilomyces* orange and of *Aspergillus* yellow but colour is not a sure guide. The insect pathogens in these genera have a very wide host range. *Aspergillus flavus* is normally saprophytic on dead insects or penetrates via wounds to become pathogenic but it may become epidemic in such situations as laboratory cultures. Leatherdale reports it attacking the bedbug, *Cimex lectularius*, in culture and an aphid, *Schizoneura lanuginosa*, in its damp and crowded galls on elm. One species of Deuteromycetes, *Verticillium lecanii*, is commercially available as a microbial agent of control for whitefly, aphids and thrips in glasshouses.

Waloff (1980) reported mortality caused by fungi to eggs of Auchenorrhyncha in the stems of rushes. Studies of *Cicadella viridis* (Cicadellidae) led to the conclusion that death of eggs from fungal infection was the key factor controlling its population density. The subaquatic eggs of *Arctocorisa germari* (Corixidae), studied by Crisp (1961) suffered a low level of mortality from fungal attack.

Sometimes dead insects are found covered with the pale, dense mycelium of *Beauveria* from which arise numerous, black, bristle-like structures. These are the stromata of *Melanospora parasitica*, a hyperparasitic fungus. It attacks a wide range of fungi pathogenic to insects. Several other fungi live in the same way but they are less common.

Laboulbeniales are non-pathogenic fungi growing on the cuticle of many terrestrial and aquatic arthropods, especially beetles. They consist of very few cells and stand erect like tiny brushes. *Coreomyces* (=*Paracoreomyces*) has about a dozen species, all confined to Corixidae, and may occur in the British Isles. A dozen more species, in several genera, are described from a small but diverse selection of Heteroptera; they are mostly confined to the tropics.

'Sooty mould' frequently develops on the honeydew voided by colonies of Sternorrhyncha. It is not pathogenic to the insects. Several genera of fungi are involved, principally *Meliola* and *Capnodium*. The filamentous, white wax produced by female Fulgoromorpha and many Sternorrhyncha might be mistaken for a mould but microscopic examination should dispel any doubts.

Protozoa

Protozoan parasites are virtually unknown in Homoptera; in European Heteroptera they affect mainly the aquatic predators. Aquatic bugs sometimes bear surface infestations of non-pathogenic stalked ciliates (Peritrichia), and an accidental infection of the water scorpion *Nepa cinerea* nymphs by a species of *Colpoda* (Holotrichia) has been reported, but the great majority of Protozoa known from Heteroptera are either trypanosomatid flagellates or Sporozoa.

Wallace (1966) gives a comprehensive synopsis of the occurrence in insects of trypanosomatid parasites, excluding those genera normally parasitic in plants or vertebrates, for which the bugs are vectors. He reports *Blastocrithidia gerridis* and *B. veliae* in Britain, from *Gerris* (Gerridae) and *Velia* (Veliidae) respectively, and also *Leptomonas jaculus* from *Nepa*. These two genera, and *Herpetomonas* as well, occur in a wide variety of aquatic bugs in Europe. *Blastocrithidia, Crithidia* and *Leptomonas* infect terrestrial

Heteroptera in many parts of the world but, from Wallace's account, they appear to be scarce in Europe and are so far unrecorded from these hosts in Britain.

Microsporidian infections of insects were summarized by Thomson (1960) and Weiser (1961). In Europe, *Nepa cinerea* is host to two species of *Nosema* and one of *Thelohahnia*. Another *Nosema* species infects *Velia*, and *Toxoglugea* species are known from *Gerris* and *Notonecta* (Notonectidae). *Nepa* is also parasitized by a coccidian, *Barrouxia ornata* (in the gut) and two cephaline gregarines of the genus *Coleorhynchus*, one in the gut and the other in the fatbody. An unidentified gregarine has been reported from a nymph of Microphysidae in Germany. Lipa (1966) made a thorough study of the protozoan parasites of a population of *Nepa cinerea* and found that seven species (a trypanosomatid, a coccidian, two gregarines and three microsporidians) were present; multiple infections of up to four species in a single individual were not uncommon.

5

NATURAL ENEMIES

Introduction

The various lifestyles of the three major divisions of Hemiptera have attracted different complexes of specialist predators and parasites to each. The relative immobility of Sternorrhyncha, coupled with their tendency to form colonies, makes them a particularly attractive food resource for many natural enemies. The impermanence of aphid colonies presents a special problem with which some of their exploiters cope by having several short-lived generations in a season and by being particularly mobile and wide-ranging as adults. Several parasitic groups, notably Strepsiptera, Dryinidae and Pipunculidae (endoparasitic in Auchenorrhyncha) and Tachinidae (in Heteroptera) are dispersed to a greater or lesser extent by their hosts, whose own dispersal flight is not impaired by the presence of small parasites. Strepsiptera and dryinid species with flightless females must be wholly dependent upon their hosts for dispersal. The strepsipteran *Elenchus tenuicornis* is one of the most widespread insect species in the world, a fact that bears impressive witness to the effectiveness of this strategy and to the power of dispersal of Delphacidae even when parasitized.

The use of the term 'parasite' requires a word of explanation. Modern workers often restrict its use to those organisms that do not cause the death of the host, at least not until it has produced viable gametes. The great majority of insects and worms that devour the living bodies of their hosts, including the parasites of eggs, are excluded by this restriction and are termed 'parasitoids' by purists. There is no corresponding term (hostoid?) for the unfortunate victims. The adjective 'protelean' is sometimes applied to those animals that are parasitic as larvae but roam freely about as adults. Almost all multicellular parasites of insects have this type of life-history. For the purposes of this handbook, the terms 'predator' and 'parasite' only are employed, in the senses in which they are widely understood: a predator usually kills one or more victims (prey) each time it feeds while a parasite is able to complete one or more stages of its development within or attached to the living body of a single victim (host) on which it feeds repeatedly or continuously. The terms 'superparasitism' and 'hyperparasitism' are also used. Superparasitism involves the presence of more than one parasite in or on a single host. Hyperparasitism (secondary parasitism) applies to the parasitization of an animal that is itself a parasite. For example, Alloxystinae (Hymenoptera: Charipidae) parasitize Aphidiidae (Hymenoptera) that are parasitic within aphids; they are primary parasites of the aphidiids but secondary parasites or hyperparasites of aphids.

There is still great scope for original research on the natural enemies of Hemiptera in Britain, even at the simplest level of recording parasite-host and predator-prey pairs. The broad outlines of the interactions of Hemiptera with other animals are probably established by now but some surprises are certainly in store, particularly as quantitive studies develop and as host or prey preferences in the field, rather than in the laboratory, are studied. The catalogues of W. R. Thompson (1950–1965) and Herting (1971–1980) facilitate entry into the world literature on the subject and Clausen's (1940) book is still unsurpassed as a guide to the biology of predaceous and parasitic insects. Waloff's summaries (1968, 1980) of the arthropod ecosystem based on Broom (*Sarothamnus* or

Cytisus scoparius) and of the natural history of grassland Auchenorrhyncha contain a wealth of detail about the natural enemies of Hemiptera and provide an introduction to the pioneering studies of Rothschild, Dempster and others on immunological analysis of the liquid gut contents of predators. Waloff & Jervis (1987) gave detailed synopses of host-parasite relations involving European Auchenorrhyncha. The parasites of Psylloidea were summarized by Jensen (1957) and those of Aleyrodoidea by Mound & Halsey (1978). Fulmek (1943) catalogued the parasites of Aleyrodoidea and Coccoidea.

Predators and parasites of Hemiptera have been used extensively in programmes of biological control in many countries. Greathead (1989) summarized the results of more than 500 such programmes directed against Sternorrhyncha and Auchenorrhyncha. His paper has a short bibliography that provides a lead into the recent literature on the subject.

Worms

Aquatic Heteroptera are sometimes intermediate hosts of cestodes (tapeworms). Kukashev (1983) reported an amabilid parasite of grebes utilizing the corixid *Sigara concinna* in this way and referred to a notocotylid from *Ilyocoris* (Naucoridae) and *Plea* (Pleidae). There are no records of Gordioidea (Nematomorpha) from Hemiptera, despite the apparent suitability of waterbugs as hosts.

Apart from one known case of a filarial infection of a tropical bedbug, all reported nematode parasites of Hemiptera belong to the families Sphaerulariidae and Mermithidae. The only British records to date are of unidentified Mermithidae from corixids.

All three sphaerulariids known from Hemiptera were described by Poisson (1933) from the haemocoels of *Aquarius* (Gerridae), *Velia* (Veliidae) and *Nepa* (Nepidae) in France. Adult worms were about 12 to 20 times as long as wide and were either small (1.6 to 2.6 mm long) and gregarious or large (7 mm) and solitary. In late summer the female worms were packed with eggs and often accompanied by male and female larvae. Details of the life cycle are unknown. In this family, maturation and sexual reproduction occur outside the host's body and the infective agent may be either the mated female or a parthenogenetically produced second-generation female.

According to Poinar (1975), Mermithidae have occurred as accidental infections of almost all the major groups of Hemiptera in Europe and North America. Young mermithids bore into the host and develop in its haemocoel into large, white larvae, looking like coiled threads of cotton many times as long as wide and several times the length of the host. They usually occur singly and the host dies when the mature larva emerges. The final change to adulthood takes place in the soil. It is usually impossible to identify immature mermithids which should, therefore, be reared to adulthood if identification is desired. Poinar (1975: 50-52) gave a key to the World genera of Mermithidae. Where identified, those from Hemiptera have been found to belong to the genera *Mermis*, *Hexamermis* and *Agamermis*.

Arachnida

Spiders, harvestmen and predaceous mites are usually generalist predators, Hemiptera featuring only incidentally in their diets. Arachnids are often so abundant that, despite this lack of specificity, they are of major importance as enemies of Hemiptera.

Precipitin tests (Rothschild, Dempster and others, summarized by Waloff, 1968, 1980) revealed widespread predation on Miridae, Psyllidae and Auchenorrhyncha by spiders of many families. Edgar (1970) recorded Delphacidae, Cicadellidae, Miridae and Anthocoridae among the prey of the common wolf spider *Lycosa lugubris* in Scotland, and Payne (1973) found that the cicadellid *Euscelis obsoleta* was taken in large numbers by three species of spiders among cordgrass (*Spartina*) in Dorset. Aphids are eaten by small spiders, especially Linyphiidae (Sunderland et al., 1986), but they are ignored by the larger ones even if they become entangled in their webs.

Harvestmen (Opiliones) of several species were major predators of nymphal *Conomelus anceps* (Delphacidae) studied by Rothschild but were less important enemies of other grassland Auchenorrhyncha studied by Tay and Solomon (summary in Waloff, 1980). Todd (1950) listed arboreal Cicadellidae among the prey of the harvestman *Leiobunum rotundum*, and the delphacid *Kelisia scotti* among that of *Oligolophus tridens*; aphids were taken by *L. rotundum* and *Mitopus morio*.

Mites belonging to more than a dozen different families have been recorded as preying on scale insects and their eggs; included among them are two tyroglyphids, *Tyrophagus putrescentiae* and *Glycyphagus domesticus*, which are usually scavengers, feeding on detritus among the colonies, and become predaceous only when their populations outstrip their normal food supply. A common specialist predator is *Hemisarcoptes malus*, which feeds on the eggs of *Chionaspis*, *Lepidosaphes* and many other armoured scales throughout Britain. Some Anystidae, including the abundant and polyphagous *Anystis baccarum*, and some Cheyletidae and Phytoseiidae include Sternorrhyncha, especially Coccoidea, in their diets.

The best known parasitic mites of terrestrial Hemiptera, and of terrestrial insects in general, belong to the families Trombidiidae and Erythraeidae. The life cycle of both families involves a parasitic, six-legged larval stage followed, after a resting phase, by an active, eight-legged nymph that enters a second resting phase before becoming adult. The larvae, when engorged, appear as small but conspicuous bright- or dark-red sacs attached to the bodies or appendages of their hosts. Both nymphs and adults are predaceous. *Allothrombidium fuliginosum* is a velvet-mite (Trombidiidae) often seen feeding on colonies of the woolly apple aphid, *Eriosoma lanigerum*, on apple trees. Its larva develops ectoparasitically upon the aphids while the other two stages prey on them. This mite is recorded as an enemy of a wide range of Aphidoidea and Coccoidea. Erythraeid larvae are reported from Heteroptera, Auchenorrhyncha, Sternorrhyncha and many other groups of arthropods; the nymphs and adults prefer inert or slow-moving prey.

Among stored products the anthocorid *Xylocoris galactinus* has been observed as an agent of phoretic dispersal for tyroglyphid and cheyletid mites. The trophic relationships among stored-products arthropods are very complex and the direction of predation between Anthocoridae and mites may vary according to the stage of the life-cycle of the participants.

Several families of water mites parasitize submerged or surface-dwelling bugs either routinely or casually; favoured alternative hosts are beetles and the larvae of Diptera. The female of the common, red water mite, *Hydrachna globosa*, inserts her eggs into the air spaces of submerged aquatic plants, a method of oviposition unique among aquatic mites to the family Hydrachnidae. On hatching, the six-legged larva seeks out a suitable host – often a corixid – to which it attaches itself by its mouthparts. Only adult insects are acceptable hosts. The larva gorges on the bug's body fluids, its dark red body becoming swollen and teardrop-shaped. When fully gorged it ceases to feed and becomes a quiescent nymphochrysalis. From the nymphochrysalis eventually emerges the eight-legged, predatory nymph. When the nymph has completed its growth, it enters a second quiescent phase, the imagochrysalis, from which the adult emerges. Males are short-lived but the predaceous females survive longer. This life history, which closely resembles

Plate 1

1, *Elasmucha grisea* (Acanthosomatidae) female guarding her newly-hatched young clustered on their empty egg shells on a Birch leaf. **2**, *Nabis rugosus* (Nabidae) preying on *Drepanosiphum platanoidis* (Aphididae) on sycamore leaf. **3**, *Graphocephala fennahi* (Cicadellidae) feeding on xylem sap from midrib of rhododendron leaf. **4**, *Aphis* sp. (Aphididae) feeding on phloem sap from a plant stem, tended by ants, *Myrmica* sp. A small parasitic wasp, *Aphidius* sp. (Aphidiidae), is laying an egg into one of the aphids. **5**, *Cimex lectularius* (Cimicidae), an aggregation of five adults and three nymphs of the bedbug.

Plate 2

Feeding effects. **1**, Bramble stem with growing point killed and leaves tattered by *Lygocoris* sp. (Miridae) feeding on the apical bud. **2**, Horehound leaves with white mottling where the cells have been emptied of their contents by *Eupteryx* sp. (Cicadellidae). **3**, Cherry leaves crumpled by effects of saliva of *Myzus cerasi* (Aphididae). **4**, Oak leaf with yellow spots caused by effects of saliva of *Phylloxera quercus* (Phylloxeridae).

those of the related terrestrial families Erythraeidae and Trombidiidae, is followed by the other aquatic families, although females of the latter lay their eggs in gelatinous masses on the surface of water plants and not into their tissues. Davids & Schoots (1975) demonstrated that parasitization of corixids by two species of Hydrachna resulted in partial or complete suppression of the hosts' egg production.

For identification of spiders, see Locket & Millidge (1951, 1953) and Locket, Millidge & Merrett (1974). For harvestmen, see Hillyard & Sankey (1990). Oudemans (1912) deals with the larvae of Trombidiidae and Erythraeidae known at that date. A compact introduction to the European Arachnida is that of Schaefer (in Tischler, 1982). Students of aquatic mites have a variety of monographic works available: Cook (1974) and Prasad & Cook (1972) for identification of the genera of adults and immatures world-wide; Sparing (1959) on development and systematics of immature stages; and Besseling (1964), Lundblad (1927), Motas (1928) and Viets (1936) for accounts of the watermite fauna of the Netherlands, Sweden, south-eastern France and Germany respectively.

Predaceous Hemiptera

In Britain, the major families of predaceous terrestrial Hemiptera are Nabidae, Anthocoridae and Miridae. All three feed extensively on Hemiptera. The semi-aquatic and aquatic families are all predators (though some doubt exists as to whether the Corixidae are exclusively so) and may engage in internecine killing and even in conspecific cannibalism as, for example, *Nepa cinerea* (Nepidae). Giller (1986) found *Notonecta* (Notonectidae) preying on *Corixa* (Corixidae), *Sigara* (Corixidae), *Plea* (Pleidae) and *Gerris* (Gerridae).

Eggs of Auchenorrhyncha and of some Heteroptera, embedded in plant stems, are located and emptied of their contents by some Anthocoridae, Nabidae and Miridae (Waloff, 1980). Only the mirid *Tytthus pygmaeus* seems to be a specialist in feeding on eggs. In the absence of eggs of Delphacidae it will prey on small nymphs and may also feed on dead or moribund arthropods of various kinds.

Most predaceous Heteroptera are governed in their choice of prey principally by its availability and size rather than its taxonomic affinity. Some Anthocoridae are an exception in that they specialize, at least in the immature stages, in feeding on one or more groups of Sternorrhyncha to the exclusion of other, apparently suitable, prey. Within the genus *Anthocoris* there is pronounced specialization in some species and an almost total lack of it in others. *Anthocoris gallarumulmi* and *A. minki* breed only in the leaf-galls of *Schizoneura* (Pemphigidae) on elm and the petiole-galls of *Pemphigus* (Pemphigidae) on poplar respectively. In addition to these aphid specialists there are psyllid specialists: *A. simulans* feeds on *Psyllopsis* species on ash while *A. visci* is restricted to mistletoe, where *Psylla visci* is the only available prey. On conifers the related genera *Acompocoris* and *Tetraphleps* prey on Aphidoidea, principally *Cinara* (Lachnidae), while *Elatophilus nigricornis* feeds on Coccoidea. *Anthocoris nemorum* and some *Orius* species are much less restricted in their feeding habits. They attack insects of several different orders, including Hemiptera, and mites as well, foraging on many different kinds of plants. Feeding rates of two *Anthocoris* species preying on the sycamore aphid, *Drepanosiphum platanoidis*, have been investigated by Russell (1970). Smith (summarized in Waloff, 1968) compared the feeding rates of *A. nemorum* on *Acyrthosiphon 'spartii'* (Aphididae) with those of three Miridae, two Coccinellidae, a chrysopid and two Diptera exploiting the same species of aphid.

Waloff (1980) claimed that Nabidae were the most important predators of grassland Auchenorrhyncha early in the season and Payne (1973) found that *Dolichonabis lineatus* was a major predator of *Euscelis obsoletus* (Cicadellidae) in a simple ecosystem based on a maritime grass. In addition to Cicadellidae and Delphacidae, the prey of nabids includes Heteroptera of several families, many Aphidoidea and a few Psylloidea. *Himacerus apterus* is the only British member of its family that habitually lives on bushes and trees. Waloff (1968) reported that it was a frequent predator of Miridae on broom, where it also fed on Psylloidea but not on the abundant aphids.

The degree of carnivory exhibited by Miridae varies greatly with the species, as does the degree of specialization in prey and the choice of habitat. *Globiceps, Megacoelum* and *Pilophorus* take a high proportion of aphids in their diet and *Pilophorus* species feed on scale insects as well. *Neodicyphus rhododendri* perhaps preys only on *Masonaphis* (Aphididae) species on rhododendrons, since it is invariably associated with colonies of the aphids. The black-kneed capsid, *Blepharidopterus angulatus* (Miridae), is unable to grow or reproduce on plant material but can be reared on a diet consisting solely of lime aphid, *Eucallipterus tiliae* (Glen, 1973). It has a wide range of arthropod prey. Stewart (1969) found that he could rear the normally phytophagous *Lygus rugulipennis* from the egg to the fourth instar on water and pea aphids alone; even when plant food was freely available, both nymphs and adults of this mirid took some aphids. Smith (summarized in Waloff, 1968) found that only fifth-instar nymphs and adults of *Heterocordylus tibialis* and of two *Orthotylus* species accepted the aphid *Acyrthosiphon 'spartii'*. They also preyed upon Psylloidea and small nymphs of Miridae, including those of their own species. Psylloidea feature in the diet of many omnivorous Miridae, including species of *Psallus, Phytocoris* and *Deraeocoris*. *Pithanus maerkeli*, which can be reared on plant material alone, will feed on small nymphs of Delphacidae and Cicadellidae.

Several other terrestrial families of Heteroptera are of minor importance as predators of Hemiptera. Turner (1984) noted Homoptera among the prey of the arboreal reduviid *Empicoris vagabundus* and Morley (1905) recorded an instance of *Coranus subapterus*, a ground-dwelling species of the same family, feeding on a rhyparochromine lygaeid. Among the small arthropods that are eaten by Microphysidae are Adelgidae, Aphidoidea and Psylloidea. The predaceous asopine shieldbugs rarely attack Hemiptera but there is a report of *Troilus luridus* preying on the cicadellid *Evacanthus interruptus*.

Hymenoptera

This huge order of insects contains many parasitic and predaceous forms. Gauld & Bolton (1988) provided an introduction to the biology of the various families. Richards (1956, 1977) keyed the British families and presented synoptic tables of parasite/host relationships.

The adults of social wasps, many ants and a few sawflies are active general predators; they probably take Hemiptera when these are available. Solitary wasps (Sphecidae) are much more specific than their social relatives (Vespidae) in their choice of prey; many species seek out Hemiptera with which to provision their nests and these are considered in some detail below. The complex symbiotic relations between ants and Hemiptera are considered elsewhere in this book.

The main hymenopteran parasites of the active stages of Heteroptera are Braconidae of the subfamily Euphorinae. Miridae are by far the commonest of their heteropteran hosts. Ichneumonidae occur in Hemiptera only as hyperparasites via euphorine braconids and are, like them, restricted to Heteroptera. There is a single record of a pteromalid parasite of a British species of Scutelleridae.

Dryinidae, though classified as Hymenoptera Aculeata, are parasitic as larvae on Auchenorrhyncha; their black or brown 'sacs' are conspicuous objects attached to the bodies of their hosts. Apart from a single species of Encyrtidae and, probably, one of Embolemidae, they are the only hymenopterous parasites of nymphal and adult Auchenorrhyncha in Britain. The ecological relationships between Auchenorrhyncha and their parasites and hyperparasites were summarized by Jervis (1980c).

Aphidoidea differ from the other three superfamilies of Sternorrhyncha in that their main parasites are the ichneumonoid family Aphidiidae whereas the other groups are parasitized mainly by two families of Chalcidoidea: Aphelinidae and Encyrtidae. A rough analysis of the host-parasite species pairs in Herting's (1972a) catalogue shows that Encyrtidae favour more mobile and less structurally modified hosts than the more sedentary and scale-like forms preferred by Aphelinidae. Thus, records of encyrtid parasitism outnumber those of aphelinid parasitism in Pseudococcidae, Psylloidea, Eriococcidae and Coccidae, by twelve to one, nine to one, three to one and two to one respectively, while Aphelinidae predominate over Encyrtidae in the ratio of four to one in Diaspididae. In Aleyrodoidea the ascendency of aphelinids over encyrtids is even greater, nearly fifty to one.

Parasitized and mummified Sternorrhyncha are a readily available food source that is exploited by several other insect groups. Many parasite larvae and pupae must fall victim to the usual predators of their hosts and overwintering mummies may be exposed to predation by birds and other animals. In addition, there are specialist hyperparasites among the Hymenoptera. Aphid mummies vacated by these are easily recognized by the jagged-edged emergence holes in place of the neat ones cut by the aphidiid primary parasites.

Thin-shelled hemipteran eggs are heavily parasitized by Mymaridae, even when embedded in plant tissues. The ovipositor is often thrust through the plant epidermis to reach the egg rather than through the oviposition slit cut by the bug. The thick chorions of the shieldbugs, coreids and reduviids are probably responsible for their complete immunity to mymarids. Instead, these exposed eggs are attacked by Scelionidae. The eggs of Psylloidea are not known to have any parasites or predators; perhaps protection here is due to their waxy coating. Trichogrammatidae, like both Mymaridae and Scelionidae, are all parasites of eggs. A few of them are recorded from those of Hemiptera. In the warmer parts of the world *Ooencyrtus* (Encyrtidae) is an important parasite of heteropteran eggs with much the same host range as Scelionidae. Eggs of terrestrial Hemiptera are occasionally attacked by Hymenoptera of the families Pteromalidae, Eulophidae and Aphelinidae.

Even the submerged eggs of aquatic Heteroptera do not escape the attentions of parasitic Hymenoptera. Poisson (1957: 16) summarized the relevant European records. *Prestwichia aquatica* (Trichogrammatidae) is a gregarious parasite of the eggs of water beetles and bugs including species of *Gerris, Velia, Nepa, Notonecta* and *Aphelocheirus*. In Mymaridae, *Caraphractus cinctus* parasitizes eggs of water beetles and *Notonecta* and *Litus cynipseus* attacks those of a hydrometrid. *Thoron metallicus*, a scelionid, has been bred from the eggs of *Nepa*. It seems likely that the closely related *Tiphodytes gerriphagus*, a Holarctic parasite of pondskaters' eggs, will also be found in Britain. It frequents the floating leaves of waterlilies and, like *Prestwichia* and *Caraphractus*, enters the water freely and swims well.

Ants as predators

Ants are believed to exert a profound influence over the composition of arthropod faunas in the areas where they forage, yet surprisingly little information is available concerning the prey of ants and quantitative research, even at an elementary level, is still very necessary. Pontin (1961) investigated the larval food of *Lasius niger*. In the brood chambers of its nests he found the remains of one specimen each of *Podops inuncta* (Pentatomidae) and *Aptus myrmicoides* (Nabidae), five cicadellids, 52 subterranean aphids and 18 other aphids. In nests of *Lasius flavus*, a species that forages only underground, all the hemipteran prey he found were root aphids. Waloff (1980) reported several species of *Myrmica* feeding on Auchenorrhyncha in grassland. The big *Formica* species can often be seen carrying recognizable pieces or whole specimens of Hemiptera back to their nests. Heads (1986) demonstrated that *Formica lugubris* had a small but significant effect in reducing the numbers of a cercopid, a delphacid and a mirid on Bracken. Whittaker (1981) found that a related ant, *F. rufa*, foraging in tree canopies, preyed commonly on one aphid species but rarely on other aphids and a psyllid that provided it with honeydew.

Sphecidae

The females of these solitary wasps provision their nests with insects of various orders and with spiders, usually stinging them to paralyse them at least temporarily during the journey back to the nest. Many species are wholly or partly dependent upon Hemiptera.

All three British species of the subfamily Astatinae take Heteroptera. The common *Astata boops* makes cells in the ground and stocks them with the nymphs of shieldbugs.

Another common ground-nesting sphecid, *Lindenius albilabris*, in the subfamily Crabroninae, provisions its nests with mirid bugs, chloropid flies or both. Most other crabronines utilize only Diptera but *Crossocerus ovalis* was once observed to take Miridae as well and a nest of *C. capitosus* in the pith of a cut twig was found to contain an unidentified cercopid or cicadellid. *C. annulipes*, which nests in rotten wood, usually provisions with typhlocybine leafhoppers but has been known to take mirids and psyllids. The common crabronine *Rhopalum clavipes* is unusual in stocking its nests with Psocoptera but it will occasionally use aphidoids, psylloids or small flies.

Most Nyssoninae, which nest in the soil, prey on Auchenorrhyncha. One common species, *Argogorytes mystaceus*, extracts nymphs of froghoppers from their froth to stock its nests. The genus *Nysson*, in the same subfamily, contains only cleptoparasites that usurp the ready-stocked nests of the related *Gorytes* species or of *Lindenius albilabris*.

The genus *Psen*, in Pemphredoninae, uses adults and, occasionally, nymphs of Cicadellidae (though rarely those of Typhlocybinae), and Issidae. *Psen equestris* and *Ps. lutarius* burrow in sandy soils but other, rarer species of the genus nest in beetle borings in dead wood. Apart from this genus and *Spilomena*, which specializes in Thysanoptera, most British Pemphredoninae take only Aphidoidea as prey. One of the three *Psenulus* species, however, confines itself to Psylloidea. Many of these little wasps nest in plant stems or in beetle borings in dead wood. Janvier (1960-1962) recorded biological data on the aphidophagous species.

Adults of British Sphecidae may be identified with the aid of Richards's (1980) handbook. Danks (1971) provided a key to the nests of Hymenoptera Aculeata occurring in bramble stems.

Dryinidae

Dryinids are perhaps the strangest of all the natural enemies of Hemiptera. They are rarely seen as adults but the larval sacs, attached to the bodies of nymphal or adult cicadellids and delphacids, are a common sight. There are four British subfamilies of Dryinidae. Aphelopinae are restricted to typhlocybine cicadellids in various habitats (Jervis, 1980c); one of the two rare Dryininae has been bred from adult Cixiidae; Anteoninae attack non-typhlocybine cicadellids on trees and in grassland; and

Gonatopodinae, in which the females are totally apterous, parasitize both delphacids and non-typhlocybine cicadellids in grassland.

Female dryinids apparently detect hosts either by sight or by odour from a distance of a centimetre or so and spring onto the victim, seizing it in the front two pairs of legs (Aphelopinae) or with the chelate tarsi of the first pair alone (other subfamilies); sometimes there is a short chase before the hopper is secured. Large hoppers are avoided or held down against the substrate but smaller ones are lifted clear. A single egg is inserted through the intersegmental membrane between the abdominal segments or between thorax and abdomen or thorax and head. The preferred oviposition site varies in different dryinid species and largely determines the site at which the larval sac will later protrude. Some Gonatopodinae (*Gonatopus, Dicondylus*) kill and eat as many hoppers as they parasitize. Anteoninae typically inflict non-fatal wounds on the prey and drink the body fluids that ooze out. The short-lived females of Aphelopinae and males of all subfamilies do not prey on hoppers at all. Males of *Gonatopus sepsoides* are unknown in Britain, where this common species reproduces asexually. Other species may lay unfertilized eggs that produce males; mated females produce about equal numbers of male and female progeny. Jervis (1979) reported an arboreal mating aggregation in *Aphelopus melaleucus*.

There are five larval instars in Dryinidae. The first is wholly internal and so, to begin with, is the second. The sac-like abdomen of the late second instar can be seen protruding from the body of its host while the head is anchored firmly inside by two elaborate tubercles. At ecdysis, the toughened integument of the larva splits along its dorsal midline but remains in situ, protecting the base of the abdomen of the third instar, the exposed part of which becomes toughened in its turn. The same procedure is followed at the next moult, so that the fourth instar's protruding abdomen is flanked by the exuviae of the second and third instars. All of the first four instars ingest only liquid body contents but the final instar, which has well developed mouthparts, forces its way into the body of the host and eats away the solid tissues, growing rapidly in bulk. As its body swells, the composite external sac splits for the last time and the fully fed larva emerges from it. It spins a double-walled cocoon on low vegetation or among litter or in the upper layers of the soil.

The larva pupates within the cocoon almost immediately in the summer generations of bi- or trivoltine species but otherwise it overwinters in the cocoon and does not pupate until the spring. *Dicondylus bicolor* overwinters as a first-instar larva within its delphacid hosts and *Aphelopus serratus*, a parasite of Typhlocybinae, may pass the winter in either of these ways. It is difficult to keep diapausing larvae alive in captivity over the winter but the summer generations of those species with more than one generation a year can be reared more easily. Some cocoons may yield specimens of *Ismarus* species (Diapriidae), showing that the dryinid larva has itself been parasitized while still attached to the hopper.

Biological and ecological details were recorded by Jervis (1980b,c) and Waloff (1974). Perkins (1976) gave keys to the British Dryinidae; Jervis (1977) provided an improved key to the species of *Aphelopus*. The family was monographed world-wide by Olmi (1984), who gave numerous records of hosts and distribution. The effects of Olmi's revision on the names used by Perkins are summarized in *Antenna* 9: 155 (1985).

Embolemidae

The small family Embolemidae is rather distantly related to Dryinidae and one species is recorded from Britain by Perkins (1976). Its may parasitize some kind of Auchenorrhyncha, since the only known host of any embolemid belongs to the infra-order Fulgoromorpha.

Braconidae

Quite often, fourth and fifth instar nymphs of Miridae are found to have the abdomen noticeably distended and shiny. These are the victims of braconid parasites of the genera *Leiophron* and *Peristenus* (subfamily Euphorinae). Leston (1961) listed 51 species of British Miridae known to be attacked by these wasps. He noted hosts in all the subfamilies except Bryocorinae; this exception is probably no more than an accident of sampling as an American species of *Monalocoris* has been found to be susceptible (Loan, 1980). Both predaceous and phytophagous species are parasitized, whether living on herbs, shrubs or trees.

Oviposition behaviour has been described by Waloff (1967), Glen (1977) and Loan (1983). The female braconid runs rapidly over vegetation and, upon encountering a mirid nymph of suitable size, which usually attempts to flee, she pursues it, seizes it between her front two pairs of legs and lifts it clear of the substrate, swiftly inserting a single, tiny, lemon-shaped egg into its underside in the region of the junction between thorax and abdomen. The immediate effect on the host varies from instant recovery upon being released by the wasp to paralysis for as long as half a minute, depending on the species of parasite. Second and third instar nymphs are the stages usually selected for attack. Later instars are often too agile and the wasp may even shy away from them. Fourth instars are sometimes attacked; they are too big to be lifted clear of the leaf and, instead, are held down against it while the egg is inserted.

Brindley (1939) described the development of *Leiophron pallipes* within the host. The whole endoparasitic period lasts about five weeks, almost equally divided between the development of the egg and that of the larva. The chorion of the egg, which lies in the host's haemocoel, is thin and elastic and there is very little yolk. A layer of polygonal cells with large nuclei forms a trophic membrane lining the chorion and completely enclosing the embryo. This membrane transports fluid and nutrients to the growing embryo. Ultimately the egg swells to six or seven times its original diameter. At hatching, the cells of the trophic membrane are dissociated and float free in the host's haemolymph; they are eventually eaten by the larva. Loan (1965) reported five larval instars in *Leiophron pallipes* but in later papers recognized only four, a figure with which other writers agree. Waloff (1967) described and figured the larvae of three species. They are white and legless and have a short tail when young. Parasitized bugs are heavier than non-parasitized ones and the efficiency of food conversion from mirid to braconid is very high (Glen, 1977). The fully fed larva cuts a slit in the side of the host's abdomen anteriorly, through which it emerges. It falls to the ground and spins a cocoon, in which it pupates, at no great depth. The host may live for a few hours, or even a day, after the emergence of the parasite; such bugs have a characteristic appearance with a broad, flat, shrivelled abdomen clearly showing the slit in the side. The instar from which the parasite emerges is usually the fifth but sometimes the adult; the host's final moult seems often to be delayed by parasitization so that, late in the season, a very high proportion of the remaining nymphs are parasitized. Only one larva develops in each host individual; if several eggs have been laid in one bug the development of all but one is suppressed (Waloff, 1967).

The overwintering stage of the braconid is the quiescent adult within its cocoon (Loan, 1983). Males emerge, on average, slightly earlier than females and do not live quite as long, so that the sex-ratio is constantly changing (Waloff, 1967; Loan & Bilewicz-Pawinska, 1973). Females may survive for as long as a month but generally live only for a week or two. There is usually a single generation each year but a few parasites of bivoltine hosts have two generations.

There are probably several dozen species of *Leiophron* and *Persitenus* attacking Miridae in Britain; most of them are undescribed. Loan (1980) reported four species of the former genus and 14 of the latter parasitizing mirids in one restricted locality in Canada. Waloff (1967) found three species of braconids attacking a complex of five mirid species on broom (*Sarothamnus*) in one locality. Host specificity seems to depend on many factors, including the relative size and agility of wasp and bug species, emergence times and habitat preferences. Some species of *Leiophron* attack Psocoptera and one unusually large and handsome *Peristenus* was described from a specimen bred from a beetle but, apart from these exceptions, all known hosts of the two genera in Europe and North America are Miridae. Rates of parasitism are generally high. Loan's (1980) Canadian study showed 21-66 per cent of susceptible hosts to be parasitized, with an average level of about 42 per cent. By no means all of the mirid species in his study area were attacked by braconids, however. Most of his braconids specialized in a single host species or several closely related species on a single host plant. *Leiophron pallipes*, which occurs in Britain, had six Canadian hosts including *Adelphocoris lineolatus*, *Capsus ater*, *Leptopterna dolabrata* and American species of *Adelphocoris*, *Labops* and *Lygus*. It was by far the most polyphagous species in his study. A key to the adults of the European species of *Leiophron* and *Peristenus* is given by Loan (1974). Rearing methods are described by Waloff (1967). Ichneumonidae of the genus *Mesochorus* sometimes occur as hyperparasites of Miridae via Braconidae.

Apart from the two genera of mirid parasites, the tribe Euphorini is represented in Britain by *Wesmaelia pendula*. In the USA this is a univoltine parasite of the males of certain Nabidae (Stoner, 1973). According to Loan (1983), females of *Wesmaelia* attack the large, later instars of these predaceous bugs, darting at the side of the abdomen with the ovipositor. He reported that this genus (like *Holdawayella*, a parasite of Tingidae) overwinters as the first-instar larva within the adult host. Larvae of a braconid parasite, tentatively identified as *Aridelus egregius* (a euphorine not yet recorded from Britain) have been found by dissection in nymphs and adults of a variety of European shieldbugs by several authors. Adults of all the genera mentioned were included in Loan's (1983) keys to North American Euphorinae and Shaw's (1985) key to world genera of the subfamily.

Aphidiidae

The most numerous parasites of viviparous aphids belong to this family, which is often regarded as a subfamily of Braconidae. No other hosts are attacked; even Adelgidae and Phylloxeridae are immune.

Adult aphidiids live for only a week or two, drinking water, nectar and honeydew. During her short life a female may parasitize 50–300 aphids or, exceptionally, more than a thousand, laying a solitary egg in the haemocoel of each. Superparasitism, resulting from more than one oviposition, may occur but the supernumeraries always fail to develop. Once laid, the small, yolkless, narrowly lemon-shaped egg develops a trophamnion – a continuous layer of nutrient cells lining the inside of the chorion – which absorbs nutrients from the body fluids of the host. At the time of eclosion the egg has swollen to several hundred times its original volume. The cells of the trophamnion separate after the larva has hatched and live independently, taking up nutrients. Their normal fate is to be consumed by the larva but if it dies they may themselves bring about the death of the host.

The larva in the first instar is a legless maggot with a short tail. Instars I and II have piercing mandibles with which they puncture aphid embryos, fatbody cells and trophamnion cells, swallowing the contents. Production of young by the aphid ceases at this time. Instar III lacks mandibles and consumes only body fluids. The fourth and last instar has well developed mouthparts and eats out all of the solid tissues of the host. It is the only larval stage with functional spiracles, since it is the only one not surrounded by liquid. When it has finished feeding, all that is left of the aphid is its dry exoskeleton, referred to as a mummy. The form and especially the colour of the mummies varies according to the different parasite genera. Those produced by *Ephedrus* and *Pseudephedrus* are black, while those of other genera are brown (e.g., *Praon*), yellow or white.

When the aphid's body is completely empty, the aphidiid larva follows one of two strategies. In most genera it cuts a small hole in the ventral surface of the mummy and attaches it to the leaf on which it stands with the secretion of the larval silk glands. It then lines the mummy internally with silk, forming a cocoon of several layers in which it pupates. In *Praon* and *Dyscritulus* (but not the closely related *Areopraon*), the mature larva emerges from the mummy by cutting a hole ventrally and spins a tent of silk attaching the mummy to the substrate. It then spins a globular cocoon inside the tent. The adult in these two genera emerges through the side of the tent but other aphidiids must cut an emergence hole in the mummy. The position of this hole is another indication of the identity of the parasite. *Ephedrus* cuts off the whole posterior end of the aphid while *Aphidius* emerges via a circular hole in the dorsum. In both of these genera the lid is not completely detached but hinges back to permit the wasp to escape.

There are usually several generations a year, each lasting a month or so. The winter is passed in the cocoon. The short generation time often enables aphidiines to exert a substantial check on the growth of aphid populations, making them important agents of biological control.

Host selection may be very specific, as in the case of *Dyscritulus planiceps*, a specialist parasite of the sycamore aphid, *Drepanosiphum platanoidis*, or hosts may be confined to those living on a particular type of plant, or the choice of host may be limited only by very general habitat requirements. A detailed catalogue of host records in this family was compiled by Mackauer & Starý (1967).

Aphelinus (Aphelinidae) and *Endaphis* (Cecidomyiidae), normally endoparasitic in aphids, may feed facultatively on aphidiid larvae if these are present in their hosts. Obligate hyperparasites of aphids, usually via Aphelinidae, are encountered in the families Charipidae, Pteromalidae, Megaspilidae and Encyrtidae.

Much biological information is to be found in the monograph by Starý (1970) and in the same author's (1966) account of the Czechoslovakian Aphidiidae, which contains keys to most of the British species and all European genera. Starý et al. (1971) dealt with the French fauna. Pungerl (1986) gave keys to the species of *Aphidius*. Starý's (1987) bibliography is a guide to the extensive literature on the family.

Ichneumonidae

Most studies of euphorine parasites of Miridae, including those of Brindley (1939), Waloff (1967) and Glen (1977) have revealed low levels of hyperparasitism by *Mesochorus* species (Ichneumonidae). Brindley found the ichneumonid egg in the haemocoel of the braconid larva and suggested, by analogy with known life-histories of other Mesochorinae, that the presence of the ichneumonid larva inhibited pupation by the braconid after the latter had spun its cocoon and that the ichneumonid pupated within the cocoon. Little is known for sure about the biology of *Mesochorus* and reliable identification of the species is not possible at present.

Charipidae

The British species of the charipid subfamily Alloxystinae are all solitary hyperparasites of aphids via Aphidiidae. The charipid egg is laid into the body of a half-grown aphidiid larva, whose development proceeds normally up to the completion of the cocoon but the presence of the hyperparasite prevents pupation. Most of the larval life is spent within the body of the aphidiid host but the last (third) instar larva emerges into the cocoon and finishes feeding externally. Alloxystinae may themselves be parasitized by tertiary parasites in the family Encyrtidae. The two British species of the subfamily Charipinae are believed to be primary endoparasites of *Psylla*. Fergusson (1986) provided keys to the adults of both subfamilies.

Megaspilidae

Species of *Dendrocerus* (Ceraphronoidea: Megaspilidae) attack aphidiids at a later stage of development than that favoured by charipids, when the host is already mummified. The egg is laid externally on the mature aphidiid larva or pupa and the *Dendrocerus* larva feeds ectoparasitically. There are four, possibly five, larval instars. The hosts of this genus include not only aphidiids but also encyrtids, aphelinids, pteromalids and charipids in various Sternorrhyncha and aphid-associated Neuroptera and Diptera of several families plus a few ants and some other insects not obviously associated with Hemiptera. *Dendrocerus laevis* has been bred from eriococcid and pseudococcid hosts but is usually hyperparasitic on aphids via Aphidiidae. *Dendrocerus serricornis* has been reared from Psylloidea and Coccoidea and from Chamaemyiidae and Cecidomyiidae associated with the latter, perhaps, in the case of the chamaemyiids, as a hyperparasite via *Melanips* (Figitidae). Identification keys and host records were provided by Fergusson (1980).

Diapriidae

Ismarus species are hyperparasitic on Auchenorrhyncha via their primary dryinid parasites, from whose cocoons they are sometimes reared. The British species of *Ismarus* were keyed by Nixon (1957: 11-12).

Platygasteridae

Allotropa mecrida is the only species of this important genus of primary parasites of mealybugs (Pseudococcidae) to have been found in the British Isles. It has been reared from *Trionymus newsteadi* on beech (*Fagus*). *Platygaster* species reared from nymphs of Psyllidae may be primary parasites but those reared from adults are probably hyperparasitic via the cecidomyiid *Endopsylla*. Similarly, *Platygaster* species from Aphidoidea are hyperparasitic via another cecidomyiid, *Endaphis*. There are no British records of *Amitus*, a genus well known to parasitize Aleyrodidae, unless Trehan's (1940) report of a species of *Isostasius* from *Asterobemisia carpini* is to be referred to this genus.

Aphelinidae

In tropical and subtropical countries, where Coccoidea and Aleyrodoidea are pests of major importance, aphelinids are of great value as agents of biological control. Most species of Aphelinidae are primary parasites of these two groups of Sternorrhyncha and may cause a very high level of mortality among their hosts. *Encarsia formosa* is available commercially as a control agent for *Trialeurodes vaporariorum*, the glasshouse whitefly.

Adult aphelinids are known to feed on their hosts' honeydew and females also take haemolymph from wounds inflicted by their ovipositors. Larvae of some genera (e.g. *Aphytis*) are ectoparasites, with functional spiracles, and live beneath the scale-like bodies of their hosts but most are endoparasitic and lack spiracles, at least until they enter the last instar. Pupation usually occurs within the host and the normal overwintering stage is the larva or pupa inside its pupal chamber. There may be several generations a year. Some species of the genus *Coccophagus* have attracted notice through the strange developmental pathways of their male larvae. Female larvae of all species of this genus are primary endoparasites of Coccoidea. Male larvae, according to species, may be primary ectoparasites of Coccoidea or hyperparasites via other aphelinids and encyrtids or even obligate ectoparasites of the immature stages of their own species within the homopteran host.

Aphelinus species are well known as primary solitary endoparasites of aphids. *Aphelinus mali* has been used successfully to control *Eriosoma lanigerum* on Apple in some parts of the world. Female *Aphelinus* feed on young aphid nymphs, piercing them with the ovipositor and drinking the fluids that exude, but do not lay eggs in them. Several aphids may be killed in this way each day by a single female. Nine or ten eggs are laid daily and a lifetime's egg production may amount to about 500 in the laboratory but usually substantially fewer in the field. Superparasitism is very uncommon even at extremely high levels of parasitization, indicating that the ovipositing female can discriminate between parasitized and unparasitized hosts. The larvae of all three instars are short and broad, lacking a tail or other appendages. Mature larvae of this genus have spiracles.

Three species of the egg-parasitic genus *Centrodora* occur in Britain. Two of them attack tettigonioid Orthoptera but the third, *C. livens*, has been reared from eggs of the cicadellid *Macrosteles sexnotatus* (Waloff, 1980: 126).

Ferrière (1965) covered the European species of the family. The works of Nikol'skaya (1952) and Peck, Bouček & Hoffer (1964) are also useful. British species of *Aphelinus* can be identified from the keys of Graham (1976).

Encyrtidae

Almost all members of the smaller of the two encyrtid subfamilies, Tetracneminae (genera *Ericydnus* to *Rhopus* in the British Check List (Fitton et al., 1978), parasitize Pseudococcidae. The larger subfamily, Encyrtinae, has a very wide host range, including Sternorrhyncha, lepidopteran caterpillars, ticks and spiders. All of the major families of Coccoidea are parasitised by encyrtines and Psylloidea are hosts to five genera (*Trechnites* to *Cercobelus* in the Check List). Most of these parasites of Psylloidea develop in the nymphs but Robinson (1961a,b) described the morphology and biology of *Sectiliclava cleone* (as *Parapsyllaephagus adulticola*), which he found attacking adult Psyllidae.

No encyrtid is a primary parasite of Aphidoidea but *Aphidencyrtus* is hyperparasitic on them via Aphidiidae, laying the egg on the surface of the late larva or pupa of the aphidiid. Encyrtids have also been bred from the immature stages of several insect families that include predators of aphids: Chrysopidae and Hemerobiidae (Neuroptera), Syrphidae, Cecidomyiidae and Chamaemyiidae (Diptera) and Coccinellidae (Coleoptera). *Heleogonatopus* is a hyperparasite of Auchenorrhyncha via Dryinidae (Hymenoptera). Some Encyrtidae are hyperparasites of Coccoidea via aphelinid or encyrtid primary parasites. Most encyrtids are internal parasites but *Microterys sylvius* is a rare exception. It is an egg predator in the ovisac of the coccid *Parthenolecanium corni* (Silvestri, 1919). The only known host of the large encyrtid *Prionomastix morio* is the membracid *Gargara genistae*, from which it was recorded in France by Granger (1944).

The egg of encyrtids is often stalked and, in some species, the larva maintains respiratory contact with the outside air via the stalk, which remains embedded in the host's body wall where the egg was laid, with the larval exuviae accumulating around it. There are usually five instars, though as few as two have been claimed for some species. The young larva has a long tail, which it progressively loses as it grows unless it is one of the few that use the egg stalk for respiration. Pupation usually takes place within the empty husk of the host but some species pupate while the host is still living and a respiratory connection is then established between the tracheal systems of host and parasite. One generation a year seems to be the norm for the British species, most of which overwinter in the pupal stage.

Noyes (1978) described *Ooencyrtus brunneipes* from Britain and suggested that it might be a parasite of heteropteran eggs. Many non-British relatives of this encyrtid develop in eggs of shieldbugs, reduviids and coreids. Trjapitzin (1979) described *O. corei* from specimens bred from the eggs of *Coreus marginatus* in the USSR.

Keys to most European Encyrtidae are to be found in the Russian-language work of Trjapitzin *et al.* (1978). The Spanish species are covered by Mercet (1921), whose nomenclature is now very much out of date, and Nikol'skaya's (1952, 1963) account of the Russian fauna is also useful. The European genera were keyed by Peck, Bouček & Hoffer (1964).

Pteromalidae

Many members of the huge family Pteromalidae develop ectoparasitically on insect larvae and pupae in enclosed situations such as leaf mines, galls and cocoons. Aphidiids in their cocoons inside or beneath aphid mummies belong to this category and accordingly are exploited by these small, metallic wasps. *Asaphes vulgaris*, *Coruna clavata* and *Pachyneuron aphidis* are the species most frequently bred from mummies. *Asaphes* and *Pachyneuron* have also been reared from psyllids in Britain, and *Pachyneuron* from *Eriopeltis* (Coccidae) as well, in these hosts probably as hyperparasites via Encyrtidae. *Pachyneuron vitodurense* is a parasite of the chamaemyiid predators of Adelgidae.

Coccoidea of several families are the only known hosts of the small pteromalid subfamily Eunotinae. The larva of *Eunotus cretaceus* is a predator in the ovisac of *Eriopeltis* (Varley in Graham, 1969).

The highly polyphagous pteromalid, *Hemitrichus seniculus*, has been reared from Lepidoptera, Coleoptera and Diptera among stored products. It was once found in Czechoslovakia as a gregarious endoparasite of the scutellerid bug *Eurygaster maurus* (Bouček, 1954).

The larval habit of burrowing in grass stems and devouring eggs of Delphacidae, as well as eggs, larvae and pupae of other insects, has been noted in two genera of Pteromalidae. *Panstenon oxylus* destroys the eggs of the delphacid *Javesella pellucida* in Sweden, but the sole British rearing was from the puparium of an agromyzid fly mining pea (*Pisum*). *Mesopolobus aequus* and *M. graminum* have also been reported feeding on the eggs of the same delphacid in Sweden and, like *P. oxylus*, have other hosts, principally the grass-feeding species of Eurytomidae and their parasites (Graham, 1969). British Pteromalidae were monographed by Graham (1969).

Eulophidae

The eulophid *Euderomphale cerris* (Euderinae) has been reared from *Aleyrodes proletella* (Aleyrodidae); other species of the genus are known as parasites of Aleyrodidae in Europe. For their identification, see Askew (1968).

Several European species of the large and diverse genus *Tetrastichus* (Tetrastichinae) have been reported as having psylloid hosts in the genus *Trioza* (Triozidae). *Tetrastichus obscuratus* is an ectoparasite living beneath the bodies of immature *Trioza centranthi* and *T. upis* is endoparasitic in *Trioza urticae* nymphs. *Tetrastichus pubescens* was once reared from an unidentified psylloid in England. Hodkinson (1973) published notes on the biology of *Tetrastichus actis*, a bivoltine parasite of the psyllid *Strophingia ericae*. Recent revisionary work on European Eulophidae (e.g. Graham, 1987) has narrowed the concept of *Tetrastichus* considerably and the tetrastichine parasites of Psylloidea have mostly been transferred to the genus *Tamarixia*.

Of two *Tetrastichus* reared from *Ceroplastes* spp (Coccidae), one is a solitary, primary endoparasite while the other is a hyperparasite via the first and also via a species of Pteromalidae: Eunotinae, *Scutellista cyanea*.

Domenichini (1968) reported one species of *Tetrastichus* from the eggs of *Capsodes* (Miridae) and Viggiani (1971) reported another from those of *Ledra* (Cicadellidae). Waloff (1980: 126) mentioned an unidentified species of *Tetrastichus* that begins its larval life as an internal parasite of an egg of *Conomelus anceps* (Delphacidae) or *Cicadella viridis* (Cicadellidae) in stems of *Juncus effusus*. When the contents of the first egg have been consumed the larva embarks on a predatory career, attacking adjacent eggs of the batch. This larval biology is exactly the same as that described by Bakkendorf (1934) for his species *Annelaria conomeli*, which he placed in Entedoninae and whose host he determined as *Conomelus limbatus*. Domenichini (1966) referred Bakkendorf's record to the British species *Tetrastichus mandanis* so it seems probable that this is the species mentioned by Waloff.

Graham (1987) commenced a revision of the European Tetrastichinae and transferred most of the species formerly placed in *Tetrastichus*, including the parasites of Hemipteran eggs, to *Aprostocetus*. In his treatment of the latter genus, he reported the British species *A. coccidiphagus* and *A. pachyneuros* from *Kermes* (Kermesidae) and *A. trjapitzini* from *Physokermes*, *Eulecanium* and *Parthenolecanium* (Coccidae). The host relationships of Tetrastichinae were catalogued by Domenichini (1966).

Eupelmidae

Two common British eupelmids have been bred from *Aleyrodes proletella* in Europe (Fulmek, 1943). Both were probably hyperparasites via Aphelinidae.

Signiphoridae

The two British Signiphoridae (=Thysanidae) belong to a small family whose members are well known as hyperparasites of Sternorrhyncha via Aphelinidae and Encyrtidae. *Thysanus ater* has been reared from Diaspididae and Aleyrodidae while *Chartocerus subaeneus* has been bred from Pseudococcidae and from the puparia of their dipterous predators, Chamaemyiidae. Both species were covered by the keys of Ferrière & Kerrich (1958).

Mymaridae

The delicate little 'fairy flies' have fascinated many naturalists. Adult Mymaridae are unique among hymenopterous egg-parasites in lacking an oblique vein running into the disc of the forewing from its costal margin. The coloration is non-metallic, yellow, brown or black; the tarsi are four- or five-segmented; the wings are stalked and they are fringed with hairs longer than the width of the posterior pair. The legs and antennae are relatively longer and more slender in this family than in almost all other Chalcidoidea.

All species develop in the eggs of other insects, their larvae living usually one to an egg, though sometimes gregariously. Choice of host species is probably governed by the mymarid's preference for searching on a particular kind of plant. Thus, cicadellid eggs in twigs of trees are mostly attacked by *Polynema* species while those in stems of grasses suffer mainly from attack by *Anagrus*. This latter genus is taxonomically difficult and recent studies have revealed that several species have been confused under one name (Walker, summarized by Waloff, 1980). Hosts of *Anagrus* include Cicadellidae, Delphacidae, Cercopidae, Miridae, Coccoidea, Lepidoptera and Odonata. *Anaphes* has a similar host range but includes also Tingidae and many Coleoptera. *Polynema* and *Gonatocerus* also have a wide range of hosts including Auchenorrhyncha, Heteroptera, Aphididae, Aleyrodidae (*Gonatocerus* only), Diptera and some other orders. The more exposed eggs of Coccoidea, Aleyrodidae and Psocoptera are favoured by *Alaptus*, which has only rarely been bred from cicadellid eggs. Some genera appear to specialize in a single host family, for example *Erythmelus* in Miridae, but such apparent specialization may be an accidental result of incomplete knowledge. Walker's researches (*op.cit.*) showed that, even within the single genus *Anagrus*, the biology of different mymarid species varies considerably. All Mymaridae overwinter as eggs or larvae in the host eggs but some of Walker's *Anagrus* species were restricted to a single host and presumably to one generation per annum while others were polyphagous. *Anagrus silwoodensis* overwintered in the eggs of *Cicadella* but hosts of its summer generation included Delphacidae.

Bakkendorf (1934) described two types of mymarid larvae. In *Anagrus*, *Erythmelus* and *Caraphractus* the first instar (known only in *Anagrus* but the others were presumed to be similar) was sacciform, divided into anterior and posterior parts by a simple constriction, and the mature larva was cylindrical with a pair of short, anterior appendages below the mouth and sometimes a similar pair posteriorly. In the remaining genera known to Bakkendorf the first instar was fusiform with a single, short, frontal process anterior to the mouth and a long, tapering tail. The main part of the body bore several rings of hairs, which were longer dorsally than ventrally. The mature larva in this group was fusiform to cylindrical with or without the frontal process and without other appendages or hairs. Bakkendorf could demonstrate only two larval instars. Mymarid larvae have no tracheal system. Pupation takes place inside the host egg.

Belgian species of Mymaridae were monographed by Debauche (1948). Kryger (1950) provided keys to most European genera. British species of some genera can be identified with the aid of Hincks (1950, 1952, 1959), Graham (1982), Matthews (1986) and Soyka (1956). Further guides to the host relations, biology and systematics of the family were presented by Schauff (1984) and Hueber (1986).

Trichogrammatidae

Trichogrammatids are of much less importance than mymarids as parasites of hemipteran eggs. The three-segmented tarsi at once set the adults apart from all other egg-parasites.

In appearance, they resemble stocky mymarids with shorter, more robust appendages and broader wings. Eggs of many insect orders, especially Lepidoptera, are attacked.

The egg is laid into that of the host and the entire larval and pupal life is spent there. Two types of first-instar larvae have been described, one simple and sacciform, the other plumply fusiform, tailed and hairy, paralleling the two types of mymarid larvae. It has been suggested that the tailed larvae of egg-parasitic Hymenoptera, which are known to be mobile, disorganize the egg contents by their activity, preventing development of the embryo. Mature larvae of Trichogrammatidae, where known, are simple, legless maggots.

The few European species bred from eggs of Cicadellidae and Miridae were mostly keyed in the genus *Oligosita* by Nikol'skaya (1952, 1963), an exception being *Paracentrobia pulchella*, described in *Monorthochaeta* by Claridge (1959), who bred it from eggs of the mirid *Capsus ater* in Britain.

Scelionidae

So far as is known, all Scelionidae develop from egg to adult in the eggs of other insects. Unlike the mymarids and trichogrammatids they appear to be unable to detect eggs embedded in plant tissue but they are undeterred by the thick eggshells of most surface-laid eggs. Very fresh eggs are usually selected for attack and one species developing in lepidopteran eggs has been shown to inject with the egg a substance that inhibits the host's embryonic development (Strand *et al.*, 1986). In Britain, the great majority of hemipteran hosts are shieldbugs but a few species have been reared from eggs of Miridae.

The first-instar scelionid larva is an hourglass-shaped creature with a few pairs of pointed, lobe-like appendages on the anterior half and sometimes a pair of bristly lobes on the posterior part of the body as well. There is a more or less complete girdle of bristles around the waist and a long, thin, sometimes forked tail. Later stages are simple and sac-like. A tracheal system is present only in the third (last) instar.

Adult scelionids are usually black in colour. They differ from Trichogrammatidae in having more than three tarsal segments and from Mymaridae in having shorter fringes around the wings and a short vein running obliquely into the disc of the forewing from its anterior margin.

Kozlov & Kononova (1983) revised the subfamily Telenominae, containing most of the parasites of hemipteran eggs, for the USSR.

Diptera

The only known dipteran parasites of British Heteroptera are Tachinidae of the subfamily Phasiinae, which are believed to have no other hosts here. Pipunculidae are similarly specific to Auchenorrhyncha. The few dipteran parasites of Sternorrhyncha belong to the widely polyphagous family Cecidomyiidae, which also contains some species with larvae predaceous on Aphidoidea and, in some parts of the world, on other Sternorrhyncha. The larvae of most hoverflies (Syrphidae) prey upon aphids and sometimes on other Hemiptera and those of Chamaemyiidae are probably all predators of Sternorrhyncha. Larvae of a few other cyclorrhaphan flies also prey on Sternorrhyncha. Predaceous adults of the dipteran families Asilidae, Empididae, Dolichopodidae and Muscidae have occasionally been known to catch and eat hemipterans but the relationship is not specific.

Tachinidae

Phasiinae attack almost all families of terrestrial Heteroptera but not the aquatic or semi-aquatic ones. Dupuis (1963) gave a compendious account of the Phasiinae in Europe and North Africa with much information on biology, host relationships, anatomy and distribution, descriptions of eggs and larvae and a comprehensive bibliography.

The little *Alophora pusilla* has been reared from Anthocoridae, small Lygaeidae and small shieldbugs; *A. obesa* from larger Lygaeidae, Miridae and a shieldbug; and *A. hemiptera*, the largest British phasiine, only from shieldbugs, which it is said to resemble in life. The known hosts of the tribe Leucostomatini belong to the families Coreidae, Rhopalidae, Lygaeidae and Nabidae. Apart from *Alophora* and the Leucostomatini all other host records for British phasiines refer to shieldbugs. Dupuis (1963) dissected thousands of bugs from France, Germany and North Africa and found that about one in ten of them contained a tachinid larva.

The eggs are laid singly into the body of the host or, in the genera *Subclytia* and *Gymnosoma*, on its surface and hatch in a few days. The young larva penetrates the cuticle, if necessary, and makes for the host's thorax. Here it moults and the second-instar larva induces the bug's haemocytes to form a respiratory siphon connecting the tracheal systems of host and parasite. If the host was attacked in the nymphal stage the parasite larva delays its second moult until the bug is adult. All three larval instars feed on body fluids but the third may also consume the fatbody and other solid tissues. Superparasitism results in the death of all but one of the parasites or sometimes all of them. The fully fed third-instar larva emerges by rupturing the intersegmental membranes of the host (which invariably dies) in the genital region and then pupates in soil or litter. The pupal stage lasts for 7–40 days. The overwintering stage is always the first- or second-instar larva, never the third. There seem usually to be two generations a year, which accounts for the records of phasiines from host species that overwinter as eggs.

Van Emden (1954) provides a key to adult Tachinidae but his classification and nomenclature are very outdated. Mr R. Belshaw is preparing a new handbook on the family.

Pipunculidae

Adult pipunculids are small flies whose eyes almost completely cover the surface of their big, globular heads. They are highly accomplished at hovering and their precisely controlled movements in very confined spaces make the larger hoverflies (Syrphidae) look clumsy by comparison. The twin qualities of acute vision and high manoeuverability enable the females to seek out motionless hosts in dense vegetation. That vision plays a prominent part in host-finding is evident from the fact that hunting females will pounce on anything of the same size and shape as their usual quarry, such as buds and other inert objects. When a suitable victim is encountered, the female lays a single egg into its body by means of the piercing ovipositor. Some species of fly lift the host bodily from its perch and inject the egg while hovering with it clasped firmly in their legs, presumably thereby avoiding injury if the hopper should try to jump in response to being attacked.

Jervis (1980a,b,c) has made a detailed study of the genus *Chalarus*, which develops in leafhoppers of the subfamily Typhlocybinae on herbs, shrubs and trees. Eggs are laid singly, though superparasitism may occur, into the anterior abdomen of nymphs of the third, fourth or fifth instar. The egg may hatch before or after the host's final moult but, in either case, the whole of the second of the parasite's two larval instars is passed within the adult, whose abdomen becomes visibly distended as the larva grows. Mature larvae emerge to pupate in soil or litter. Some species are univoltine, others bivoltine; the overwintering stage is usually the pupa but in one species it may be either this stage or the first-instar larva.

Other genera that have been investigated have broadly similar life histories to *Chalarus* but attack different hosts. *Verralia*, which overwinters as a pupa, parasitizes Cercopidae and oncopsine Cicadellidae; hunting females attack adults of both groups and nymphs of the cicadellids, but avoid the protective froth of the cercopid nymphs. Two species of *Cephalops*, a genus that parasitizes Delphacidae and Cixiidae, are known to overwinter as first-instar larvae within their hosts, as is one species of *Eudorylas*. This latter genus and the genera *Tomosvaryella*, *Pipunculus* and *Dorylomorpha* have all been reared from non-typhlocybine Cicadellidae. Most species overwinter as pupae in the ground but some *Dorylomorpha* that frequent damp habitats pupate on vegetation, where the danger from winter flooding is less. Coe (1966) keys the adults of the British species.

Cecidomyiidae

Two genera of this family are endoparasitic as larvae in the British Hemiptera: *Endopsylla* in psylloids and *Endaphis* in aphidoids. Lal (1934) studied the natural history of *Endopsylla* in Scotland. He reported that the female midge laid one to three yellow eggs on the forewing of the adult host, adjacent to the veins, in the period June–August. After 8–13 days the eggs hatched and the young larvae fed externally for 3–4 days. They then entered the host's haemocoel via the intersegmental membrane and fed internally for 6–10 days before emerging again. After spending a day or two on the leaves of the psylloid's host plant, they dropped to the ground to pupate. The orange-coloured adult midges emerged the following summer. Lal figured the egg, larva, pupa and adult and reported rearing a number of Hymenoptera from the same host species. One of these was a species of *Platygaster*; as Platygasteridae are almost invariably parasitic on Diptera, especially Cecidomyiidae, it was probably hyperparasitic on the psylloid via *Endopsylla*. *Endaphis* is much less common than the hymenopteran parasites of Aphidoidea and also much less frequent than their cecidomyiid predators. Its biology is probably similar to that of *Endopsylla*; the eggs are laid among aphidoids though not necessarily on their bodies. The reddish larvae can be easily seen through the body wall of pale-coloured hosts. Like *Endopsylla*, it may be attacked by Platygasteridae. These lay their eggs in the eggs of the cecidomyiids and it is supposed that some degree of polyembryony occurs as several hyperparasites are usually produced from each parasitized aphid.

Predaceous cecidomyiid larvae have been reported in a number of genera. Harris (1973) revised the aphidophagous ones, reducing the British representatives to four species in two genera. The small, yellow or orange maggots, only 3–4 millimetres long at maturity, are frequently present among aphid colonies on both herbaceous and woody plants throughout the summer but are often overlooked. They differ from the larvae of both Syrphidae and Chamaemyiidae in that their posterior spiracles are flush with the body wall, and not carried on spiracular processes. There is a small, dark-coloured 'sternal spatula' visible in ventral view beneath the mouthparts. The commonest of these predaceous species is *Aphidoletes aphidimyza*. It has several generations in the course of a year. Pupation takes place in the soil, in a cocoon; the overwintering stage is the prepupa inside this. The only recorded prey of *A. abietis* is a species of Adelgidae on Spruce but records are too scarce to establish whether or not this is a specific association. Despite its name, *Monobremia subterranea* is not particularly associated with subterranean aphid colonies. Harris figured the larvae of the three commonest aphidophagous cecidomyiids and gave prey lists for all of them. In a later paper (Harris, 1982) he reviewed the usefulness of *A. aphidimyza* in the biological control of aphids. Baylac (1986) observed cecidomyiid larvae, which he referred to the genus *Lestodiplosis*, preying on the eriococcid *Cryptococcus* on Beech in northern France. This is the only European record of any cecidomyiid predator of Coccoidea. Harris (1968) revised the known cecidomyiid predators of Coccoidea throughout the world.

Syrphidae (Hoverflies)

Large colonies of aphids seldom escape predation by the larvae of hoverflies. Aphidophagous species have been known to feed on Coccoidea, Psylloidea, Aleyrodoidea and even on young nymphs of Cicadellidae but such instances are uncommon. Female hoverflies show a strong preference for depositing their eggs, which are usually about a millimetre long, white and banana-shaped, on plants infested with aphids, the more highly infested plants receiving more eggs (Dixon *in* Waloff, 1968). The larvae are blind, legless, carrot-shaped maggots that progress by undulating movements of the body. Their posterior spiracles, in the last of the three instars, are borne on a single projection resulting from the fusion of the two separate processes present in the first two instars. The slender, anterior end of the body is very flexible and extensible; it casts from side to side as the larva moves forward. When the larva meets an aphid it plunges its hooked mouthparts into the latter's body and hoists it clear of the substrate so that its legs wave uselessly in the air and it is unable to escape. Pupation, as in all the higher Diptera, involves the conversion of the skin of the third larval instar into a puparium. This is commonly formed among the aphids of the colony on which the larva has fed. It is pear-shaped and has a glassy appearance; the narrow, posterior end still bears the characteristic 'tail' with its two spiracles.

There are usually several generations a year though some hoverflies have only one. The polyvoltine species are not particularly selective as regards the species of prey or host-plant, following the shifting fortunes of the aphid fauna as a whole in their chosen habitats. Where active colonies of aphids persist through the winter, as on some cabbage (*Brassica*) crops, syrphid larvae may be found among them in the coldest months but it is usual for most species to overwinter as diapausing pupae. Not all syrphids are opportunists. Some specialize in subterranean prey, others in gall-formers such as *Pemphigus* and *Tetraneura*, galling poplar and elm respectively. The waxy secretions of *Eriosoma* on apple and of Adelgidae on conifers seem to confer some degree of protection against most syrphid larvae but *Neocnemodon*, which often feeds on gall-formers (themselves wax-producers), is not deterred.

Adult hoverflies feed on honeydew, nectar and pollen, the last providing protein for the production of eggs.

Most syrphid larvae that prey on aphids belong to the large subfamily Syrphinae, few of whose members feed on anything else; the remainder belong to the tribe Pipizini, in the subfamily Milesiinae. There is an admirable identification guide to adult British hoverflies by Stubbs & Falk (1983). Eggs of some aphidophagous species were keyed by Chandler (1968). Larvae of the commoner species were described by Dixon (1960); later papers describing additional species were summarized by Rotheray (1987). Goeldlin de Tiefenau (1974) gave much ecological information and descriptions of immature stages.

Chamaemyiidae

Careful searching among dense colonies of aphids, particularly in such places of concealment as leaf-axils, may reveal the larvae and puparia of *Leucopis* species. They are smaller than most Syrphidae and the posterior spiracles are borne on separate, widely spaced projections in all three larval instars and, of course, the puparium. Mature syrphid larvae have the spiracular processes fused together and cecidomyiids lack them altogether. Adult chamaemyiids are small, greyish acalypterate flies. Most species probably undergo several generations a year. Much remains to be discovered about their taxonomy, distribution, natural history (especially of the adults) and ecological preferences. It is possible, nonetheless, to make some generalizations about the food of the larvae. *Leucopis* (s. str.) probably all feed on viviparous aphidoids on herbaceous plants (including thistles, nettles and *Phragmites*), shrubs and trees. *Leucopis* (*Leucopomyia*) species have been reared from egg-masses of *Eriopeltis* scales. *Leucopis* (*Neoleucopis*) *obscura* and *Lipoleucopis praecox* attack Adelgidae on conifers. All of these belong to the subfamily Leucopinae; the remaining genera all belong to Chamaemyiinae and their prey, where known, are all mealybugs living in the leaf-sheaths of grasses and sedges. This way of life is undoubtedly the reason they are seldom encountered except as adults.

Smith (1963) keyed and described the adults and the known larvae of the species recorded as British at that date. More up-to-date accounts than Smith's of the British and Palaearctic species of *Chamaemyia* are, respectively, those of Collin (1966) and Tanasiychuk (1970, 1986). McAlpine (1960) published a key to the world genera, including *Acrometopia*, which was not included by Smith; its biology is still unknown.

Other predaceous larval Diptera

The subterranean larvae of *Thaumatomyia notata* and *Chloropisca glabra* are predaceous, unlike most members of their family, Chloropidae. They have been observed feeding on the lettuce root aphid, *Pemphigus bursarius*, by Dunn (1960). *Thaumatomyia notata* sometimes forms vast swarms in the autumn. Larvae of *Acletoxenus formosus* (Drosophilidae) prey on Aleyrodidae, including *Siphoninus phillyreae* and *Aleurotrachelus jelinekii*. Séguy (1934) included both chloropids in *Thaumatomyia*; this work also includes *Acletoxenus*. Disney (1983:6) in a summary of the feeding habits of the larvae of British Phoridae, cited *Megaselia rufa* as a predator of Coccoidea and their eggs and *Phora holosericea* as a predator of root aphids. Larvae of *Phaonia trimaculata* (Muscidae) have been found feeding on aphids at the roots of cabbages as well as on larvae of the cabbage root fly.

Predaceous adult Diptera

Predaceous adult flies of several families have been reported as taking hemipteran prey, though all were general predators, not specialists in taking bugs.

Laurence (1952) studied the insects carried as nuptial gifts by a swarm of male *Hilara litorea* (an empidid that normally takes small Diptera from the air) in the shade of a wood. He found only three specimens of dipteran flies being carried, together with one typhlocybine leafhopper, six Psyllidae of three different species and large numbers of alate aphids among which 26 species could be identified. Poulton (1906) reported typhlocybine cicadellids and a psyllid among the prey of two other species of Empididae in the genus *Hybos* and Prior (1972) recorded another empid, *Platypalpus notata*, preying on deltocephaline cicadellids of the genus *Macrosteles*.

Dioctria atricapilla (Asilidae) darts at its victims from a perch among grasses. It has been noted preying on the ground-dwelling lygaeid bug *Peritrechus geniculatus*. The similarly named *Machimus atricapillus*, in the same family, frequents woodland, where it hunts from tree-trunks or foliage. Several ground-dwelling Cicadellidae (species of *Euscelis* and *Aphrodes*) have been listed among its prey. The related *Neoitamus cyanurus* was reported by Poulton (1906) to prey on *Anthocoris nemorum*.

In the family Dolichopodidae, *Neurigona suturalis* has been known to capture a small cicadellid, *Typhlocyba cruenta*. Allen (1964) observed *Coenosia lineatipes* (Muscidae) preying on the same species.

Coleoptera

Both adult and larval ladybird beetles (Coccinellidae) are well known as major predators of aphids and other Sternorrhyncha. Some have been used in biological control of their prey. One species of Derodontidae has been introduced into Britain for the same purpose. One genus of the otherwise phytophagous family Anthribidae feeds on scale insects. Strepsiptera, which used to be considered as a separate order, are currently treated as a suborder of Coleoptera; three British species parasitize Auchenorrhyncha. Apart from these specialists, some beetles that are generalist predators are known to take Hemiptera and may have an important impact in some habitats.

Keys to the families of adult beetles (except for Derodontidae, which had not been introduced at that date) were given by Crowson (1956). Marshall, Jessop & Hammond (in prep.) provide keys to the families and some subfamilies of the larvae and summaries of the larval biology of each group.

Coccinellidae (Ladybird beetles)

Ladybirds lay their usually yellow or orange eggs singly or in small groups among colonies of all kinds of Sternorrhyncha. Their long-legged larvae can be distinguished from those of Neuroptera by their much shorter antennae and mouthparts.

There are four larval instars, the last of which usually pupates on the host-plant without making any special provision for concealment. Probably all stages, at least of the larger species, are distasteful to predators, making concealment unnecessary. The prey of coccinellids includes mites, the eggs of chrysomelid beetles and occasionally Auchenorrhyncha and Heteroptera, but these alternative foods are probably of only minor importance to all but a few species. Trials with larvae of of various aphidophagous species have revealed a complex mosaic of predator-prey relationships: aphids acceptable to larvae of one species may be toxic to another. Adult ladybirds are predaceous but also feed on honeydew, nectar and pollen. One British species is phytophagous throughout its life. There is usually a single generation per year but some aphidophagous species regularly have two. Adults overwinter, sometimes in huge aggregations, and sometimes large swarms of them are seen migrating. On occasions, countless thousands of their bodies have been washed up on the strandlines of lakes or on the seashore. The impact of ladybirds on their prey species may be considerable. A big larva may consume up to 100 aphids in a day and it is not uncommon for colonies to be completely wiped out.

The commoner species, especially the two-spot ladybird (*Adalia bipunctata*) and the seven-spot ladybird (*Coccinella septempunctata*), are known to be widely polyphagous both as adults and as larvae, feeding on Aphidoidea, Psylloidea, Coccoidea and mites on a variety of plants. Most of the large, colourful ladybirds that prey on Aphidoidea are known to take Psylloidea as well. Young cicadellids are sometimes eaten but the older nymphs and adults are too agile to be caught (Wratten, 1976). Two familiar conifer-haunting species, *Aphidecta obliterata* and *Anatis ocellata*, feed mainly on adelgids. *Chilocorus* and its relatives *Exochomus* and *Platynaspis* prey on Coccoidea but may also eat some Aphidoidea, especially the wax-producing kinds. Aleyrodidae rarely figure among the prey of British ladybirds but the little *Clitostethus arcuatus* specializes in them. It is found sometimes on ivy, where it feeds on *Siphoninus immaculatus*; *S. phillyreae* and *Aleyrodes proletella* are also recorded as prey. A related species, *Scymnus* (*Pullus*) *auritus*, specializes in attacking *Phylloxera glabra* on oak. Other *Scymnus* species prey on adelgids on conifers and yet others on scale insects.

The bionomics and ecology of ladybirds were reviewed by Hodek (1967). Wratten (1973) conducted a study of the impact of *Adalia bipunctata* on a population of the aphidoid *Eucallipterus tiliae* and Smith (summarized in Waloff, 1968) measured the feeding rates of its various instars on another species of aphid. Pope (1973) gave details of the prey and British distribution of *Scymnus* species. Keys to adults were provided by Pope (1953); van Emden (1949) keyed the larvae of some species.

Anthribidae

Anthribus species (often placed in the genus *Brachytarsus*) are predators of scale insects of the genera *Parthenolecanium* and *Eulecanium*, though a few records refer to *A. nebulosus* attacking *Physokermes*, *Quadraspidiotus*, *Asterolecanium*, *Kermococcus* and even the adelgid *Gilleteella cooleyi*. Up to half the scales in a colony may be attacked, though some eggs laid by the attacked scales may still survive to hatch. According to Clausen (1940), the female *Anthribus* feeds extensively on the scales and their eggs. In April she lays a single egg into each scale selected for oviposition, inserting her ovipositor through the wound caused by feeding to reach the egg chamber. The plump, curled larva has a well developed head and thirteen body segments. Each thoracic segment bears a pair of fleshy tubercles in place of the jointed limbs of most beetle larvae. The larva feeds on eggs and the body of the scale and eventually pupates beneath the remains of its dead host. Adults of the new generation emerge in June or July, biting large, circular exit holes in the dorsum of the scales. There is only one generation of beetles a year. They overwinter in crevices in the bark of trees, especially of conifers. Morris (1990) deals with the two British species.

Derodontidae

This family of beetles is not native to Britain. One species, *Laricobius erichsoni*, has been introduced in an attempt to control adelgids on conifers, principally *Adelges piceae*. An introduction into Kent failed but by 1982 the beetle was widely established on conifers in Suffolk, apparently having arrived there independently (Hammond & Barham, 1982). Early larvae of *L. erichsoni* feed only on the eggs of the adelgids but later instars consume the active stages as well. The adult beetles consume adelgids and their eggs and also fungal hyphae and spores.

Other predaceous beetles

The big, predaceous beetle families Carabidae and Staphylinidae may have considerable impact on populations of Hemiptera living at or near ground level or in the upper layers of the soil. Aphids, shieldbugs and grassland Auchenorrhyncha have all been recorded as prey of Carabidae. Adults of a few species of this family (e.g. some *Dromius*) ascend trees and may include arboreal Hemiptera in their diet. In cereal crops, various Carabidae, especially *Agonum dorsale* and *Demetrias atricapillus*, take substantial numbers of aphids (Sunderland & Vickerman, 1980). Staphylinidae are known to prey on on aphids, psyllids, mealybugs, armoured scales, shieldbugs and Auchenorrhyncha but only a few species are involved and, as with Carabidae, none is believed to be solely dependent on Hemiptera.

Adult Cantharidae inhabit tall, herbaceous plants, shrubs and trees, where they prey on a wide range of soft-bodied insects, among them aphids and mirids. Schmutterer (1952c) noted adults of two Cantharidae, *Cantharis obscura* and *C. fusca*, and an elaterid, *Prosternon tessellatum*, feeding on the honeydew, eggs and adults of some scale insects.

The main predaceous waterbeetle family is Dytiscidae. Both adults and larvae are known to feed on aquatic Heteroptera; the larvae may themselves fall victim to Notonectidae.

In the smaller beetle families a few instances are known of some degree of dependence on Hemiptera. Larvae of two Phalacridae, *Olibrus affinis* and *Stilbus atomarius*, feed at least sometimes on aphidoids; and adults of *Enicmus minutus* (Lathridiidae) have been found feeding on the beech scale, *Cryptococcus fagi*. Both adults and larvae of the genus *Cybocephalus* (Nitidulidae) live as predators in colonies of scale insects and, sometimes, of whiteflies in many parts of the world, including Europe, but the genus has not yet been found in the British Isles. *Carpophilus mutilatus*, which does occur in Britain and which belongs to the same family, has been reported feeding on aphids in the West Indies. Some species of Melyridae, like those of the common British genus *Malachius*, are general predators that probably sometimes take Hemiptera. In North America, some species of the related genus *Collops* are known to feed voraciously on many different kinds of aphids and on other soft-bodied Hemiptera (including immature leafhoppers and Heteroptera) on herbaceous plants and also on the eggs of some Pentatomidae and Lygaeidae.

Joy (1932) provided identification keys to adults of all British Coleoptera; Balfour-Browne (1953), R. T. Thompson (1958) and Lindroth (1974) supplied more modern keys to Dytiscidae, Phalacridae and Carabidae respectively. Van Emden (1942) keyed many genera of world carabid larvae.

Strepsiptera

Strepsiptera are probably best known to British naturalists through the familiar textbook example of *Stylops*, a parasite of solitary bees. Indeed, the term 'stylopization', which was originally coined for the condition of infertility and genital malformation induced by this genus in its hymenopteran hosts, is generally applied to the parasitization of any host by any strepsipteran. The modern consensus on the systematic position of these strange little parasites is that they constitute a group of uncertain affinities within the Coleoptera. Three species attack Hemiptera in Britain.

Elenchus tenuicornis can probably utilize any species of Delphacidae as a host. Its life history in Germany has been worked out by Baumert (1958, 1959); his study was more detailed than Hassan's work on a British population of the same species. The sac-like female of *Elenchus* remains all her life embedded in the body of the planthopper host with only her head and a few associated structures (the button-like 'cephalotheca') protruding from the side of the host's abdomen. Each female produces 1450 to 1500 oval eggs about 0.07 mm long. These hatch within her body and the first-instar larvae, called 'triungulins', escape via three genital openings on their mother's sterna II-IV into the brood pouch, a cavity between the female's body and her puparium.

The body of the triungulin larva is about 0.18 mm long and is shaped rather like a silverfish (*Lepisma*), with a large head bearing two large, dark eye-patches with a few scattered ocelli on each and a pair of short, bristle-like antennae; three pairs of well developed, single-clawed thoracic legs; and a ten-segmented, bristly abdomen whose tenth segment bears two very stout bristles about one-third as long as the body. The triungulins crawl out of the brood pouch, which opens to the outside world via a single, large pore below the female's mouth, emerging onto the body of the planthopper, and embark upon the hazardous business of finding new, nymphal hosts for themselves. Their free-living existence lasts only a short while, from a few hours to a few days, and many must perish without finding a suitable host. Those triungulins that are successful in their search bore into the host's body and begin to feed on its haemolymph. The fully fed triungulin becomes plump and curled, assuming a 'scarabaeidoid' shape (Hassan, 1939). The second and succeeding larval instars are legless, glabrous and have a less well differentiated head than the triungulin. They live in the host's haemocoel, in which they are capable of wandering about, drinking the haemolymph. By the fourth instar, the sexes can be distinguished morphologically.

The late fifth-instar larva bites a hole in the side of the host's abdomen through which the cephalotheca protrudes. The cut edges of the hole rapidly heal, sealing the cephalotheca in position. In a few hours the pupal moult takes place. It is succeeded very rapidly by the adult moult in the female but the male does not emerge from his puparium for one to three weeks. Eruption of the female larva occurs only in adult hosts but male puparia typically are formed in nymphs, particularly of overwintered hosts, though in some species in summer they develop in adults. Males emerge from their puparia in the late morning or afternoon and live for only a few hours or, if unmated, for up to three and a half days. They seek out adult hosts containing mature females and copulate once, very briefly; the whole process of copulation occupies from one to eleven seconds, after which the male soon dies.

In southern England, Japan and the Berlin area there are two generations annually but in Finland only one. In England, triungulins of the summer generation are liberated in late July and those of the overwintering generation in early October. The species overwinters as third or fourth instar larvae within nymphal hosts. Although numerous triungulins may penetrate a single host it is unusual for two larvae to survive to maturity and only rarely have three or more puparia been found in one planthopper. Nymphs containing male puparia invariably fail to survive to adulthood but if eruption does not occur until after the final moult the hoppers may live for some time. Hosts containing females live for several weeks, during which time the triungulins are slowly released. Effects of stylopization on the host vary from greatly reduced fertility to, most usually, total sterility involving suppression of production of gametes and abnormalities in the shape of the genitalia (Baumert & Behrisch, 1960a,b). Hassan (1939) found *Elenchus* to be the most frequent parasite of Delphacidae; the incidence of parasitism in the nine species he studied varied from 1.5 to 28.3 per cent.

Halictophagus silwoodensis seems to be a monophagous parasite of *Ulopa reticulata* in southern England. Waloff (1981) published an account of its morphology and life history, with figures of all stages. Its biology is similar in most respects to that of *Elenchus*. It is univoltine with a partial second generation in mild autumns. Triungulins of the main generation are liberated in July and August and penetrate hosts of any instar, including adults. There are six larval instars; the sexes are recognizable from the third instar onwards. Late sixth instars erupt through the abdominal wall of adult hosts only. Males pass through a pupal instar but female sixth instar larvae moult directly to adults. Waloff found rates of parasitism of the host ranging from 0 to 38 per cent in nymphs and to 65 per cent in adults, averaging about 14 and 29 per cent respectively.

Halictophagus curtisi was described from a male caught in Britain. It appears to be rare and its host is unknown. Crowson (1976) found *Eupelix cuspidata* to be stylopized in south-western Scotland and southern England and conjectured that the parasite might be *H. curtisi*.

Triungulins of Strepsiptera may enter insects only distantly related to their normal hosts, as in the case of an aphid mentioned by Prior (1976). They invariably fail to develop even as far as the second instar and their presence cannot therefore be taken as establishing a true host record. However, several species are known to develop in Heteroptera (Pentatomoidea and Coreoidea) in the tropics.

Neuroptera

Larvae of the familiar green lacewings (Chrysopidae) feed mainly on Aphidoidea and Pseudococcidae. Some other families of Coccoidea, Aleyrodidae and a few Psylloidea are also attacked. Very occasionally, probably in the absence of their preferred prey, they have been known to eat other Hemiptera and even insects of other orders. Brown lacewings (Hemerobiidae) have similar preferences except that they show little liking for Aleyrodidae or Diaspididae. Larvae of powdery lacewings (Coniopterygidae) feed on the more sedentary or immobile Sternorrhyncha: adelgids, *Phylloxera*, Coccidae (s. str.) and Diaspididae have all been recorded as prey, but Herting's (1978) catalogue contains no record of predation by them on Aphidoidea. Mites also feature prominently in the diets of Coniopterygidae.

Eggs of Chrysopidae are shaped like short sausages and are carried at the tips of hair-like stalks several times as long as the eggs themselves. Some species lay them singly; others lay them in groups, sometimes with the stalks twisted together. The unstalked eggs of the other two families lie on their sides. The larvae are long-legged and mobile, like those of ladybirds, from which they differ in having antennae much longer than the head and, usually, long, slender, curved, forwardly-directed mandibles. The small, pyriform larvae of powdery lacewings are unusual in having short mandibles, conspicuous, club-shaped labial palps and long hairs on the antennae. Some chrysopid larvae cover their bodies with debris. Each tarsus bears a single, trumpet-shaped arolium between the claws in chrysopids and first-instar hemerobiids; later instars of hemerobiids have much reduced, pad-like arolia. There are three larval instars (four reported in a coniopterygid). Pupation takes place in a silken cocoon and the overwintering stage is usually the mature, diapausing larva within this cocoon. *Chrysopa carnea* is unusual in overwintering as an adult. There are typically two generations in the course of a year but only one in some Chrysopidae and three or more in Coniopterygidae.

The potential of Chrysopidae and Hemerobiidae as agents of biological control was reviewed by New (1975). Killington's (1936) monograph provides useful biological information. Adult lacewings may be identified with Fraser's (1959) handbook, where some larvae are also figured.

Other insect predators

Almost any terrestrial or aquatic predator or omnivore, unless it is particularly highly specialized in its food or habitat preferences, is likely to feed on some kind of hemipteran at some time.

Rothschild (1966) found that a springtail identified as *Sminthurus* sp. (Collembola) preyed on small nymphs of a delphacid, *Conomelus anceps*, in the laboratory and in the field.

Staddon & Griffiths (1968) identified remains of three species of Corixidae and a notonectid among meals of a dragonfly nymph, *Aeshna juncea*.

In a saltmarsh studied by Payne (1973) the bush cricket *Conocephalus dorsalis* fed mainly on the flower spikes of a grass but also on the cicadellid *Euscelis obsoletus*.

Earwigs, *Forficula auricularia* among others, are often found sheltering in leaves rolled and crumpled by aphids but no longer tenanted by them; presumably the earwigs have devoured the former occupants. *Forficula auricularia* was shown to be a major predator of cereal aphids by Sunderland & Vickerman (1980).

Thysanoptera are mainly fungivores or plant-feeders but a few are predaceous. *Aeolothrips intermedius* has been recorded as a predator of a cicadellid and a delphacid in Europe and in some other parts of the world Thysanoptera play a part in controlling populations of various Sternorrhyncha. Palmer (1986) found an abundance of *Haplothrips subtilissimus* on oak and stated that it was believed to prey on scale insects and other arthropods.

Larvae of snakeflies (Raphidioptera: Raphidiidae) have been recorded as predators of Coccoidea, Adelgidae and Aphidoidea.

Vertebrates

Jansson (1986) cited predation by fish as a major determinant of habitat preferences of Corixidae and it is probably also responsible for maintaining the correlation of the different colour forms of some corixids with the colour of the bottom sediments of the ponds where they occur. No doubt amphibians, reptiles and small mammals also feed on Hemiptera if they have the opportunity.

Birds undoubtedly have the greatest impact of any vertebrate group on terrestrial Hemiptera, yet few quantitative studies have been made. The gross effect of predation by birds can be demonstrated easily by caging some of the insects' host-plants with netting of a mesh size that will exclude birds but permit the passage of insects. By comparing the fate of aphid populations on broom bushes inside such cages with that of populations on uncaged plants, Smith (summarized in Waloff, 1968) was able to demonstrate that at times of peak aphid populations birds were more important predators than all the invertebrates together. Blue tit and chaffinch were the main insectivorous birds in the study area; whitethroat and hedge sparrow were judged to be less important. Blue tits and other titmice are known to feed extensively on scale insects, especially in winter.

6

DEFENCE

Almost every aspect of an insect's physical form, physiology and behaviour has some bearing on its susceptibility to or immunity from attack by natural enemies. Mechanical protection may be afforded by a thick body wall, as in Pentatomidae and Lygaeidae, for example. Adults of some of the larger Hemiptera, especially Pentatomidae, Coreidae and Membracidae, have spine-like lateral projections of the pronotum that may injure birds that attempt to eat them or may make them too wide to be swallowed whole. Even if the spines can eventually be broken off, they may so increase the time taken to eat the bug that the bird finds it more economical to pursue alternative prey. Pentatomids and some of the water-surface bugs are compactly built, with few chinks in their armour that might enable a predator that cannot swallow them whole to tear them apart. Many Coccoidea, being slow-moving or immobile for much of their lives, rely for protection on a coating of wax which may assume various forms: powdery, filamentous, plate-like and sheet-like waxy coatings are all known. Diaspid scales make their protective roofs from a silk-like secretion rather than a waxy one. Size is another factor in defence. Some insects may be so small that they are overlooked by large predators while others may be too large for small predators to tackle.

The size ratio between predator and prey may determine the efficacy of defence mechanisms. Evans (1976) found that large aphids kicked out at small anthocorids that approached them but small aphids had a better chance of escape from large anthocorids if they dropped from their host plants. At intermediate size ratios the aphids attempted to flee on foot.

Behavioural responses to predators, besides running away, dropping from the plant and kicking, include leaping, characteristic of most Auchenorrhyncha and adult Psylloidea. Escape by leaping is often continued in a short flight, putting extra distance between prey and predator. Flight alone is not a common means of escape except in adult whiteflies and Miridae. The ability to shed limbs when seized by a predator is developed to a high degree in mirids, which are often encountered with less than the full complement of legs. Heteroptera of the predaceous families Reduviidae, Notonectidae and Naucoridae defend themselves against humans who try to handle them and, presumably, against other enemies, by biting painfully, injecting toxic saliva into the wound. Even some aphids, including the British species, *Pemphigus spyrothecae* (family Pemphigidae), attack intruders into their colonies by biting and kicking.

Chemical secretions may function to protect insects in several ways. Gregarious Heteroptera, seeking safety in numbers, secrete a pheromone that serves to attract others of their kind. When attacked, they emit a repellent fluid that both deters predators and causes aggregations of bugs to disperse. The acrid secretion of the metathoracic glands of the adults can be squirted out in a jet, at least in some species. It probably functions mainly against invertebrate enemies. Schmuck (1987) found that the first two dorsal abdominal glands of nymphs of the pyrrhocorid *Pyrrhocoris apterus* secreted a substance that attracted other members of the species while the third produced a noxious substance similar to that found in the metathoracic gland of the adult. The whole topic of the chemical attractants and defences of Heteroptera was reviewed by Aldrich (1988). Aphids, too, produce 'alarm pheromones' which, when secreted from the siphunculi of one of their number under attack by natural enemies, cause their colonies to disperse. A few brightly coloured Heteroptera are believed to be distasteful to vertebrate predators that hunt by sight which, in general, means birds. The clustering behaviour of *Pyrrhocoris apterus* lends support to this theory. The bright colours of the bugs serve as a means of easy identification for birds that have experienced their unpleasant flavour so that, when one bug has been attacked and found to taste unpleasant, others in the vicinity will be left alone. Shieldbugs of the genus *Eurydema* also have warning coloration (largely bright red or metallic green), suggesting that they may pick up a disagreeable flavour from feeding on their usual host plants in the family that includes Cabbage, Mustard and Hedge Garlic. Alternatively, they may be able to synthesize mustard oil glycosides for their own protection and so possess the metabolic mechanism necessary to deal with these same compounds in their similarly protected food plants.

Most bugs are coloured similarly to the background on which they normally rest. Rose aphids, for example, are green or pink and many grassland cicadellids and mirids are grass-green or straw-coloured. Some Corixidae occur in pale and dark forms and tend to resemble the colours of the substrates at the bottom of the waters in which they live, particularly if fish are present. Pondskaters and their relatives are strongly countershaded, appearing dark brown or black when viewed from above the dark waters on which they live but silvery when viewed from below against the sky. Inconspicuous coloration is a simple case of the phenomenon called crypsis. A more elaborate kind of crypsis is found in the water scorpions *Nepa* and *Ranatra*, where the body shape, as well as the coloration, is involved in giving these bugs the appearance of a leaf and a stick respectively. Perhaps this resemblance to inanimate objects also helps them to approach potential prey undetected. Among the cicadellids, *Ledra* provides another example of an elaborate form of crypsis, with its resemblance to the broken base of a twig.

Cercopidae (froghoppers) have some of the most elaborate defences of any Hemiptera. Their nymphs are protected from all but a few enemies by the froth they secrete around themselves, while adults of *Aphrophora*, *Philaenus* and *Cercopis* show three different uses of colour for defensive purposes. The grey-brown bodies of *Aphrophora alni* and some of the forms of *Philaenus leucophthalmus* incorporate pale wedges that serve to break up their outlines and render them less conspicuous. Such patterns are classified as disruptive coloration. *Philaenus leucophthalmus* adults are long-lived and it would be relatively easy and economically effective in the use of searching time for a bird to specialize in searching for this species if all individuals looked alike. However, adult *P. leucophthalmus* exist in several strikingly different colour patterns, which make it difficult for an insectivorous bird to form a 'searching image' of the species as a whole. The colour polymorphism is probably kept stable by predators preferentially taking the commonest morphs available at any time. Despite the substantial differences in appearance between the different forms, all of them are cryptic to some extent. In areas where the vegetation is blackened by industrial pollution the darker forms predominate although they form only a relatively small proportion of the population in unpolluted places. Unlike *Philaenus*, adults of *Cercopis* live for only a few weeks. With rare exceptions they are strikingly coloured, with oval red spots on a black background. It has been suggested (Cloudsley-Thompson & Sankey, 1958) that they are avoided by birds because of the strong resemblance they bear to Burnet moths, which contain poisonous cyanogens (precursors of cyanides), although the froghoppers are not themselves poisonous or even distasteful.

Defensive mimicry, involving the protective resemblance of an edible organism to a distasteful or aggressive one (as in the supposed case of *Cercopis* and the Burnet moths), is widespread in insects. Mimicry depends for its existence on the presence of an easily recognised organism, the model, and another that is able to perceive and respond to specific signals (such as shapes and colours). In the case of protective mimicry, the mimic is usually less abundant than the model as the former would not gain protection from its resemblance to the latter unless the predators involved quickly learnt to avoid animals of that general form. Ants are almost ubiquitous and shunned by many insectivores, so it is not surprising that many insects, particularly those that normally walk rather than fly, like the Hemiptera, have come to resemble them. Miridae of the tree-dwelling genus *Pilophorus* prey on aphids that are often attended by ants. The mirids avoid the ants by their alertness and agility and so coexist with them. Both as nymphs and as adults they bear a resemblance to ants that is believed to afford them protection against other predators. Adults of two other mirid genera, *Systellonotus* and *Hallodapus*, and nymphs of the alydid *Alydus* and the nabid *Aptus* all run on the ground and resemble ants in various ways though only the mirids have a close association with ants.

7
BIORHYTHMS

Four regular environmental rhythms affect living organisms: the annual, monthly (lunar), daily and tidal cycles. Tidal cycles are relevant only to the few Hemiptera that live on the sea shore and the influence of the lunar cycle, which strongly affects the nocturnal flight activity of many tropical bugs, is scarcely detectable in those of temperate regions. It is the daily and annual cycles that most profoundly affect the latter.

Daily fluctuations occur most noticeably in light intensity but temperature also varies on a daily basis, albeit in a less regular way. Even on an overcast day, the minimum temperature is likely to occur around dawn and the maximum in the late afternoon. Wooded habitats are rather equable with respect to temperature, both on the shady ground and in the well ventilated canopy, cooled by transpiration in the heat of the day. The most extreme variations occur on the surface of dry, bare or thinly vegetated ground, which may easily register a temperature range of 30 degrees on a cloudless day. The microclimate of the litter layer of dry grassland is surprisingly mild in winter, due partly to the ease with which the upper layers of dead vegetation warm up in the sunshine and partly to the insulating properties of the air trapped in it. Many Hemiptera become torpid at low temperatures and can operate efficiently only in the warmth of the daytime. *Alydus calcaratus* (Alydidae) is an outstanding example of such dependence on temperature, flying only in warm sunshine. On dull days it can be found only by carefully searching its host plants. Predators that hunt by sight, like Saldidae, are naturally dependent on daylight to see their prey and cease hunting at night. On the other hand, sap-sucking insects are largely unaffected by the alternation of day and night. It is known that aphids, for example, feed at a constant rate (given uniform temperature conditions) regardless of the time of day. They are clearly not unaware of the alternation of light and dark, since winged aphids fly only at certain times of day.

The lives of insects in temperate regions are dominated by the annual cycle. Almost all of them are able to survive the rigours of winter in only one stage of the life cycle and they have a variety of mechanisms that ensure that the majority of individuals are in the right stage as winter approaches.

Because insects are 'cold blooded', the metabolic processes that govern their growth, maturation, digestion and so forth are largely dependent on the ambient temperature. Mid-winter temperatures are often too low to allow development to proceed at all, but the unpredictable nature of British weather is such that occasional warm spells, particularly in early winter, might easily lead to precocious development unless some mechanism existed to prevent it. Usually, the overwintering stage enters a special physiological state of arrested development, termed diapause. In those species that overwinter in the egg state, diapause involves a suspension of embryonic development. In those overwintering as adults, it is usually sexual maturation that is delayed, the gonads being incapable of producing gametes until diapause is broken. A few instances are known in the families Miridae, Cicadellidae, Aleyrodidae and Coccidae where sexual maturation and pairing occur in the autumn and only the mated females survive the winter, with the egg rudiments not developing fully until the spring. Nymphal diapause is commonplace in Fulgoroidea and occurs sporadically in other groups, including some Coccoidea and three genera of Heteroptera: *Micronecta* (Corixidae), *Pentatoma* (Pentatomidae) and *Odontoscelis*

(Scutelleridae). The puparia of those Aleyrodidae that live on deciduous trees overwinter on the fallen leaves but winter diapause affects the younger nymphal stages of their relatives that live on the leaves of evergreens.

Diapause is often obligatory and must be entered by the insect at the appropriate stage of its development regardless of the environmental conditions. In many instances, however, it is induced by the conditions experienced by the insect as it develops, or by its mother as she develops. In those aphids in which the factors controlling the production of sexuales have been studied, it has been found that the shortening of autumn days is the determining factor. Sexuales can only produce winter eggs, which have to undergo diapause. In an unpredictable climate, daylength is a much more reliable indicator of the approach of winter than is temperature. Aphids are unusual in going through many generations in the course of the year. Most insects of cool temperate regions are univoltine (that is, they pass through a single generation annually) or bivoltine (two generations). Parker (1975) studied populations of the same species of Anthocoridae in northern and southern Britain and found that the northern population was univoltine and the southern one bivoltine. He showed that the differences between the populations were genetically determined and were not simply physiological responses to the different length of the growing season in the areas they inhabited. Samples from the northern population brought south and reared alongside the southern ones entered obligatory diapause after one generation while the southern population embarked on a second generation as normal without entering diapause. In a species of Miridae studied by Dolling (1973), the daylength experienced by the late nymphs determined whether they would diapause as adults or not. By manipulating the daylength it was possible to rear successive generations of either form as well as alternate diapausing and non-diapausing forms. Such environmental control of diapause makes it possible for the species to respond appropriately to different environmental conditions. In a cold year, development may proceed so slowly that the late nymphs of the first generation experience the short days of autumn and thus become diapausing adults without risking a second generation before the winter. In a good year, however, they will experience the long days of high summer and will produce a second and perhaps even a third generation.

Diapause may be ended by any of a number of factors. In some cases, it may simply be that a period of dormancy must elapse before the interrupted developmental processes are resumed. Alternatively, a cold spell followed by a warmer one may be the necessary stimulus. Possibly, eggs embedded in living plant tissues respond to some chemical present in the sap or to the increased mechanical pressure as turgor increases with the spring sap flow. Daylength is also a factor. In the case of the mirid mentioned above, adults remained in diapause if the daylength was kept the same as that under which they were reared but came out of it if the daylength was increased. More interestingly, if they were exposed to the very short daylengths typical of midwinter and then to only slightly longer daylengths, less than those under under which they had been reared, they also came out of diapause. They must, therefore, have been responding to perceived increases in the hours of daylight rather than to the actual daylength.

It should be remembered that diapause and overwintering are not synonymous and that an overwintering insect may not be in diapause at all. This seems to be the case with the soil-dwelling nymphs of the froghopper *Cercopis vulneraria*, which grow steadily throughout the winter. Conversely, diapause is not only a winter phenomenon. Another froghopper, *Philaenus leucophthalmus*, has a winter egg diapause and also a summer reproductive diapause. The adults often appear by midsummer but they do not become fertile until the early autumn. Adults of some tree-dwelling aphids have a summer diapause. Aestivation (the summer equivalent of hibernation) seems to be a response to the reduced quality of the sap in summer. It is usually the parthenogenetic adult females that aestivate, resting quietly beneath the leaves and feeding but not reproducing, but in the genus *Periphyllus* (Chaitophoridae) specialized resting nymphs, called aestivales or dimorphs, occur.

8

DISPERSAL

A population of insects going about their everyday lives will tend to disperse more or less evenly through an area of suitable habitat. There will be no pressure on them to leave the area and travel across inhospitable terrain unless conditions at home become unsuitable for one reason or another. The most frequent of these reasons are the non-availability of suitable food and seasonal changes in the environment. Dispersal behaviour may occur in response to the conditions prevailing at the moment or spontaneously, triggered either by an internal physiological mechanism or by environmental cues in advance of actual need.

Whatever the cause, dispersal exposes the insects to a range of physical hazards and predators not encountered in the original habitat. For this reason, physical inability to fly away from a deteriorating habitat may actually favour an insect's chances of survival. Hence the high incidence of brachyptery among ground-dwelling Delphacidae, Cicadellidae and Lygaeidae that must historically have been limited to small islands of grassland and herbaceous plants amid a sea of uninhabitable forest. Suitable habitats for aquatic Hemiptera are also few and far between and, as might be expected, brachyptery is frequent among them as well. Tree-dwellers, on the other hand, would normally have had forests all around and thus little chance of being lost amid inhospitable terrain. Moreover, most insects living on deciduous trees, unless they overwinter as eggs, need to seek shelter on the ground or among evergreen plants in the winter. For these reasons, most tree-dwelling bugs are fully able to fly. Coccids are the main exception. Tree-dwelling species of this group maintain perennial colonies, overwintering on the bare twigs or stems, and rely for dispersal on chance transfer of the active first-instar larvae. Most normally flightless species (but not bedbugs or female coccids) produce at least a few fully winged individuals that are able to take on a pioneering role in searching for new habitats. A complex mixture of genetic and environmental factors determines the proportion of fully winged individuals in populations of such species.

Dispersal may be a routine part of a bug's life-cycle, happening in the absence of population pressure or habitat deterioration. Young (1966) found that in several species of Corixidae there was an obligatory dispersal flight in early spring before the gonads matured. Even flightless individuals rose to the surface of the water and seemed to be trying to take off at this time. At other times of year the corixids took to the wing only in response to deterioration in their living conditions. The non-overwintering generations of these bivoltine bugs did not have an obligatory pre-maturation dispersal flight. A similar difference between summer and winter generations was shown by Iheagwam (1977) in *Aleyrodes proletella*. The summer morph of this whitefly took off only when disturbed and always flew downward, away from the light, settling immediately. The overwintering morph, on the other hand, would start to fly spontaneously and climbed upwards, towards the light, usually remaining airborne for 1–9 minutes but sometimes much longer, up to two hours or more. The ability to fly does not seem to be restricted to any particular phase of adult life. Female Cicadellidae and Delphacidae are able to fly at both pre- and post-maturation age, up to the time of death.

The ability to remain aloft for long periods can lead to very long-distance dispersal that might properly be termed migration. Cheng & Birch (1978) sampled the insects floating on the surface of the sea in the English Channel and Hardy & Cheng (1986) netted those flying over the central North Sea. These two samples contained one or two representatives each of the families Corixidae, Lygaeidae, Tingidae, Miridae, Nabidae, Piesmidae, Cicadellidae and Delphacidae, dozens of Psylloidea and hundreds of Aphidoidea. These figures speak for themselves.

Many Aphidoidea have life-cycles that exploit highly seasonal sap-flows. Several totally flightless generations may occur in quick succesion until, in response to overcrowding, a decline in the quality of the sap or simply the length of time since the colony was founded, winged emigrants are produced. Local climate and weather conditions combine to synchronise emigration by these individuals into massed flights.

Although flight is the prevalent means of dispersal for most Hemiptera a few rely on other methods. Freshwater species must sometimes be dispersed by downstream drift or by adherence of their eggs to waterfowl. Pondskaters may be swept across the whole width even of large lakes by the wind as they rest on the surface. An even more unusual mode of dispersal was encountered by Foster (1975) in his study of the saltmarsh aphid *Pemphigus trehernei*. He was unable to find evidence for dispersal by flight in this aphid, which produced only a small number of winged individuals that were often crippled by immersion in seawater. Dispersal over longer distances than could be covered by walking was effected by tidal action, with the first-instar nymphs floating on the surface of the sea.

Plate 3

Galls. **1**, Pocket-galls of *Trichochermes walkeri* (Triozidae) on edges of leaves of purging buckthorn. **2**, Chambered, cone-like gall of *Adelges* sp. (Adelgidae) distorting shoot of spruce. **3**, *Pemphigus bursarius* (Pemphigidae) gall on petiole of poplar leaf. **4**, *Pemphigus spyrothecae* (Pemphigidae) gall on petiole of poplar leaf.

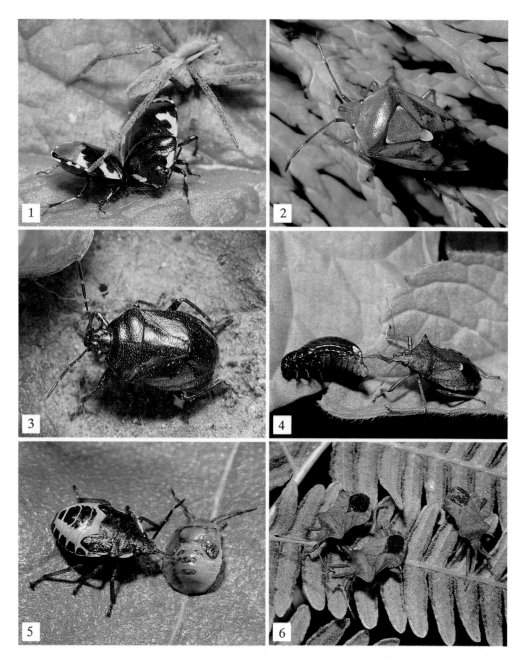

Plate 4

Heteroptera. **1**, *Sehirus bicolor* (Cydnidae) courting. The smaller male is transmitting the vibration of his stridulation to the female by bodily contact. A passing spider, *Pisaura mirabilis*, ignores the unpalatable bugs. **2**, *Elasmostethus tristriatus* (Acanthosomatidae) on cypress. 3–5, Asopine Pentatomidae. **3**, *Zicrona caerulea* among leaf-litter on the ground. **4**, *Picromerus bidens* preying on a leaf-beetle grub. **5**, *Troilus luridus* nymph preying on nymph of *Elasmostethus interstinctus* on birch. **6**, Group of dock bugs, *Coreus marginatus* (Coreidae), on bracken frond in Autumn.

9

DISTRIBUTION

The British Isles lie wholly within the geographical unit known to biogeographers as the Palaearctic Region. This huge area embraces Europe, northern Africa and the temperate and colder parts of Asia. Few species or even genera of Hemiptera are shared between the Palaearctic and the neighbouring but mainly tropical Afrotropical (subsaharan African) or Oriental (southern Asian) Region. Many genera and some species, however, have natural ranges that include at least the cooler parts of the Palaearctic and climatically similar areas of the Nearctic Region, which comprises Canada, the U.S.A. and most of Mexico. Human activity, particularly the horticultural trade, has further increased the faunal resemblance of these two regions through an interchange of species between Europe and North America that shows no sign of slowing down despite modern plant quarantine procedures.

Factors affecting distribution

The distribution of living organisms is strongly influenced both by climate and by vegetation, which itself has reciprocal influences on climate and soil. Vegetation types all over the world were described by Eyre (1963). He characterized the original vegetation of most of the British Isles, before the advent of agriculture, as 'deciduous summer forest'. This formation extended virtually unbroken eastwards across Europe, including the extreme south of Sweden, to Poland and southwards to the Alps and Pyrenees. Beyond these mountains, the same type of forest was found in northern Spain, northern Italy and parts of south-eastern Europe as well as in the central part of European Russia. Not surprisingly, many Hemiptera are found throughout these areas or in large parts of them, despite the destruction of much of the original vegetation by human activity. Mixed deciduous and coniferous forest, now greatly reduced in extent, occured in the highlands and islands of Scotland, where the only native conifers are scots pine and juniper. Remnants of this formation, termed 'mixed boreal and deciduous forest' by Eyre, are also found on the south coast of Norway, in southern Sweden (apart from the extreme south), in northern Poland and the Baltic States of the U.S.S.R. and thence eastwards to the vicinity of Moscow, with outliers in the Alpine and Carpathian mountain systems. These forests have a distinctive hemipteran fauna, particulary characterized by species associated with the coniferous trees. The central highlands of Scotland retain an isolated fragment of the high-altitude and high-latitude tundra and alpine vegetation system. In the western areas of Ireland and Scotland blanket-bog alternates with deciduous summer forest and mixed boreal and deciduous forest respectively.

Southwood (1957) presented a classification of distribution types of British Heteroptera, based on an earlier scheme by Deville. He recognises five main types: arctic-alpine (associated with high-altitude and high-latitude vegetation), Siberian (mixed forest, moorland and bog), European (deciduous summer forest), Mediterranean and Atlantic. Species of the Mediterranean element are most at home among the drought-resistant evergreen vegetation of southern Europe and North Africa but can survive also in dry, sunny spots in more northerly latitudes. Typical habitats are heaths, downs, breckland

and coastal areas, especially dunes. Ground-dwelling insects like Lygaeidae and Coreidae are well represented in this element of the fauna. The Atlantic type of distribution is characteristic of species that can tolerate cool summers and mild, damp winters and is, perhaps, best regarded as a subordinate category of the European type.

The highland areas of Britain are situated chiefly in the north and west. They are characterized by lower temperatures and higher rainfall than the lowlands. Proximity to the ocean has the effect of evening out the annual extremes of temperature so that western winters are milder than eastern ones. Large chalk and limestone masses have a similar effect on the local climate. It is difficult to disentangle the separate and inter-related effects that these factors have on the composition of the fauna. Undoubtedly, there is a strong correlation between the average annual temperature of a locality and the number of species of insects to be found there. The mild winters of the west seem to be less important than the warm summers of the south and east. Many species found quite widely in England to the south of a line from the Bristol Channel to the Wash are scarce or absent to the north of it.

The distribution of Heteroptera is better known than that of the other major groups of Hemiptera and some fairly reliable comparisons can be drawn between the heteropteran faunas of different parts of the British Isles. There are very few species breeding in Ireland or Scotland that are not also native to England. For Ireland, these probably amount to no more than three: the corixid *Sigara fallenoidea*, which is not found in Great Britain but occurs from northern Scandinavia across Siberia and in North America; the large pondskater *Limnoporus rufoscutellatus*, widespread in Europe but unaccountably absent from Britain except as an occasional migrant; and the mirid *Notostira erratica*, which was once established in the Dublin area, possibly as the result of a chance introduction, and may no longer persist there. Scotland's list of specialities is not much longer: one aradid, one lygaeid and three mirids, associated with the mixed boreal and deciduous forest, and a corixid. About 50 species of Heteroptera have a northern or north-western distribution in Britain or at least are less common in the southern and eastern lowlands. Of these 22 are Miridae, mainly associated with the wet, upland heaths or the remnants of mixed boreal and deciduous forest, 10 are Corixidae and 7 are Saldidae. The bleak countryside of north-western Scotland is very poor in species. Only about 150 of the 520 British Heteroptera occur north of the Great Glen (the diagonal gash across Scotland that includes Loch Ness) and only 30 or so of these belong to the northern or north-western element; the rest are hardy species, widespread in the British Isles but more abundant towards the south.

The aquatic, semiaquatic and moisture-loving Nepomorpha, Gerromorpha, Leptopodomorpha and Dipsocoromorpha show little preference for the south. If anything, a slight preference for the cooler but wetter north can be discerned. Of the 86 British species in these groups, 68 are reported from Scotland, 61 from Ireland and only 60 from England's most south-easterly and best worked county, Kent. North of the Great Glen, these groups account for about one-third of the total heteropteran fauna but in Kent they amount only to one-seventh. The warmth-loving and predominantly tropical groups Pentatomoidea, Coreoidea, Tingoidea and Reduvioidea have 89 British species in all, of which only 30 are Irish and 24 Scottish, with only 10 of them crossing the Great Glen. Lygaeidae and Berytidae are only slightly less sensitive to climate but Miridae are more tolerant, with 105 of the 200 British species in Ireland, 123 in Scotland and 63 reported from beyond the Glen.

The effect of one animal's presence on the distribution of another is more difficult to demonstrate than that of climate or vegetation but a few instances are known where competitive displacement seems to be occurring among British Hemiptera. Woodroffe (1958) found that the niche usually occupied by the mirid *Dicyphus stachydis*, on

Woundwort, was occupied by its relative *D. constrictus* in the extreme south-west, where *stachydis* does not occur. In areas where both species are present, *constrictus* does not usually feed on this plant. Displacement may also be occurring in the aquatic family Corixidae. Jansson's (1986) maps show that *Corixa iberica* is confined to two widely separated areas: the southern half of the Iberian Peninsula and the Atlantic coasts of the British Isles from Connemara to Shetland. It is difficult to see any factor that these two areas have in common except for the absence of the widespread and common *Corixa punctata*, the limits of whose known distribution extend to within a few miles of the areas occupied by *iberica*. It is tempting to speculate that *punctata* is in the process of replacing *iberica* and has already out-competed it and driven it to extinction in most of Europe.

The distribution of Heteroptera was summarized by Massee (1955) on a county basis for England and Wales, with presence in Scotland and Ireland noted. Halbert (1935) published detailed Irish records for Heteroptera, Auchenorrhyncha and Psylloidea. Identification guides usually give at least an indication of distributions and many local lists and individual records are published in British journals devoted to entomology or natural history. The *Victoria County Histories of England*, which cover most counties, often have species lists of Heteroptera and sometimes of Auchenorrhyncha. Some indication of the distribution of Aphidoidea can be gathered from comparing the annotated lists of species from Scotland (Shaw, 1964) and Kent (Wood-Baker, 1980).

Distributional data on many groups of British insects are now being organized on a systematic basis by mapping schemes under the supervision of the Biological Records Centre, Monks Wood Experimental Station, Abbots Ripton, Huntingdon, Cambridgeshire. At the time of writing there are active schemes covering both terrestrial and aquatic Heteroptera and Auchenorrhyncha.

Changes in the fauna

Historical factors have played a major part in shaping the British hemipterous fauna. At the last glacial maximum all but the most hardy bugs must have been banished from the tundras of southern England, while nothing could survive in the ice fields further north. As the climate became milder, forest vegetation with its associated insect inhabitants spread northwards and about 8000 years ago the sea level, fed by melting ice, had risen so high that Britain and Ireland became islands. The Irish Sea and the English channel, although narrow, have almost certainly been responsible for checking the spread of brachypterous insects, at least. A striking example is the absence of the lygaeid subfamily Geocorinae from Britain despite the presence of flightless species of *Geocoris* on the dunes of Holland and Scandinavia. Within the British Isles, *Saldula scotica* and *S. c-album* (Saldidae) have similar habitat requirements – shingle and boulders at the margins of rivers – and similar distributions in the western and northern parts of Britain. The macropterous *S. scotica* is widespread in Ireland but its flightless relative has never been found there.

A very large proportion of the land in the British Isles once occupied by deciduous summer forest has been converted to pasture or to arable or horticultural use. Such woods and forests as exist in this area are largely planted and consist mainly of introduced conifers or sweet-chestnut coppice. Almost all remnants of native forest bear at least the marks of past management. Most of the mixed boreal and deciduous forest of Scotland has been converted to sheep-walks, grouse moor or timber production using non-native conifers. Drainage of the remaining blanket bogs and, in Ireland, the large-scale extraction of peat is proceeding apace. In view of this near-total destruction of the native vegetation, it is surprising that so few native plants and animals have been completely lost. Many species are, of course, much rarer than in former times. On the credit side, introduced plants provide a habitat for formerly non-native insects and the open aspect of the countryside has resulted in the species characteristic of grassland becoming very widespread whereas, formerly, they must have been confined to dunes, cliffs, marshes, windswept hilltops, landslips, lake shores and other scarce natural breaks in the forest cover.

No fauna is ever static. At any locality, some species are close to the edge of their range. As climatic conditions fluctuate, a site may become more or less able to support such marginal species, which may become locally extinct in some years and re-invade later from neighbouring areas. One such marginal species is *Liorhyssus hyalinus* (Rhopalidae), one of the most widespread Hemiptera in the world. Occasional migrants of *Liorhyssus* reach Britain or Ireland from southern Europe and may survive to breed here (nymphs have been found more than once), though it is doubtful if the species could survive a winter of average severity. *Lygaeus equestris* (Lygaeidae) has occasionally reached Britain but there is no evidence that it has ever bred here. As the years go by, records of such strays accumulate and unwary biogeographers may wrongly conclude that such species have become extinct here when, in fact, they have never been established. Nevertheless, there probably have been some extinctions of long-established species. Examples are *Eurygaster austriaca* (Scutelleridae), once found in several places on the coast of Kent but no longer to be found there or in Denmark or the lowlands of Belgium, where it used also to occur, and *Chlorochroa juniperina* (Pentatomidae), which has disappeared from the now greatly depleted stands of juniper that used to support it in southern England. Shirt (1987) evaluated the status of the less common British Heteroptera. He classified 6 species as probably extinct, 6 as vulnerable, 14 as endangered and 53 more as rare. Britain is not home to any species that is in danger of global extinction but an endemic subspecies of Piesmidae, *Piesma quadratum spergulariae*, is known only from the Scilly Isles where it might conceivably be in danger from human interference with its seashore habitat. Among the Auchenorrhyncha, the continued survival of the only British cicada, *Cicadetta montana*, must be seriously in doubt. It maintains a precarious hold in a single locality in Hampshire.

Despite a few extinctions the hemipteran fauna is growing larger. Species new to Britain are being reported at the rate of about one a year. Moreover, some native species are increasing their ranges. A berytid, *Berytinus hirticornis*, once restricted to a small area in Devon, is now widespread in southern England and *Metatropis rufescens*, in the same family, is apparently becoming more common in Ireland. The juniper-feeding acanthosomatid *Elasmostethus tristriatus*, unlike the locally extinct juniper-feeding pentatomid *Pitedia juniperina*, has extended its host range to include cultivated junipers and cypresses in parks and gardens, and is now more widespread and abundant than ever.

Ischnodemus sabuleti (Lygaeidae), on wetland grasses, *Eysarcoris fabricii* (Pentatomidae), on woundwort, and *Deraeocoris olivaceus* (Miridae), on hawthorn, have all spread out recently from small foci, a pattern that suggests in each case recent arrival in Britain due to chance introduction or, just possibly, long-range dispersal.

As well as additions like these to the fauna associated with native host plants, a number of Hemiptera are now established on non-native hosts. The widespread planting of conifers for forestry and amenity has led to two developments in the British fauna. First, some species associated with scots pine in its native habitat in Scotland have been able to extend their ranges into England, Wales and Ireland and, secondly, the large number of alien conifer species now grown in the British Isles has led to a great diversification in the conifer-associated hemipteran fauna, mainly through accidental introductions. Of the dozen or so Adelgidae now established in Britain, only *Pineus pini* is able to complete its life cycle in the absence of alien conifers. Similarly, of the more than two dozen species of the conifer-feeding Cinarinae (Lachnidae) listed as British, only five live on native host plants.

Among widely planted alien broadleaved trees sweet chestnut, sycamore and robinia have all acquired an alien aphid fauna and a psyllid has become established on laburnum. *Rhododendron ponticum* is a particularly interesting plant in this context. A native of southern Europe and Asia Minor, it is now widely established in many parts of the British Isles, having escaped from parks and gardens. Despite its Old World origins, it supports a hemipteran fauna derived largely from North America, where other species of *Rhododendron* grow wild. Among these North American bugs are a tingid, *Stephanitis rhododendri*, a predaceous mirid, *Neodicyphus rhododendri*, an aphid, *Illinoia lambersi*, and a cicadellid, *Graphocephala fennahi*. *Rhododendron* species in Britain also support another North American aphid, a European cicadellid, *Placotettix taeniatifrons*, and two Aleyrodidae probably of Old World origin. There is also a rhododendron-feeding race of *Kleidocerys resedae* (Lygaeidae) of unknown origin; this species usually feeds on birch. Alien herbaceous plants, too, have been steadily colonized by Hemiptera, especially aphids. The more recent aphid arrivals include *Acyrthosiphon auriculae* on 'Auricula' species of *Primula* (Martin, 1981), *Impatientium asiaticum* on *Impatiens parviflora* (Blackman, 1984), *Uroleucon erigeronis* on canadian fleabane (Blackman, 1984) and *Macrosiphon albifrons* on various kinds of lupin (Stroyan, 1981; Carter & Fourt, 1984).

Artificial heating in domestic and industrial buildings and in greenhouses provides additional opportunities for insects to extend their geographical ranges. Bedbugs could not survive frosty winters without artificial warmth; nor, probably, could the reduviid *Reduvius personatus* and several *Xylocoris* species (Anthocoridae) that prey on small arthropods in mills, warehouses and food processing plants. The great majority of introduced species of Hemiptera that prosper under artificial conditions, however, are those that feed on crops and ornamental plants in glasshouses. Most of the diaspid scale insects recorded in Britain have been found on tropical or subtropical glasshouse plants such as palms, orchids, bromeliads, aroids and various ferns. *Planococcus citri* (Pseudococcidae), *Saissetia oleae* (Coccidae) and *Icerya purchasi* (Margarodidae) can be troublesome on *Citrus* and many other plants under glass. *Spilococcus cactearum* (Pseudococcidae) is a nuisance on cacti. The most widespread of all alien Hemiptera in greenhouses and on house plants is the glasshouse whitefly, *Trialeurodes vaporariorum*. It may move out of doors in the summer months but survives in winter only under cover, doing extensive damage to fuchsias, tomatoes, cucumbers and other edible and ornamental plants. Some Hemiptera occurring under natural conditions in southern Britain, such as *Coccus hesperidum* and *Parthenolecanium corni* (Coccidae), are found under glass outside their native ranges and a number of warmth-loving aphids also thrive indoors during the winter.

10

THE BRITISH HEMIPTERA AS A SAMPLE OF THE WORLD FAUNA

British bugs in a World context

All currently recognized families of bugs are listed in Table 3 together with estimates of the number of species known to science. In all, there are nearly 150 families and rather more than 80 000 species. About half of the families are represented in Britain but only two per cent of the species. Some idea of how representative the British bug fauna is as a sample of that of the whole world can be had from Table 2. Here, for each major group, the actual number of British species is compared with two per cent of the number of species that have been described world-wide. Some allowance must be made for the fact that the tropical and Southern Hemisphere faunas are much more poorly known than those of the temperate parts of the Northern Hemisphere and that big insects are generally better known than small ones. The most striking points to emerge from a consideration of this table are: the great richness of Britain's fauna of Aphidoidea; the very poor representation of the mainly tropical auchenorrhynchan superfamilies Cicadoidea, Membracoidea and, to a lesser extent, Cercopoidea; and the complete absence of two small groups, the suborder Coleorrhyncha and the heteropteran infraorder Enicocephalomorpha. At the family level, other discrepancies are noticeable: in particular, many large, mainly tropical families of Fulgoromorpha and Pentatomoidea are not found in the British Isles and only about one in a thousand species of Reduviidae breed here.

Perhaps the most characteristically British species is the marine bug, *Aepophilus bonnairei*, which is found only on the Atlantic coasts of Europe from the British Isles to Spain, Portugal and possibly Morocco, and is the only member of its family (Aepophilidae) in the world.

Major groups not found in Britain

The suborder Coleorrhyncha contains a single family of about two dozen small, flattened and rather tingid-like insects associated with mosses in the southern forests of South America, Australia, New Zealand and Norfolk Island. They have a very short gular bridge closing the head capsule behind the mouthparts. Their short, stubby antennae, concealed by the broad, flattened head in dorsal view, suggest (probably falsely) a relationship with Nepomorpha. Fossil evidence shows that Coleorrhyncha were numerous and widespread in the Northern Hemisphere in the early Jurassic period, about 180 million years ago, and a few have been found in British rocks.

The 260 known species of Enicocephalidae, the only family of the heteropteran infraorder Enicocephalomorpha, resemble small Reduviidae but lack the prosternal stridulatory groove. Their wholly membranous fore wings are completely divided into cells by longitudinal, peripheral and cross-veins. They mainly inhabit leaf litter and some tropical kinds dance in swarms like midges. The family is very poorly represented in the temperate regions of the Northern Hemisphere but some species are known from subantarctic islands, so their virtual absence from the temperate North cannot be entirely attributable to the climate.

The largest family of Dipsocoromorpha is Schizopteridae. These are tiny and often beetle-like insects living in leaf litter, including the fibrous vegetable matter that accumulates in and around epiphytic plants on the trees of tropical forests. The distribution of the family is rather like that of Enicocephalidae, though there are no subantarctic species. There must be many hundreds of undescribed species but only about 160 have been named so far.

The warmer waters of the world support about 150 species of the predaceous Belostomatidae (Nepomorpha). Some are very large, as much as 12 cm long, and have received the names of giant water bugs, electric light bugs (because they are attracted to light in the tropics) and 'toe biters' (from occasional attacks on swimmers). One of the larger species, *Lethocerus patruelis*, lives in south-eastern Europe and adjacent areas to the east and south. Adults of this species have sometimes been captured swimming in the salt water of the Adriatic Sea.

In Pentatomomorpha, the 100 or so species of Largidae are sometimes treated as a separate family closely allied to Pyrrhocoridae, with 300 species. Some members of both of these mainly tropical families are at least occasionally predaceous. The cotton stainers (*Dysdercus* species, Pyrrhocoridae) are well known both as pests of cotton and related crops and as laboratory animals. In the key to families, adult Pyrrhocoridae are keyed out on the basis of the characters of the single British species, which is on the extreme edge of its range in southern Britain.

The remaining large pentatomomorphan families not represented in Britain are all plant-feeding shieldbugs (Pentatomoidea). Urostylidae (100 species) are found in both tropical and temperate parts of eastern Asia. They are the only Pentatomoidea, apart from a few genera of Cydnidae, with a claval commissure. Tessaratomidae (250 species) and Dinidoridae (115 species) not infrequently have only four antennal segments. Both families are almost restricted to the tropics of Asia and Africa. Tessaratomids are large bugs, bigger on average than most Pentatomidae. Plataspidae (500 species) are also largely confined to the tropics and there are no American species of the family. A few species of *Coptosoma* occur in Europe. The scutellum of plataspids is very large and the elongate hemelytra are transversely folded in the middle when stowed away beneath it. Most species feed on plants of the pea family.

One very small family, Polyctenidae, deserves a mention because of its unusual biology. All 31 species are permanently ectoparasitic on bats, feeding on their blood. They are completely eyeless and various parts of their bodies bear ctenidia (comb-like rows of stout setae) like those of fleas. Uniquely for Hemiptera, their tarsi are 4-segmented. Polyctenids are viviparous and the developing young receive nourishment through a sort of placenta. They are born at an advanced stage of development and pass through fewer than the normal five instars. These developmental characteristics parallel those of some blood-sucking flies.

The great majority of British Fulgoromorpha belong to the family Delphacidae, which has about 1300 described species around the world. Most other fulgoromorph families are predominantly tropical in distribution. Three of them have about 1000 species each but of these Britain has only a dozen Cixiidae, two Issidae and no Flatidae at all. Flatids are broad-winged and often brightly coloured insects with rich, branching venation. When set with the wings spread they somewhat resemble butterflies. Four species, in two genera, have been reported from southern Europe. Derbidae often have long, parallel-sided fore wings although more normally proportioned forms also occur in the family. They are found throughout the tropics and a few species of the genus *Malenia* live as far north as southern Europe. Their immature stages may feed on fungi. Another group believed to have fungus-feeding habits is the family Achilidae. These rather flattened bugs are encountered under bark and in rotten logs in most parts of the world and in Europe their range includes Scandinavia. All seven European species belong to the genus *Cixidia*. The large and colourful Fulgoridae, which often have bizarrely shaped heads, suck sap from the trunks and branches of trees throughout the tropics. The rather similar Eurybrachidae have normal heads except for the very broad frons and are more restricted in distribution, being absent from the American tropics. Dictyopharidae often have their heads prolonged in front of the eyes and their wings are usually narrow and transparent. Eight genera and nearly 40 species of the family occur in Europe and the range of *Dictyophara europaea* extends from the Mediterranean countries northwards to Belgium and Germany. Like Tropiduchidae and Nogodinidae, the overwhelming majority of them live in the tropics and subtropics of both the Old World and the New. Three species, in two genera, of Tropiduchidae live in central or southern Europe and a single species of Ricaniidae reaches the south-eastern fringes of the continent. No Lophopidae are European but the family is represented, albeit very poorly, in North Africa and Iran. Both Ricaniidae and Lophopidae are almost wholly confined to the Old World tropics and subtropics but a few species of both are reported from South America. A single species of the small family Meenoplidae has been found in the Mediterranean countries of Europe.

Cercopoidea are rather poorly represented in Britain. The largest subfamily, Cercopinae, has about 1400 species in the world, but only one in Britain, while Aphrophorinae has only 800 world species, nine of them British.

Cicadelloidea are well represented in Britain by Cicadellidae but the related family Eurymelidae, with about 100 species, is absent.

The only non-British family of Sternorrhyncha with more than 100 species is Greenideidae, a group of mainly tropical Asian aphids with hairy siphunculi. They live on broadleaved trees, without host-alternation. Some species are found as far north as Japan and a few are Australian but the family has no European, African or American representatives.

Table 2 Comparison of British and world faunas of the major groups of Hemiptera.
(Comparison of the actual number of British species in each group with the numbers expected if each group were represented in Britain by two per cent of its known world species.)

Group	2 per cent of world species in group	Total British native species
Coleorrhyncha	0.4	0
Enicocephalomorpha	5	0
Dipsocoromorpha	5	3
Nepomorpha	33	43
Gerromorpha	26	20
Leptopodomorpha	6	22
Cimicomorpha	277	295
Pentatomomorpha	299	154
Fulgoroidea	162	85
Cercopoidea	48	10
Cicadoidea	40	1
Cicadelloidea	400	255
Membracoidea	50	2
Psylloidea	46	80
Aleyrodoidea	23	12
Adelgoidea	2	4
Aphidoidea	76	500
Coccoidea	134	84

11

MORPHOLOGY

Most Hemiptera display the usual tripartite division of the insect body into the head, bearing the mouthparts, antennae and eyes; the thorax, bearing the organs of locomotion; and the abdomen, concerned mainly with digestion, excretion and reproduction. In some Sternorrhyncha, especially immature and adult female scale insects (Coccoidea) (Figs 156, 157, 160–164) and immature whiteflies (Aleyrodoidea) (Figs 98, 99), the gross morphology shows little sign of such differentiation. Adult Hemiptera (except Aleyrodidae) are more readily identifiable than immatures and the information in this chapter is concerned with adults unless otherwise stated.

Head

The most fundamental and characteristic feature of the order is the modification of the mouthparts into a *rostrum* (Figs 2, 4). No other insects, even those with sucking mouthparts, have a similar structure. Palps are completely lacking and the mandibles and maxillae are modified into very slender, strongly sclerotized stylets (Fig. 106). The *maxillary stylets* interlock to form a tube with salivary and food canals running between them. The functional mouth is a small hole or a slit at or near the tip of this tube. The lateral surfaces of the maxillary stylets are usually ridged and grooved to form guides for the *mandibular stylets*, whose inner surfaces are so shaped as to make a sliding fit with them. The mandibular stylets are toothed and often barbed apically. During penetration of a food item, the mandibular stylets make alternate (or, in some groups, simultaneous) thrusts into the host tissues and the maxillary tube is advanced between them towards the feeding site. In several groups of plant-feeding Hemiptera that feed for long periods at one site, the saliva may harden to form a 'salivary sheath' around the stylets inside the plant tissues and a 'salivary cone' enveloping them for a short stretch where they are exposed between the plant epidermis and the tip of the labium. The *labium* is modified into an often jointed tube, open dorsally and enveloping the bundle of stylets, which can be protruded from its tip. Dorsally, its base is roofed by the *labrum*. Because the rostrum is usually directed posteriorly, its morphologically dorsal surface faces ventrad. The labium varies from a long, four-segmented tube, as in most Heteroptera, to a short, unsegmented cone in some Coccoidea (Fig. 163). In Heteroptera its base is frequently flanked by a pair of longitudinal, erect plates called *bucculae* (Fig. 2), arising from the under-surface of the head. Some of the plant-feeding members of this suborder can hinge the second and third segments of the labium backwards, away from the stylets, so that the latter emerge from the apex of the first segment and enter the base of the fourth, bridging the gap between them. This procedure enables the stylets to penetrate the host plant to an additional depth equal to the combined lengths of the two by-passed segments of the labium. Some Pentatomomorph bugs can disengage the stylets from the labium altogether, allowing very deep penetration of the host plant's tissues.

Not only does the presence of a rostrum define the order but its position is diagnostic of each of the three suborders. In Heteroptera there is a bridge of cuticle, the *gula* (Fig. 2), closing the head capsule ventrally behind the rostrum, which therefore appears to arise from the anterior end of the head. There is no gula in the other two suborders. In Auchenorrhyncha, the rostrum clearly arises from the base of the head (Fig. 78) but in Sternorrhyncha it is displaced posteriorly so that it appears to originate between the anterior coxae (Fig. 154).

The sclerites of the head are not well defined in most Sternorrhyncha but they can more easily be made out in Psylloidea, Heteroptera (Figs 2, 3) and Auchenorrhyncha. The *lora*, or *mandibular plate*, is situated between the antennal insertion and the labrum. If they are visible in the dorsal or frontal aspect, the lorae flank the clypeus and their dorsal or frontal surfaces are termed *paraclypei* or, in Heteroptera, *juga*, with the term lora being restricted to the component visible in lateral or ventral view. The main part of the side of the head is the *gena*. A ventral *maxillary plate* may be separated from the main part of the gena by a suture in Heteroptera. The labrum is attached basally to a median dorsal sclerite, the *clypeus*, lying between the mandibular plates. The clypeus may be divided by a transverse suture into *anteclypeus* and *postclypeus*. The latter bears the attachment for the cibarial muscles, which operate the food-sucking pump. In many Auchenorrhyncha the postclypeus bears a series of conspicuous chevron-shaped marks that correspond to the muscle attachments on its inner surface. The anterior part of the clypeus in Heteroptera, lying between the juga, is termed the *tylus*. In most Heteroptera the tylus and juga are visible in dorsal view but in most Auchenorrhyncha and Sternorrhyncha the facial area is inclined ventrally. Heteroptera show little differentiation of the dorsal region of the head: even the boundary between clypeus and frons is not marked by a suture and it is impossible to separate frons and vertex externally. The facial area of Auchenorrhyncha varies somewhat between the families. An *epistomal suture* across the lower part of the face in the fulgoromorph families marks the posterior limit of the postclypeus; the upper part of the face is occupied by the *frons* (Fig. 78). In the cicadomorph families the postclypeus is very large, a reflection of the great development of the cibarial muscles, which are inserted on this sclerite, at the expense of the pharyngeal muscles, inserted on the frons. Cicadellidae lack an epistomal suture, so the main facial sclerite is a compound structure involving both postclypeus and frons and termed the *frontoclypeus*. The top of the head is equivalent to the *vertex* in Fulgoromorpha but in Cicadomorpha the frons often extends onto this area. In Cercopidae the small frons is entirely dorsal and very clearly defined by sutures (Fig. 86). The head has a broad, membranous connection to the thorax in Sternorrhyncha and Auchenorrhyncha. Heteroptera typically have a well developed *occipital region* at the back of the head that is retracted into the anterior collar of the thorax. *Cervical sclerites*, in the neck membrane, are present only in Cicadidae.

Paired *compound eyes* are usually present and conspicuous, with many facets, the facets being the lenses of *ommatidia*, the units of which the eye is compounded. In Aphidoidea there is an *ocular tubercle* or *triommatidium* (Fig. 4), closely applied to the posterior margin of the eye and bearing a group of three ommatidia. In some subterranean and gall-making aphids the eyes are greatly reduced and only the triommatidium is present in nymphs and even in wingless adults (Fig. 137). The eyes are also greatly reduced in Coccoidea, among which compound eyes occur only in male Ortheziidae and Margarodidae. Males of other coccoid families have four widely spaced ommatidia. Female Coccoidea have either a pair of ommatidia or, sometimes, no eyes at all. Primitively, three *simple eyes* or *ocelli* are present, in addition to the compound eyes. Two are situated on the vertex or on the genae and one on the frons. All three are present in Psylloidea, winged Aphidoidea, Cicadidae and Cixiidae but the frontal ocellus is lacking in most Auchenorrhyncha and all Heteroptera. Issidae, some Cicadellidae, some Heteroptera (notably Miridae and the submerged aquatic families), wingless Aphidoidea and all immature Hemiptera lack ocelli altogether. The lateral ocelli are usually situated on top of the head but in some Cicadellidae they are on the face and in Fulgoromorpha they are on the sides of the head,

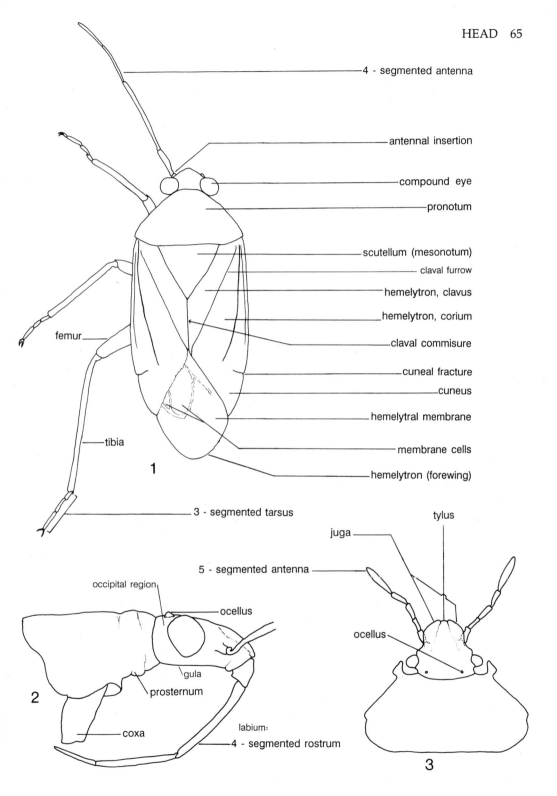

Figs 1–3. *Anatomy of Heteroptera.* 1, Lygus (Miridae), dorsal view; 2, Nabis (Nabidae), lateral view of head and prothorax; 3, Podops (Pentatomidae), dorsal view of head and prothorax.

low down in front of the eyes. In Coccoidea, the males have a pair of ocelli whereas females and nymphs have none.

The *antennae* vary greatly in length and in their position of insertion. They are often inserted on low tubercles named *antennifers* or *antennal tubercles* (Fig. 4). The first antennal segment, or *scape*, has a single condyle (point of articulation) at its base, allowing great flexibility of movement. Its articulation with the second segment, or *pedicel*, has two condyles, so mobility at this joint is restricted to simple back-and-forth motion. Since neither of these condyles is aligned with the basal one, the pedicel effectively has a universal joint with the head via the scape. The part of the antenna beyond the pedicel, the *flagellum*, articulates with the pedicel by a simple annular joint without condyles and is itself usually divided into segments by similar articulations. Most Heteroptera have only four antennal segments, including the scape and the pedicel, but adult shieldbugs (Pentatomoidea) have five, due to division of the pedicel at the last moult, and some adult Hebridae have three flagellar segments. In some aquatic families the flagellum is not subdivided. Small *ring-segments* or *annuli* may be present between the main segments of the antennae in some Heteroptera but they are not counted as true segments. The flagellar segments in this suborder may be about as thick as the scape and pedicel or distinctly thinner. In both cases they bear numerous, specialised, sensory hairs. Sometimes, as in Coreidae and Tingidae, these sensilla are restricted to the last segment, which is then usually ovoid or shortly spindle-shaped and thicker than the preceding segment. The aquatic Nepomorpha have very short antennae that are confined in bubbles of air in depressions below or behind the eyes, where they serve to indicate to the bugs their attitude in the water. The antennal flagellum in Auchenorrhyncha is bristle-like and usually short. In most Fulgoromorpha it is unsegmented. In a few Cicadellidae it is expanded apically into a leaf-shaped *palette*. The sense-organs on the antennae of Aphidomorpha provide important diagnostic characters. These *placoid sensilla*, or *rhinaria* (Figs 4, 115), are round or oval pits floored with a thin membrane. They are of two types. The *primary sensilla* occur singly or in small groups near the apex of the penultimate segment and part way along the last segment. The primary sensillum (or sensilla) of the last segment marks the beginning of the slender and often long *terminal process*, whose length relative to the basal part of the segment is useful in identification. The *secondary sensilla* that are often present on various antennal segments are also useful in this regard.

Other features of the head that provide taxonomic characters are the prominent keels of the face and vertex in Fulgoromorpha, the two spots (*thyridia*) at the transition between face and vertex in Cicadellidae, the *genal cones* of many Psylloidea and the *median frontal prominence* between the antennae of some aphids.

Thorax

The three leg-bearing segments of the body constitute the thorax. The first of these, the *prothorax*, bears only a pair of legs while the *meso-* and *metathorax* bear the fore and hind wings as well. Consequently, the prothorax is the smallest in volume. Each thoracic segment is roofed by a *notum*, walled laterally by *pleura* and floored by a *sternum*.

The *pronotum* (notum of the prothorax) of Heteroptera frequently has a distinct anterior *collar*, which may be continuous around the pleura and sternum and marks the region of the prothorax into which the occipital region of the head can be retracted. In Heteroptera and many Auchenorrhyncha there is a *posterior lobe* of the pronotum, extending backwards to cover part of the mesonotum like a roof. In Membracidae this lobe extends posteriorly in a spine-like process that reaches as far as the abdomen (Fig. 84). Its posterolateral angles are often produced laterally into spreading lobes or spines in Membracidae, Pentatomidae and Coreidae. The area of the heteropteran pronotum between the collar, if any, and the posterior lobe often bears a pair of large, transversely oval or oblong *calli* (raised or sunken areas, Fig. 32) and is consequently termed the *callar region*. Although this small area is the true central region, or disc, of the pronotum the term disc is usually taken to refer to the central part of the much larger posterior lobe.

The *mesonotum* is usually much the largest of the three nota, as the mesothorax houses the main flight-muscles. During flight its main sclerite, the *mesoscutum*, deforms along certain well-defined lines of weakness which, in Heteroptera and Auchenorrhyncha, are covered and protected from damage by the posterior lobe of the pronotum. In these groups the only part of the mesonotum that is normally visible is the *mesoscutellum*, filling the triangular space between the pronotal lobe and the fore wings when the latter are folded down in the resting position. It is usually simply termed the *scutellum* (Fig. 1). In some Pentatomoidea the scutellum is greatly expanded posteriorly, covering nearly all of the abdomen and the folded wings (Figs 29, 30). The rest of the mesonotum and the whole of the *metanotum* are not usually visible and consequently have rarely been used to provide diagnostic features. In both segments there are four basic notal sclerites which are named, from front to back, *praescutum*, *scutum*, *scutellum* and *postnotum* (with prefixes meso- and meta- as appropriate). The base of the fore wing in Fulgoromorpha and Psylloidea is protected by a small, separate, flap-like sclerite, the *tegula* (Figs 74, 90). Psylloidea have an additional, tubercular or flap-like sclerite of similar size, called the *paryptera*, in front of the tegula.

The lateral sclerites of the thoracic segments, the pleura, are usually each divided by a vertical or oblique suture, the *pleural suture*, into an anterior *episternum* and a posterior *epimeron*. The pleural suture is a line of strength rather than weakness. It bears an articulation for the coxa at its lower end and, on the wing-bearing segments, a wing articulation at its upper end. In Heteroptera, there are usually downwardly directed outgrowths from both epimeron and episternum, enclosing the coxa in an *acetabulum* (Fig. 35). The suture between the two elements of the acetabulum is the *coxal cleft*. The episternum may be divided into an upper anepisternum and a lower katepisternum and the epimeron may be similarly divided into anepimeron and katepimeron. There are two thoracic spiracles, lying in the intersegmental membranes between the thoracic segments. In Heteroptera and Auchenorrhyncha the first of these cannot usually be seen without separating the prothorax from the mesothorax. The second spiracle, too, is often not visible externally, its position being indicated by a narrow slit between the meso- and metapleura. A conspicuous pit often visible high on the mesopleuron is the external manifestation of an internal muscle-attachment and should not be confused with a spiracular opening. The metathorax in adults of almost all Heteroptera houses a large gland that opens either by a single median aperture on the sternum (Fig. 68) or by paired apertures on the metapleura (Fig. 36). In the latter case, the metepisternum in the area around the aperture, the *peritreme*, is modified into an *auricle* or spout surrounded by a dull patch of elaborately microsculptured cuticle. This dull patch, which may extend onto the mesopleuron as well, is termed the *evaporatorium* or *evaporative area* in the probably mistaken belief that it retains the secretion from the gland and allows it to evaporate slowly. It is more probable that it is an unwettable region, protecting the bug from the corrosive effect of its own secretion.

The sterna are often continuous laterally with the episternal elements of the pleura. Apart from the occasional presence of various keels and sulci, they present few characters of taxonomic value.

The legs are inserted between the sterna and the pleura. The basal segment of the leg, the *coxa*, is partly sunken within a *coxal cavity* excavated from the sternum and often

68 MORPHOLOGY

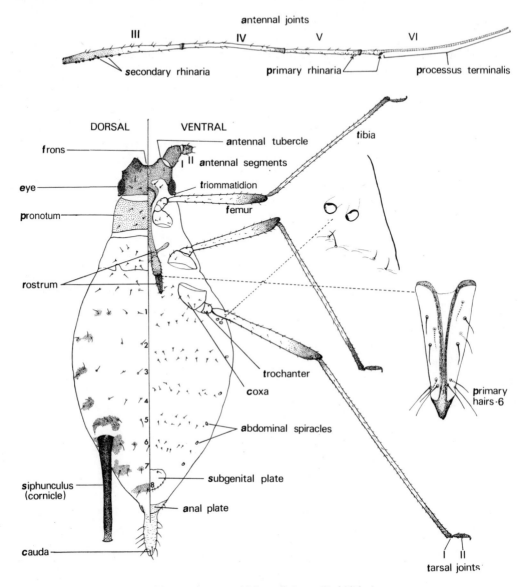

Fig 4. *Anatomy of* Macrosiphum *(Aphididae).*

partly also from the pleuron. These cavities may be *closed* or they may be *open* in either of two ways: the members of a pair may not be fully separated one from the other by the sternum; or the posterior pair of cavities may not be surrounded posteriorly by it. The position of insertion of the coxae has been used as a diagnostic character for higher taxa. For example, the mid coxal articulations are widely separated and the coxae themselves are elongate like the fore coxae in Fulgoromorpha but in Cicadomorpha they are short and inserted close together. There was once a fashion for dividing the families of Heteroptera into two groups on the basis of the coxal types, with most Cimicomorpha and some Nepomorpha characterised as *pagiopodous*, with a long, *cardinate* (hinged) coxa with a limited range of movement, and the other families characterised as *trochalopodous*, with a shorter and more freely mobile coxa with a *rotatory* articulation. Much less emphasis is placed on this distinction now. Some, perhaps all, Hemiptera have a small sclerite, the *trochantin*, situated anterolaterally in the coxal articulation, hinged at one end to the coxa and at the other to the pleuron or, sometimes, only to the latter. The second segment of the leg is the *trochanter*. It is short and usually simple but in Miridae it is divided into two apparent segments by a suture at which the leg can be broken off and the wound automatically sealed without bleeding or the need to form scar tissue. The *femur* and the *tibia* are the largest segments of the leg. The femur houses the powerful muscles operating the knee joint and is, therefore, almost always noticeably thicker than the tibia. Both femora and tibiae may bear outgrowths of multicellular origin (spurs); the posterior tibia of Delphacidae bears an articulated spur apically but leg spurs in other Hemiptera are immobile. The shape of the tibia is often important diagnostically at higher taxonomic levels. It may be simply cylindrical, longitudinally sulcate (grooved) or longitudinally carinate (keeled). The *tarsus*, or foot, consists of one, two or three segments. Very rarely it is absent or fused with the tibia. Distally, the tarsus terminates in a group of structures collectively termed the *pretarsus*. The most conspicuous of the pretarsal elements are the strongly sclerotized *claws*, which are usually paired but are invariably single in Coccoidea and sometimes single in the fore or hind legs of certain water bugs. Inserted into the end of the tarsus, between and below the claws, is the *unguitractor plate*. The claws articulate with this plate and with the sides of the terminal aperture of the tarsus. The effect of contraction of the muscle pulling on the tendon attached to the plate is to draw the claws together and downwards. At its apex, the unguitractor plate usually bears a pair of socketed bristle-like (rarely flap-like) structures, the *parempodia*. The membrane above the unguitractor plate may bear outgrowths called *arolia*. Any ventral appendage to a claw is by definition a *pulvillus*. In the pentatomomorphous Heteroptera, the pulvillus is a stalked pad, the stalk being termed the *basipulvillus* and the pad the *distipulvillus*. In Miridae the pulvillus, if present, is a simple pad- or leaf-like structure attached directly to the claw. The pretarsal structures have been used extensively in the study of the systematics of the Heteroptera but have received less attention from students of Auchenorrhyncha and Sternorrhyncha.

The general shape of the leg varies greatly according to the use to which it is put. In bugs that leap, the hind femora are usually enlarged to accommodate the increased musculature but in a few aphids it is the anterior coxae that are so modified. Some predaceous Heteroptera have enlarged anterior femora against which the curved tibiae fold; the opposing surfaces are frequently armed with spurs, spines or denticles to help retain the prey. Many members of the predaceous families Reduviidae, Nabidae and Anthocoridae have a wedge of thin, adhesive cuticle, the *spongy fossa*, set into the apical ventral part of the fore tibia, which also helps in retention of the prey. The fore legs of many Lygaeidae appear to be of the raptorial type (though they have no spongy fossa), yet the bugs feed on seeds and not on animal prey. Many aquatic Heteroptera have strikingly modified legs. The fore legs are modified, sometimes strongly, for catching the prey. Corixidae (Fig. 73) have a usually scoop-shaped 'hand' (the *pala*), which bears a long fringe of hairs for sorting small food items from detritus. In male Corixinae the pala also bears a row of tubercles (the *palar pegs*) that serve to hold the female during copulation. The hind legs of the submerged swimmers, especially those of Corixidae and Notonectidae, provide the main propulsive force and are flattened and fringed with hairs to enhance their effectiveness. The mesothoracic legs of the pondskaters (Gerridae) are strikingly elongated but not flattened and are used to row the insect across the surface of the water.

There are usually two pairs of wings, inserted between the nota and pleura of the meso-

and metathorax. They develop as posterior extensions of the nota of these two segments, becoming equipped with basal articulations only at the final moult. The wings may be reduced in size or absent altogether in various groups. The fore wings are larger than the hind wings (except in Aleyrodoidea) and often also distinctly tougher. Those of Cicadellidae, Cercopidae and a few Psylloidea are opaque and fairly evenly strengthened throughout, with the cells between the veins thickened. Such uniformly toughened fore wings are termed *tegmina*. In Cicadellidae the apical cells are often more membranous and transparent than the others, paralleling the condition in Heteroptera. The fore wings of other Auchenorrhyncha and Psylloidea derive strength from a thickening of the veins only, the cells remaining transparent and glassy. In Heteroptera, usually only the basal two-thirds or so of the fore wings are toughened and opaque between the veins, while the apical part (the *membrane*) is thin, flexible and often transparent like the hind wings (Figs 1, 32, 43–45). These part-thickened fore wings are called *hemelytra* or *hemielytra*, by analogy with the elytra of beetles. The fore wings of the heteropterous groups Dipsocoromorpha and Gerromorpha are not modified in this way, though the venation of the apical part of the fore wing is often weaker than that of the basal part or lacking altogether.

At rest, Heteroptera hold the wings flat over the back, with the membranes overlapping. Auchenorrhyncha, Psylloidea and Aleyrodoidea hold them in a tent-like attitude with little or no apical overlap of the fore wings, though the hind wings may overlap widely. Male Coccoidea and some Aphidoidea hold them flat over the back, while typical Aphidoidea, in which the fore wings are twice as long as the abdomen, have a modification of the tent-like position with the apical parts of the wings pressed flat together in the vertical plane.

Certain structures in the wings, especially the anterior pair, give flexibility along predetermined lines when the insect is in flight. The boundary between the coriaceous basal part and the membranous tip of the heteropteran hemelytron is such a structure. In Miridae, Anthocoridae and some other Heteroptera there is a fracture in the costal (anterior) margin of the fore wing, called the *costal fracture* or *cuneal fracture* (Fig. 1). The coriaceous area between this fracture and the membrane is termed the *cuneus*. In Auchenorrhyncha there is often a *nodal line* of fairly closely aligned cross-veins towards the apex of the fore wing, sometimes with associated breaks in the longitudinal veins. The nodal line allows the wing tip to behave differently from the rest of the wing, as in Heteroptera, but there is never a costal fracture in Auchenorrhyncha. There are two major longitudinal lines of weakness in the fore wing. They are most easily seen in the coriaceous hemelytra of Heteroptera (Fig. 1) and tegmina of various Auchenorrhyncha. The *claval furrow* cuts off the posterior basal area, the *clavus*, from the rest of the wing, the *remigium*. In Heteroptera, the coriaceous basal part of the remigium is called the *corium*. Where a cuneus is present, it is not reckoned as being part of the corium. At rest, the clavi lie alongside the scutellum and usually then meet one another along the midline of the body for the rest of their length. The line along which they are in contact is the *claval commissure*. Among Sternorrhyncha, only Psylloidea have a separate clavus. The second longitudinal line of weakness in the fore wing is the *medial fracture*, which runs from the base of the wing to about the level of the cuneal fracture, if there is one, and never reaches the wing margin, unlike the claval furrow. In Anthocoridae there is another line, of unknown function, anterior to the medial fracture and not quite reaching the cuneal fracture. This line is the *emboliar groove* and it marks the posterior border of a narrow, costal strip of the hemelytron, termed the *embolium*. A number of flexion lines can be made out in the hind wings. Two of these can be homologized with the medial fracture and claval furrow of the fore wing. The claval furrow may fork or it may be duplicated. The most conspicuous of the flexion lines in the hind wing is the *vannal fold*, which allows the posterior basal field, the *jugal lobe*, to be folded under the rest of the wing when the insect is not flying.

In Heteroptera that part of the underside of the costal margin of the hemelytron that is exposed at rest is thickened into a *hypocostal lamina* or *epipleur*. Towards the outer end of the costal margin of the transparent forewings of Cixiidae (Figs 5, 6), Psylloidea (Fig. 96) and Aphidoidea (Fig. 134) there is a pigmented and sclerotized area called the *pterostigma*, or simply the *stigma*. Oval patches of a white, powdery substance on the ventral surface of the subcostal area of the tegmina in most Cicadellidae, and particularly conspicuous in Typhlocybinae, have been termed *wax-patches* or *waxy areas*. It is now known that they are composed not of wax particles but of brochosomes, small spheres of an excretory product expelled from the anus with the liquid faeces.

Not all adult Hemiptera are *macropterous*; that is, fully winged and able to fly. Female Coccoidea are invariably *apterous* (completely lacking wings) and the males of this group are sometimes so, though they normally have functional fore wings. The hind wings of male Coccoidea are reduced to short, rod-like structures, sometimes linked to the basal angle of the fore wings by up to three small hooks. Totally apterous females are usually present at some stage in the life-cycle of Aphidoidea and apterous males sometimes occur. Complete aptery is rarely encountered among other groups of Hemiptera. There are several different kinds of abbreviated wings between the two extremes of the apterous and macropterous types. They do not form a smooth continuum, though intermediates do occur between some of them. In almost all British Heteroptera and most Auchenorrhyncha that are habitually flightless, occasional macropters have been found. It is unusual for more than one flightless form to occur in the same species. All of the flightless forms between the extremes are loosely called brachypters but the following types can usefully be distinguished in Heteroptera and most of these terms are applicable also to other groups:

(a) *submacropterous* Both pairs of wings slightly reduced in length but otherwise unmodified; hind wings distinctly shorter than fore wings but not by much; flight muscles reduced or not; flitting flight may be possible. (In all the other forms flight muscles are reduced and flight is impossible.)

(b) *subbrachypterous* Clavus and corium more or less normal, membrane with area of overlap often greatly reduced; hind wings greatly abbreviated.

(c) *brachypterous* Fore wings reaching beyond apex of scutellum, with a claval commissure but membranes not overlapping; hind wings greatly abbreviated.

(d) *micropterous* Fore wings not reaching beyond second or third abdominal tergite, usually undifferentiated (Fig. 70) but sometimes with a rudimentary membrane.

(e) *staphylinoid* Fore wings abruptly truncate, lacking membrane and without claval furrow, resembling elytra of staphylinid beetles (Fig. 58).

(f) *coleopteroid* Fore wings undifferentiated, usually convex, meeting beyond scutellum without overlap, commissure longer than claval commissure of macropterous form (Fig. 42).

It is difficult to standardize the nomenclature of the various insectan wing-veins and the cells determined by them, even within the Hemiptera. Different authors give different names to the same structures and differ in their interpretations of homologies betwen fore- and hind wings of the same insects. In an attempt to establish a ground-plan for the Heteroptera, Wootton & Betts (1986) discussed some of the problems. These authors also addressed the question of the nomenclature of the fractures, furrows and folds and the fields that they delimit (see above).

Wootton & Betts (1986) believed that, in Heteroptera, the costal margin (leading edge) of the fore wing may be traversed by an unbranched *Costa* (C) or an unbranched *Subcosta* (Sc) or both. Behind the leading edge, usually some distance from it, run the *Radius* (R) and *Media* (M), which are united basally until they reach the level of the medial fracture, where they diverge more or less abruptly. Sometimes, the first branch of R (R1) arises close to the point where R and M diverge and runs obliquely towards the costal margin, as in *Nabis* (Nabidae) for example. Usually, a cross-vein (r-m) unites R and M at the apical margin of the corium, thus enclosing the *first discal cell*. Between the medial fracture and the claval furrow runs the *Anterior Cubitus* (CuA). Two cross-veins (1m–cu, 2m–cu) connect M and CuA, enclosing the *second discal cell*; 2m–cu is situated on tha apical margin of the corium, in line with r–m. What happens when the longitudinal veins enter the corium is unclear. Probably R, M and CuA each fork once; certainly the last of the three does. Venation in the membrane is often weak and variable and the courses of the tracheae are a poor guide to the homologies of the veins there. Wootton & Betts denied the existence of a Posterior Cubitus (PCu), claiming that it has become obliterated by the claval furrow. They recognized only the *First and Second Anal veins* (1A, 2A) in the clavus. The trachea of 1A always crosses the claval furrow near its apex and enters the posterior branch of the Cubitus (CuA2) in the membrane. In the heteropteran hind wing the vein at or closest to the leading edge is C+Sc or, more usually, C+Sc+R. M runs separately from R but its basal region is more or less fully obliterated by the medial fracture, which ultimately crosses it to run between R and M for a short distance. A cross-vein, r–m, stops the median fracture from running beyond about the middle of the wing. In Cimicomorpha there is no cross-vein since R and M fuse in the middle of the wing, usually diverging again distally. CuA runs separately from M and is connected with it by m–cu. Typically, therefore, the anterior basal quarter of the hind wing contains a single cell that is traversed by the medial fracture and part of M and is bounded anteriorly by R (or C+Sc+R), posteriorly by CuA and apically by r–m (if present), a short segment of M and m-cu. The section of M that lies within this cell (the only cell in the hind wing) is termed the *hamus*. In the broad hind wings of some Pentatomoidea and Coreidae, CuA is strongly curved, with posterior convexity. A non-tracheate vein may then be present running parallel to the claval furrow and just in front of it, uniting apically with CuA where the latter comes close to the furrow. This secondary vein is called the *glochis* or *antevannal vein*. The area between the two branches of the claval furrow, or the two claval furrows if there are two, sometimes has two non-trachete, secondary *interclaval veins*. The jugal lobe, marked off from the rest of the wing by the vannal fold, contains only the *Third Anal vein* (3A).

The wings of Auchenorrhyncha (Figs 5–7) often have much richer venation than those of Heteroptera. There are usually more than the heteropteran maximum of three cross-veins in the fore wing and two in the hind wing and hence more cells. Either or both wings may have a *Peripheric Vein* uniting the apices of all branches of the longitudinal veins. There may be a narrow, veinless area (called, in the fore wing, the *appendix*) between the peripheric vein and the apical margin of the wing. The veins in the basal part of the fore wing can fairly easily be compared with their counterparts in the Heteroptera but the question of the homologies of the veins in the distal part of the wing may never be settled to everyone's satisfaction. LeQuesne's (1960, 1965b, 1969; Le Quesne & Payne, 1981) handbooks dealing with Cicadellidae employ a unique and unambiguous system for naming the features in this area, recognizing an apical and a subapical series of cells. The hind wings of Auchenorrhyncha usually have more than three anal veins. The venation of Sternorrhyncha is simpler than that of the other Hemiptera, due largely to

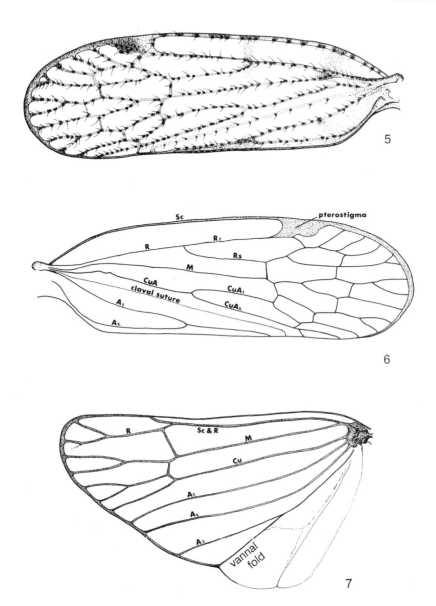

Figs 5–7. *Wings of* Cixius *(Cixiidae). 5, tegmen (fore wing); 6, map of tegmen with main veins labelled; 7, hind wing with main veins labelled.*

fusion of the longitudinal veins. The junction between the wing and the thorax is an area of great complexity, with a variety of small sclerites, folds and flexion-lines whose arrangement is determined by the need to control the curvature, attitude and motion of the wing in flight and to position it neatly and securely at rest. The characters afforded by the structures of the wing-base are difficult to observe and are not normally used in identification.

Abdomen

There are ten abdominal segments plus a post-segmental region in all Hemiptera. Studies of the embryonic and postembryonic development do not support the idea, often advanced, that an eleventh segment is present. The abdominal segments unaffected either by the basal attachment of the abdomen to the thorax or by the presence of structures surrounding the genital openings are simpler than the thoracic ones. Each has a dorsal *tergum* or *tergite* and a ventral *sternum* or *sternite*. There are no pleura but the outer parts of the terga and sterna may be cut off by longitudinal sutures (Figs 37, 38), in which case these outer parts are termed *laterotergites* and *laterosternites* (or *paratergites* and *parasternites*) respectively. Most pentatomomorphan Heteroptera have two series of laterotergites, the inner ones being narrower than the outer ones, and no laterosternites. The row of laterotergites and, where present, laterosternites are collectively called the *connexivum*.

Most immature Heteroptera have from one to four median *dorsal abdominal glands* with conspicuous single or paired apertures opening between the tergites Figs (25–27). The scars of these gland apertures persist in the adults (Figs 37, 38) and the glands themselves may sometimes still be functional. Most Aphidoidea bear a pair of conspicuous *siphunculi* or *cornicles* (Figs 4, 127) on the dorsum of the fifth or sixth abdominal segment. These vary in shape from simple pores or low cones to long tubes and have a secretory function. Many Sternorrhyncha, including aphids, and Fulgoromorpha produce wax and other secretions from abdominal glands. The greatest diversity of these glands is found among the Coccoidea, where they are extensively used in identification.

The base of the abdomen in male and often also in female Auchenorrhyncha carries a pair of sound-producing *tympanal organs*. These consist of a pair of usually inconspicuous plates (*tymbals*) on the first tergum, which are vibrated by muscles anchored to inwardly projecting *apodemes* or *phragmata* of the first and second sternites and second and third tergites. The shape of the tymbals themselves is rarely used in identification but the apodemes supply important characters at the species level.

The maximum number of abdominal spiracles is eight pairs. The ninth and tenth segments never bear spiracles. If there are fewer than eight pairs, those most likely to be lost are the first and last pairs, since they are situated in the most highly modified regions of the abdomen. In the most extreme case of reduction, many Coccoidea lack abdominal spiracles altogether, relying only on the two thoracic pairs. Spiracles are usually situated near the outer edges of the sternites or on the laterosternites if these are present or, rarely, on laterotergites.

The male copulatory apparatus is carried on the ninth abdominal segment. It is frequently of great diagnostic value at the species level, particularly in Psylloidea, Cicadellidae, Delphacidae, Miridae and Lygaeidae. The eighth segment is sometimes strongly modified as a consequence of its proximity to the ninth. In pentatomomorph Heteroptera it forms a short tube, usually without any trace of spiracles, that enables the ninth segment to be retracted by 'telescoping' into the end of the abdomen. Where the eighth segment is strongly modified, the seventh is also slightly modified as well. Very rarely (Dipsocoromorpha) the segments anterior to the ninth have articulated appendages, probably modified from the laterosternites.

In most male Hemiptera (Figs 8–10, 12–17) the ninth sternum is enlarged into a more or less cup-like *genital capsule*, *pygofer* or *pygophore*, which typically bears a pair of *claspers* or *parameres* and a median intromittent organ, the *aedeagus*, *phallus* or *penis*. Dorsally, the ninth tergum may or may not be apparent as a separate sclerite. The tenth segment and the post-segmental region most usually form a single structure, the *anal tube* or *proctiger*, which often covers the aedeagus completely when viewed from above. In Cicadellidae

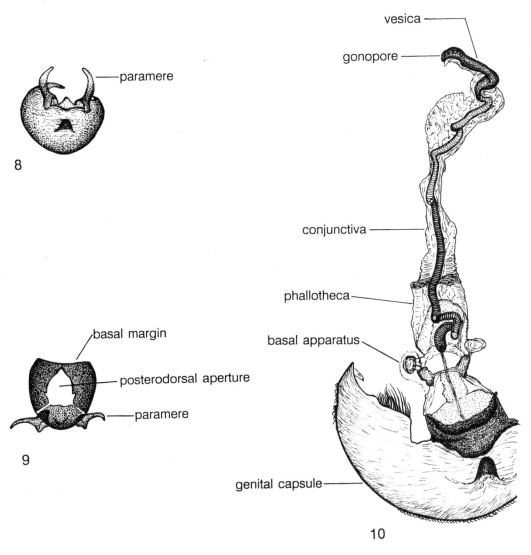

Figs 8–10. *Male genitalia of* Nysius *(Lygaeidae). 8, posterior view of genital capsule and parameres; 9, dorsal view of same; 10, posterior view of part of genital capsule and fully inflated aedeagus.*

(Figs 18, 19) the pygofer appears to be derived from the much enlarged ninth abdominal tergum; the sternum is small and triangular and is termed the *genital valve*. It bears a pair of broad *genital plates*, which enclose the aedeagus ventrally and are derived from parts of the parameres. The claspers in this family are therefore not fully homologous with the parameres of other Hemiptera and are usually termed *styles*. Unfortunately, the same term is also applied to the true parameres of other Auchenorrhyncha. The parameres are connected, tenuously in Heteroptera but by a direct articulation in most Auchenorrhyncha, with the *basal plate* of the aedeagus, otherwise known as the *basal apparatus* or, in Auchenorrhyncha, the *connective*. This structure may be stirrup- Y- or T-shaped.

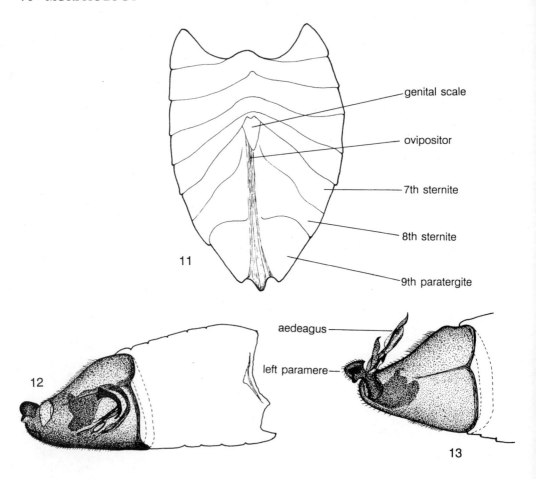

Figs 11–13. *Genitalia of* Psallus *(Miridae). 11, ventral view of female abdomen; 12, dorsal view of male abdomen and genital capsule (shaded), showing aedeagus in repose by transparency of capsule; 13, male genital capsule with aedeagus protruded.*

The phallus, which articulates with the basal plate or connective, is often an elaborate structure. It may be wholly sclerotized or partly to wholly membranous. In the latter case its apical part may be eversible. The heteropteran aedeagus (Fig. 10) typically has a basal sclerotized region, the *phallotheca*, into which the membranous *conjunctiva* is retracted in repose. The gonopore may be borne at the tip of a sclerotized tube, the *vesica*, or it may open directly from the conjunctiva. In all groups the distal part of the aedeagus may bear a variety of sclerotized or membranous appendages that provide useful characters for species identification.

A ridge arising from the ventral wall of the genital capsule may lie between the aedeagus and the parameres. In Delphacidae (Fig. 17) this ridge forms a high wall, the phragma, which almost reaches the dorsal wall of the capsule (pygophore), leaving a small aperture, through which the tip of the aedeagus may be projected, ventral to the anal tube.

Slight asymmetry of the male genitalia is frequent in Hemiptera. The greatest degree of asymmetry is encountered in those Anthocoridae in which one paramere is missing and

ABDOMEN 77

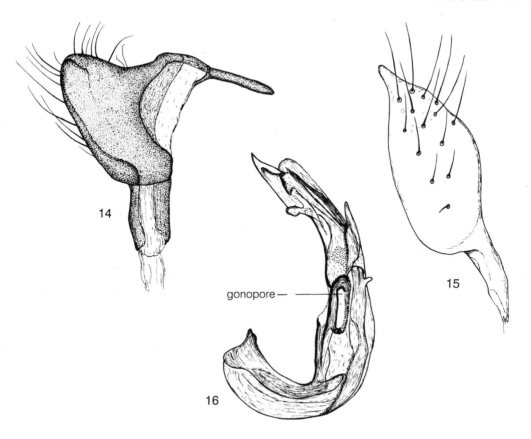

Figs 14–16. *Genitalia of* Psallus *(Miridae). 14, left paramere; 15, right paramere; 16, aedeagus (14–16 greatly enlarged).*

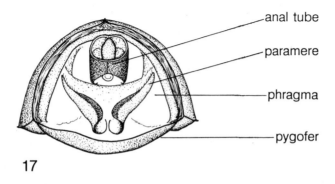

Fig 17. *Male genitalia of* Criomorphus *(Delphacidae), posterior view.*

the other is elaborately developed. The aedeagus and parameres of Miridae, too, are strongly asymmetrical. In the mirid subfamily Phylinae the phallotheca is fused with the genital capsule.

In male Psylloidea the ninth sternum is referred to as the subgenital plate. Dorsally in this group there is a large proctiger, presumably incorporating the ninth tergum as well as the tenth segment. It may be secondarily two-segmented. The male genitalia of other Sternorrhyncha offer rather few characters and are not routinely used in identification although they may be of use in identifying whiteflies.

In female Heteroptera (Fig. 11) and Auchenorrhyncha the eighth to tenth tergites are usually discernible as separate sclerites and the last pregenital sternite is the seventh. In forms with a long ovipositor, the seventh sternite may be elongate and medially cleft and several preceding sternites modified into chevrons to accommodate it. The lateral regions of the eighth and ninth tergites normally extend laterally down the sides of the body so that they are visible in ventral view. They may be marked off from the main part of the tergite by longitudinal sutures and are then termed *paratergites* or *ventral laterotergites*.

The eighth and ninth sternites together form the *ovipositor*. Each sternite is divided longitudinally into two halves and each half is divided into a basal *valvifer* or *gonocoxa* and an apical *valvula* or *gonostylus*. The valvulae on each side lock together by means of grooved, sclerotized rods, the *rami*, to give a sliding fit. The eggs pass between the valvulae as they are laid. As well as guiding the eggs to the oviposition site, the valvulae may be provided with serrations with which to cut an oviposition slit in plant tissue for the reception and protection of the eggs. Usually the gonocoxae of the ninth segment bear a second pair of appendages in addition to their gonostyli. These are the *gonoplacs, third valvulae* or *ovipositor sheath*. They enclose the other two pairs of valvulae at rest. A small, median *genital scale* is often visible ventrally at the base of the ovipositor.

Female Psylloidea usually have only five well developed abdominal tergites and four sternites visible in the pregenital abdomen. The ovipositor in this group is protected

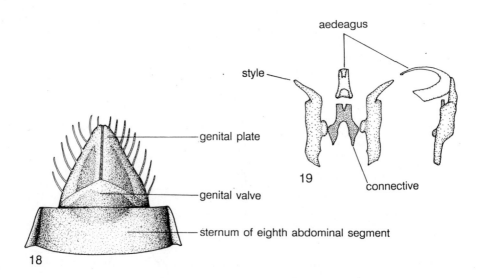

Figs 18, 19. *Male genitalia of* Euscelis *(Cicadellidae). 18, ventral view of abdominal apex; 19, dorsal and left lateral views of connective, aedeagus and styles.*

ventrally by a large *subgenital plate* (the seventh sternite), laterally by the so-called *palps* (gonoplacs) and dorsally by a large, triangular plate, the *proctiger*, which bears the anus dorsally near its anterior end.

In other groups of Sternorrhyncha the ovipositor is usually absent or rudimentary and of little use in identification. The anal region of these groups, however, often has features of diagnostic value. In aphids (Figs 4, 122–126 etc.), the ninth tergite is modified into a *cauda* of often characteristic shape and below the genital and anal openings there are, respectively, the *genital plate* or *subgenital plate* and the *anal plate* or *subanal plate*. The anus of Aleyrodidae is carried at the tip of the *lingula*, a tube lying in the characteristic dorsal *vasiform orifice*, where it is partly or wholly concealed by the overlying flap-like *operculum* (Figs 98–100). The anus of female Coccoidea may be surrounded by a ring whose setae and general form provide useful diagnostic features (Fig. 154). In this superfamily it may also be flanked by a pair of plates (Fig. 157) with characters of diagnostic value.

A concise introduction to the genitalia of Hemiptera may be found in chapters 27 (Homoptera, pp 179–190) and 28 (Heteroptera, pp 190–209) of Tuxen (1970).

12

CLASSIFICATION

Ideas about the higher classification of Hemiptera have undergone rapid changes in recent years. It has been usual to divide the order into two suborders, Heteroptera (with the head capsule closed ventrally behind the rostrum) and Homoptera (with the head capsule open behind the rostrum). Heteroptera were then divided into land-dwellers (Geocorisae), water-surface dwellers (Amphibicorisae) and fully aquatic bugs (Hydrocorisae), while Homoptera were split into Auchenorrhyncha (with 3-segmented tarsi and with the rostrum clearly arising from the back of the head) and Sternorrhyncha (with 1- or 2-segmented tarsi and the origin of the rostrum displaced posteriorly between or behind the anterior coxae). The modern view is to give Heteroptera, Auchenorrhyncha and Sternorrhyncha equal status as suborders, along with a fourth, very small suborder, Coleorrhyncha, which is now restricted to the Southern Hemisphere. Some people even treat these four groups as separate orders but this seems inappropriate in view of the great similarity of structure of the three groups, particularly in the form of the mouthparts. Below the level of suborders, infraorders are increasingly being used, with the termination -morpha to distinguish them from the next level down, superfamilies, with the suffix -oidea. Family names end with the suffix -idae.

A synopsis of the classification of the whole order Hemiptera is given in Table 3. Groups with no indigenous British species are marked with an asterisk (*). The expression 's.lat.' means that the family name is used in a broad sense and that such families are divided into two or more families by some authorities. The number following each family name indicates the approximate number of described species. Vernacular names in English are given opposite their Latin equivalents where they have attained reasonable currency. Some are inappropriate to the British members of the groups involved or are mainly used in other countries, principally the USA. These are cited in quotation marks.

Hypotheses about the relationships among the various groups of Hemiptera are constantly being tested and there is no reason to suppose that the classification presented in this synopsis will either be universally accepted or endure for very long. The accepted limits of some families have already changed since the publication of some recent identification guides and will probably change again before long.

Table 3 Classification of the families of Hemiptera.
Non-British families are marked with an asterisk (*). The approximate number of species known in the world is indicated for each.

*Coleorrhyncha		
*Peloridiidae	20	
Heteroptera		
Pentatomomorpha		
Aradoidea		
Aradidae	1800	Flatbugs, Barkbugs
*Termitaphididae	9	

Table 3 Classification of the families of Hemiptera – *continued*

Pentatomoidea		
*Urostylidae	100	
Acanthosomatidae	200	Shieldbugs
Cydnidae	400	Burrower Bugs
*Plataspidae	500	Shieldbugs
*Canopidae	8	Shieldbugs
*Megarididae	16	Shieldbugs
*Cyrtocoridae	15	Shieldbugs
*Phloeidae	3	Shieldbugs
*Lestoniidae	3	Shieldbugs
*Tessaratomidae	250	Giant Shieldbugs
*Dinidoridae	115	Shieldbugs
*Serbanidae	1	Shieldbugs
Scutelleridae	400	Shieldbugs
*Aphylidae	2	Shieldbugs
Pentatomidae	5000	Shieldbugs
Idiostoloidea		
*Idiostolidae	4	
Piesmatoidea		
Piesmatidae	40	Beetbugs
Lygaeoidea		
*Colobathristidae	60	
Berytidae	160	Stiltbugs
*Malcidae	20	
Lygaeidae	3000	Groundbugs, Seedbugs
Pyrrhocoroidea		
Pyrrhocoridae	300	Firebugs, 'Cotton Stainers'
*Largidae	100	
Coreoidea		
Stenocephalidae	40	Spurgebugs
*Hyocephalidae	2	
Coreidae	2000	'Squashbugs'
Alydidae	250	
Rhopalidae	150	
Cimicomorpha		
*Thaumastocoroidea		
*Thaumastocoridae	12	
*Xylastodoridae	3	Palm Bugs
*Joppeicoidea		
*Joppeicidae	1	
Tingoidea		
Tingidae	1820	Lacebugs
*Vianaididae	5	
Miroidea		
Miridae	6000	Capsids or Leaf Bugs
Microphysidae	25	Minute Bugs
Cimicoidea		
*Velocipedidae	4	
*Medocostidae	2	
Nabidae	370	Damsel Bugs
*Plokiophilidae	13	
Anthocoridae	500	Flower Bugs
Cimicidae	80	Bedbugs
*Polyctenidae	31	Batbugs

Table 3 Classification of the families of Hemiptera – *continued*

Reduvioidea		
*Pachynomidae	16	
Reduviidae	5000	Assassin Bugs
Leptopodomorpha		
*Omaniidae	4	
Aepophilidae	1	Marine Bug
Saldidae	260	Shorebugs
*Leotichiidae	2	
*Leptopodidae	30	
Dipsocoromorpha		
Dipsocoridae	23	
Ceratocombidae	45	
*Hypsipterygidae	3	
*Schizopteridae	155	
*Stemmocryptidae	1	
Gerromorpha		
Mesoveliidae	30	Pondweed Bugs
Hydrometridae	112	Water Measurers, 'Marsh Treaders'
Hebridae	150	Sphagnum bugs
Paraphrynoveliidae	2	
Macroveliidae	3	
Veliidae	500	Water Crickets
Gerridae	500	Pondskaters, Wherrymen, 'Water Striders'
Hermatobatidae	10	
Nepomorpha		
Nepidae	200	Water Scorpions
*Belostomatidae	150	Giant Water Bugs
Corixidae	500	Lesser Water Boatmen
*Gelastocoridae	80	Toadbugs
*Ochteridae	32	
*Potamocoridae	2	
Naucoridae	300	Saucer Bugs
Aphelocheiridae	30	
Notonectidae	300	Water Boatmen, 'Backswimmers'
Pleidae	40	
*Helotrephidae	20	
*Enicocephalomorpha		
*Enicocephalidae	260	
Auchenorrhyncha		
Fulgoromorpha		
Cixiidae	1000	
Tettigometridae	70	
Delphacidae	1300	Planthoppers
*Eurybrachyidae	180	
*Fulgoridae	700	Lantern Flies
*Achilidae	350	
*Tropiduchidae	330	
Issidae	1000	
*Lophopidae	120	
*Ancaloniidae	81	
*Ricaniidae	360	
*Flatidae	1000	
*Derbidae	800	
*Achilixidae	9	

Table 3 Classification of the families of Hemiptera – *continued*

*Meenoplidae	80	
*Kinnaridae	42	
*Dictyopharidae	540	
*Gengidae	2	
*Nogodinidae	140	
*Hypochthonellidae	2	
Cicadomorpha		
Cercopoidea		
Cercopidae s.lat.	2400	Froghoppers, Cuckoo-spit Insects, 'Spittlebugs'
Cicadoidea		
Cicadidae s.lat.	2000	Cicadas
Cicadelloidea		
Cicadellidae	20000	Leafhoppers
*Eurymelidae	100	
Membracoidea		
Membracidae	2400	'Treehoppers'
*Aetalionidae	50	
*Biturritiidae	10	
*Nicomiidae	15	
Sternorrhyncha		
Psylloidea		Jumping Plant Lice
Liviidae	25	
*Homotomidae	72	
Psyllidae	1050	
*Spondyliaspididae	300	
*Calophyidae	33	
*Carsidaridae	37	
Triozidae	650	
Aleyrodoidea		
Aleyrodidae	1156	Whiteflies
Adelgoidea		
Adelgidae	47	Conifer Woolly Aphids
Phylloxeridae	69	Phylloxeras
Aphidoidea		Aphids, Plant Lice
Lachnidae	347	
Chaitophoridae	141	
Callaphididae	488	
Aphididae	2229	
*Greenideidae	127	
Anoeciidae	32	
Thelaxidae	28	
Hormaphididae	130	
Phloeomyzidae	5	
Mindaridae	9	
Pemphigidae	266	
Coccoidea		Scale Insects
Ortheziidae	81	'Ensign Coccids'
Margarodidae	250	'Ground Pearls'
*Phenacoleachiidae	2	
*Stictococcidae	15	
Pseudococcidae	2000	Mealybugs
Coccidae	1200	Soft Scales
*Aclerdidae	53	

Table 3 Classification of the families of Hemiptera – *continued*

*Dactylopiidae	9	Cochineal Insects
Kermesidae	69	
Eriococcidae	500	Felted Scales
*Apiomorphidae	62	Pegtop Coccids
*Kerriidae	65	Lac Insects
*Beesonidae	3	
*Lecanodiaspididae	74	
*Cerococcidae	64	
Asterolecaniidae	200	Pit Scales
*Conchaspididae	23	
*Phoenicococcidae	1	Red Date Scale
*Halimococcidae	17	Palm Scales
Diaspididae	2500	Armoured Scales

13

KEY TO SUBORDERS OF BRITISH HEMIPTERA

1. Head capsule closed ventrally behind rostrum by a gula (Fig. 2) .... HETEROPTERA
— Head capsule not closed ventrally behind rostrum or rostrum absent 2
2. Rostrum clearly arising from ventral surface of head (Fig. 78); tarsi of adult always with three segments .... AUCHENORRHYNCHA
— Rostrum apparently originating between anterior coxae (Figs 4, 102, 106, 154, 155) or, rarely, absent; tarsi of adult never with more than two segments .... STERNORRHYNCHA

14

HETEROPTERA

Of the three suborders of Hemiptera, only Heteroptera possess a gula. This is an area of cuticle closing the head capsule ventrally behind the mouthparts. The effect is to bring the rostrum forwards, allowing it greater freedom of movement than is possible in Auchenorrhyncha and Sternorrhyncha. The greater versatility of the mouthparts enables Heteroptera to exploit a greater variety of foodstuffs, including animal tissues. In most adult Heteroptera the fore wings, which are held flat over the body at rest, comprise a tough, opaque, basal part (the corium) and a thin, transparent or translucent apical membrane in the area where the two fore wings overlap. The transition between the two areas is abrupt.

Scent glands with conspicuous external openings are characteristic of Heteroptera, although they are lacking in some aquatic forms. In the nymphs they are situated intersegmentally on the abdominal dorsum and in the adult there is a metathoracic gland opening by a median aperture at the front of the metasternum or by a pair of lateral apertures on the metapleura.

All truly aquatic Hemiptera and all predaceous ones belong to this suborder but the majority of species are terrestrial and phytophagous. Eggs are laid either exposed on the surfaces of leaves or twigs, or inserted into plant tissues or crevices in such substrates as moss and bark. The form of the ovipositor reflects these differences in egg-laying habits. The valves are short and lobe-like in those species whose eggs are laid superficially, and long and pointed in those laying eggs in concealed situations. Most species have five nymphal instars but a few with strongly abbreviated wings as adults have only four.

There are more than 500 British species of Heteroptera. Of these, 43 live submerged in fresh water, 20 live mainly on the surface film, one is active below the surface of tidal pools and a few more can survive immersion by the sea water in tidal marshes. The most recent comprehensive identification guide is that of Southwood & Leston (1959), replacing the monographs of Douglas & Scott (1865) and Saunders (1892). Stichel's (1955–1962) keys cover the entire heteropteran fauna of Europe, and Kerzhner's and Jaczewski's contributions to Bei-Bienko (1964) are also useful in this context. Wagner (1952, 1966, 1967) dealt with the German terrestrial bugs. Savage (1989) covers the aquatic groups in Britain and Poisson's (1957) work on the French fauna is also relevant. Butler (1923) gave detailed biological information and figures the immature stages of many species. Cobben's (1968, 1978) works provide a very useful introduction to the embryology, and many aspects of the anatomy and morphology, of the suborder.

The terrestrial Heteroptera most often encountered in general ecological surveys, apart from the readily recognizable shieldbugs and lacebugs, belong to four families, macropterous specimens of which can be separated on a rough-and-ready basis as follows:

— Body soft; ocelli absent; membrane venation consisting of two closed cells, one of them very small; cuneal fracture present (Fig. 1) .... MIRIDAE

— Body soft; ocelli present; membrane venation rich, with closed cells and many free veins; cuneal fracture absent (Fig. 55) .... NABIDAE

— Body soft; ocelli present; membrane venation poor, with no closed cells and a few longitudinal veins; cuneal fracture present (Fig. 56) .... ANTHOCORIDAE
— Body hard; ocelli present; membrane venation poor, with one or no closed cell and with a few longitudinal veins; cuneal fracture absent (Figs 44, 45) .... LYGAEIDAE

Key to families of British Heteroptera (adults)

1. Antennae longer than head, easily visible from above when insect is active but sometimes capable of being folded beneath body at rest; insects living on land or on the surface of water, not immersed 2
— Antennae much shorter than head, usually concealed in dorsal view; insects living submerged in water 33

2. Scutellum at least half as long as distance from hind margin of pronotum to posterior end of body. ('shield bugs', always macropterous and all species normally with five antennal segments.) 3
— Scutellum distinctly shorter than half distance from hind margin of pronotum to posterior end of body. (Antennae with four segments except for *Hebrus*, which is less than 2 mm long and has the scutellum only about one-sixth as long as distance from hind margin of pronotum to posterior end of body.) 6

3. Tibiae with strong spines (Fig. 28) as well as fine setae; if scutellum large, insect almost wholly glossy black .... CYDNIDAE
— Tibiae without strong spines or, if with fairly strong spines (*Odontoscelis*), scutellum greatly enlarged and almost wholly concealing membrane and insect not glossy black 4

4. Scutellum concealing at least three-quarters of hemelytral membrane at rest AND pronotum without projections at anterolateral angles (Fig. 29) .... SCUTELLERIDAE
— Scutellum concealing at most one-quarter of hemelytral membrane at rest (Fig. 32) OR, if concealing three-quarters (*Podops*, Fig. 30), anterolateral angles of pronotum with distinctive projections just behind eyes 5

5. Tarsi with 2 segments. Base of abdomen ventrally with anteriorly projecting median acute process whose apex lies alongside ventral thoracic keel (Fig. 23) ...... ACANTHOSOMATIDAE
— Tarsi with 3 segments. Base of abdomen ventrally either without a median process or, if such a process is present, its apex does not reach median thoracic keel (Fig. 31) ... PENTATOMIDAE

6. Head very long and narrow (Fig. 66). (Living on surface of still waters.) .... HYDROMETRIDAE
— Head less than twice as long as wide 7

7. Antennae with 5 segments or with segment 4 constricted in middle. (Living on surface of still waters.) .... HEBRIDAE
— Antennae with 4 segments, the last not medially constricted 8

8. Totally apterous and largely green OR macropterous with scutellum visible and metanotum raised in a tongue-like process behind it (Fig. 65). (Living on surface of still waters.) .... MESOVELIIDAE
— If totally apterous, then largely black; if macropterous then without tongue-like process behind scutellum (but very occasionally with a spine behind the scutellum, which itself is spine-like) 9

9. Macropterous or brachypterous with pronotum extending posteriorly so that scutellum is completely concealed OR totally apterous 10
— Scutellum not completely concealed by pronotum; never totally apterous 12

10. Claws of anterior tarsi apical; surface of pronotum and hemelytra areolate (*Agramma*) or reticulate (Fig. 48). (Terrestrial.) .... TINGIDAE
— Claws of anterior tarsi subapical (Fig. 67); surface of pronotum and hemelytra neither areolate nor reticulate. (Living on surface of still or flowing water.) 11

KEY TO FAMILIES OF BRITISH HETEROPTERA (ADULTS) 87

11.	Middle pair of legs attached to body much closer to hind pair than to fore pair and obviously longer than hind pair (Fig. 68)	GERRIDAE
—	Middle pair of legs attached to body about equidistant between fore and hind pairs and about as long as hind pair (Fig. 67)	VELIIDAE
12.	Ocelli absent (or very indistinct)	13
—	Ocelli present	21
13.	Tarsi with 2 segments	14
—	Tarsi with 3 segements	15
14.	Very flat, leathery, black or brown insects with all antennal segments stout (Fig. 20)	ARADIDAE
—	Small, rather delicate, brachypterous insects, not flat and leathery; hemelytra without obvious venation, resembling beetle elytra (Fig. 58)	(most females of) MICROPHYSIDAE
15.	Head with a transverse furrow between or just behind eyes. Part of head behind furrow either globular or coarsely punctured and hairy and not capable of retraction into thorax (Figs 46, 52). (Narrow-bodied insects with very slender legs and antennae.)	16
—	Head without a transverse furrow. If a furrow can be seen behind the eyes, it marks the limit to which the head can be retracted into the thorax and the region behind it is neither globular nor punctured and hairy	17
16.	Scutellum short, with an apical spine, metanotum with a similar spine visible behind it; first antennal segment and femora not clavate; fore legs raptorial, their femora stouter than those of the other two legs and with numerous, slender spines beneath (Fig. 52). (Subfamily Emesinae.)	REDUVIIDAE (part)
—	Scutellum sometimes short, with an apical spine, but metanotum never with a spine; first antennal segment and femora abruptly clavate at apex; fore legs not raptorial, their femora no thicker than those of the other legs and without spines (Fig. 46). (Ocelli present but often difficult to see.)	BERYTIDAE
17.	Pronotum with disc black and anterior, posterior and lateral margins red. (One rare species in Britain; British examples almost all brachypterous, without hemelytral membrane; a robust insect with a characteristic red and black colour pattern, Fig. 42)	PYRHOCORIDAE
—	Pronotum rarely red and black and then not patterned as above	18
18.	Head coarsely sculptured, with dense punctures or granular appearance. (Ocelli present but difficult to see; membrane of hemelytra, unless reduced, with four or five more or less parallel, longitudinal veins reaching, or almost reaching, the margin.)	LYGAEIDAE
—	Head smooth, without coarse puncturation or granulation	19
19.	Micropterous, posterior margins of hemelytra sinuate (Fig. 62). (Ocelli present but very difficult to see.) In the intertidal zone of rocky sea shores	AEPOPHILIDAE
—	If micropterous, then posterior margins of hemelytra truncate or slightly convex	20
20.	Pronotum with lateral flanges which are particularly conspicuous at the anterior angles. (Always micropterous. Infesting houses, birds' nests and bat roosts.)	CIMICIDAE
—	Pronotum smoothly rounded into propleura or with a more or less distinct keel laterally but these keels, if present, never developed into flanges at anterior angles (Fig. 59)	MIRIDAE
21.	Rostrum short, stout, curved, apparently 3-segmented, its apex reaching only to a minutely cross-striate furrow on anterior part of prosternum (Fig. 51); head dorsally with a transverse furrow behind eyes (Fig. 50). (Femora not clavate.)	REDUVIIDAE
—	Rostrum usually longer, prosternum lacking a cross-striate furrow; if rostrum short then head without a transverse furrow between eyes. (Berytidae, with a short, straight rostrum and a furrow between the eyes, have femora and first segment of antennae abruptly clavate at apex, Fig. 46.)	22
22.	Tarsi with 2 segments	23
—	Tarsi with 3 segments	24

HETEROPTERA

23. Hemelytra heavily punctured, insects resembling Tingidae; pronotum with two or three longitudinal keels .... PIESMIDAE
— Hemelytra not punctured; pronotum without keels; delicate insects, females usually brachypterous, hemelytra lacking a membrane. (Anthocoridae may key out here if the tarsal segmentation is indistinct and has been wrongly counted as two instead of three; compare 3-segmented rostrum of Anthocoridae, Fig. 56, with 4-segmented rostrum of Microphysidae, Fig. 58) .... MICROPHYSIDAE

24. Antennae with segments III and IV very slender and twice as long as segments I and II taken together (Fig. 61) .... CERATOCOMBIDAE and DIPSOCORIDAE
— Antennae with segments III and IV stout or slender but never twice as long as segments I and II taken together 25

25. Rostrum apparently 3-segmented. (Cuneal fracture often present.) 26
— Rostrum obviously 4-segmented. (Cuneal fracture absent.) 27

26. Eyes small (Fig. 56); if brachypterous, hemelytra truncate or convex apically. (The intertidal *Aepophilus*, family Aepophilidae, would key out here if its small ocelli were not overlooked; its hemelytra lack a membrane and are apically sinuate: Fig. 62) .... ANTHOCORIDAE
— Eyes large (Fig. 63) .... SALDIDAE

27. Rostrum slender, curved, not appressed to underside of body at rest (Fig. 53); ventral surface of anterior tibia with a double row of spines that are usually small and peg-like (Fig. 54) .... NABIDAE
— Rostrum straight, appressed to underside of head and thorax at rest; anterior tibiae not minutely spined 28

28. Legs and antennae slender, femora and segment I of antennae abruptly clubbed at apex (Fig. 46); if indistinctly clubbed then antennal segment I at least twice as long as head .... BERYTIDAE
— Legs and antennae relatively robust, femora and segment I of antennae not abruptly clubbed; antennal segment I never twice as long as head 29

29. Membrane of hemelytra with 4 or 5 roughly parallel veins reaching or almost reaching margin (Figs 44, 45); if hemelytron reduced, insect less than 6 mm long .... LYGAEIDAE
— Membrane of hemelytra with more than 5 veins reaching or almost reaching margin (Figs 39, 43); if hemelytron reduced, insect elongate and over 6 mm long 30

30. Antennae and legs with dark and pale annulations; head anteriorly bilobed as juga meet in front of tylus (Fig. 41) .... STENOCEPHALIDAE
— Antennae and legs without dark and pale annulations; head anteriorly trilobed as juga do not extend beyond tylus 31

31. Posterior femora strongly dentate AND antennal segment IV longer than any other segment, curved (Fig. 39) .... ALYDIDAE
— Antennal segment IV not the longest OR posterior femora not dentate 32

32. Corium opaque and punctured throughout and never black and red. Metathoracic gland orifices visible in lateral view (Fig. 36); anterior margins of abdominal tergites 5 and 6 emarginate, the scars of the two dorsal abdominal glands thus both displaced posteriorly (Fig. 37) .... COREIDAE
— Corium with at least some of the areas between the veins impunctate, glassy and transparent or, if not, whole insect red with bold black markings. Metathoracic gland orifices not visible in lateral view (Fig. 35); abdominal tergite 5 emarginate both anteriorly and posteriorly, the scars of the two dorsal abdominal glands thus approximated (Fig. 38) .... RHOPALIDAE

33. Apex of abdomen with a long, slender respiratory siphon, its two halves sometimes separating after death (Fig. 69) .... NEPIDAE
— Apex of abdomen without siphon 34

34. Each pair of legs much modified in structure and clearly different from the others (Fig. 73). Tarsi of middle pair of legs one-segmented, with very long claws .... CORIXIDAE
— With at most one pair of legs (first or last pair) much modified in comparison with the other two pairs. Tarsi of middle pair of legs 3-segmented (basal segment very small) 35

35. Fore legs strongly raptorial, their tarsi 1-segmented and without claws (Fig. 71)   NAUCORIDAE
— Fore legs not strongly raptorial, their tarsi 3-segmented (basal segment very small) and with a pair of claws 36

36. lattened, oval insect, almost invariably brachypterous, living under stones in running water (Fig. 70) .... APHELOCHEIRIDAE
— Deep-bodied insects, always macropterous, usually living in still waters 37

37. Body length over 1 cm; hind tarsus densely fringed with hairs (Fig. 72) .... NOTONECTIDAE
— Body length less than 3 mm; hind tarsus with sparse hairs not arranged in dense fringes .... PLEIDAE

Key to instars of Heteropteran nymphs

The following key applies to nymphs destined to become macropterous or sub-macropterous adults. Separation of the first two instars is not as easy as couplet 4 implies.

1. Meso- and metanota with wing-pads projecting posteriorly at least as far as the length of those segments 2
— Meso- and metanota at most with posterior margin lobed at the sides, the lobes not as long as the segments 3

2. Mesothoracic wing-pads longer than metathoracic pads, their tips covering the tips of metathoracic pads .... Fifth instar
— Mesothoracic wing-pads about as long as metathoracic pads, not covering their tips .... Fourth instar

3. Posterior margins of meso- and metanota distinctly produced into posteriorly projecting lobes at each side .... Third instar
— Posterior margins of all thoracic segments not lobed 4

4. Mesonotum longer than metanotum .... Second instar
— Mesonotum and metanotum about the same length .... First instar

Key to families of British Heteroptera (nymphs)

In general, the characters of the immature stages are less well developed and more difficult to see in the earlier instars. The following key should be used with caution, particularly for nymphs of the first to fourth instars.

1. Insects living submerged in water; antennae short, concealed under head in dorsal view 2
— Insects terrestrial or living supported on the surface of the water; antennae long, easily visible 7

2. Hind tarsus with a single claw; openings of dorsal abdominal glands present between segments 3–4, 4–5 and 5–6; mid femur distinctly longer than hind femur .... CORIXIDAE
— Hind tarsus with two claws; abdomen with a single dorsal gland (between segments 3–4) or with none at all; mid and hind femora almost equal in length 3

3. Fore tarsus with paired claws 4
— Fore tarsus without claws, ending in a single, acutely pointed tarsal segment 6

4. Body flattened, disc-shaped; paired dorsal abdominal gland openings visible between segments 3–4 .... APHELOCHEIRIDAE
— Body strongly convex dorsally; dorsal abdominal gland opening single, difficult to see (Pleidae) or absent (insects swim with dorsal surface directed downwards) 5

5. Rostrum 4-segmented; hind legs with dense fringes of swimming hairs; body length, except in first instar, more than 2.5 mm .... NOTONECTIDAE
— Rostrum 3-segmented; hind legs with sparse, long hairs; body length at most 2.5 mm .... PLEIDAE

6. Dorsal abdominal gland with paired openings; anterior femur not grooved; respiratory siphon absent .... NAUCORIDAE
— Dorsal abdominal glands completely absent; anterior femur with groove into which tibia folds at rest; respiratory siphon present at apex of abdomen but shorter and broader than in adult .... NEPIDAE

7. Insects living supported on surface of water (among moss and debris in winter); underside of body clothed with dense, silvery pubescence 8
— Insects terrestrial or in marginal aquatic habitats; underside of body without dense covering of silvery hairs 12

8. Claws subapical (Fig. 67; difficult to see in small specimens; dorsal abdominal glands absent) 9
— Claws inserted at extreme apex of tarsi (Fig. 65) 10

9. Middle pair of legs distinctly the longest; hind femora reaching beyond apex of abdomen by at least half their own length; middle pair of legs attached to body much closer to hind pair than to fore pair .... GERRIDAE
— Middle and hind pairs of legs almost equal in size; hind femora not, or only slightly, reaching beyond apex of abdomen; the three pairs of legs evenly spaced .... VELIIDAE

10. Head at least twice as long as wide; abdominal glands absent .... HYDROMETRIDAE
— Head less than twice as long as wide; abdominal glands present 11

11. Antennal segment I at most equal to inter-ocular width; colour of insect mainly reddish or yellowish brown .... HEBRIDAE
— Antennal segment I longer than inter-ocular width; colour of insect largely greenish brown .... MESOVELIIDAE

12. A single dorsal abdominal gland present with one opening or a pair of openings, situated between segments 3–4 13
— Two or three, single or paired, dorsal abdominal gland openings present 15

13. Each abdominal tergite with several sclerotized plates; insects flattened, living under bark (Fig. 22) .... ARADIDAE (part)
— Abdominal segments without sclerotized plates 14

14. Head dorsally with 3 pairs of trichobothria; femora without trichobothria; rostrum appearing 3-segmented (Fig. 64). Mostly black or grey; the intertidal *Aepophilus*, family Aepophilidae, which is brown, also keys out here .... SALDIDAE
— Head without trichobothria; mid and hind femora each with several trichobothria; rostrum clearly 4-segmented (Fig. 60) .... MIRIDAE

15. Two dorsal abdominal glands present (trichobothria present on abdominal sternites of all except Tingidae) 16
— Three dorsal abdominal glands present 23

16. Dorsal abdominal gland openings in intersegmental areas 3–4 and 4–5 17
— Dorsal abdominal gland openings in intersegmental areas 4–5 and 5–6 18

17. Paraclypei projecting freely on each side of clypeus (Fig. 47); trichobothria present near margins of abdominal sternites 5 and 6. (Head not spiny dorsally.) .... PIESMIDAE
— Paraclypei not projecting freely beside clypeus; abdomen without trichobothria; head very often with dorsal spines (Fig. 49) .... TINGIDAE

18.	Antennae and legs slender, threadlike, antennal segment I and femora usually slightly clavate apically	BERYTIDAE
—	Antennae and legs shorter and stouter	19
19.	The two dorsal abdominal gland openings both indented into segment 4, one anteriorly and one posteriorly (Fig. 38)	RHOPALIDAE
—	Each dorsal abdominal gland opening indented into the segment behind it (Fig. 37)	20
20.	Insect ant-like; head narrowing evenly from anterior border of eyes to apex; antennal segment IV long, gently curved (Fig. 40)	ALYDIDAE
—	Insect not strikingly ant-like; head often truncate anteriorly or with strongly projecting antennifers; antennal segment IV not curved	21
21.	Paraclypei, in the later instars, projecting beyond clypeus and contiguous in front of it (as in Fig. 41)	STENOCEPHALIDAE
—	Paraclypei not projecting beyond clypeus	22
22.	Antennae inserted on or below an imaginary line connecting tip of clypeus and middle of eye in lateral view (Fig. 34); trichobothria of abdominal sternites 5, 6 and 7 more than three times as far from abdominal midline as from abdominal margin	LYGAEIDAE (part)
—	Antennae inserted above an imaginary line connecting tip of clypeus and middle of eye in lateral view (Fig. 33); trichobothria of abdominal sternites 5, 6 and 7 less than three times as far from abdominal midline as from abdominal margin	COREIDAE
23.	Abdominal sternites with trichobothria (Fig. 24)	24
—	Abdominal sternites without trichobothria	29
24.	Trichobothria of abdominal sterna 3 and 4 close to midline, those of 5, 6 and 7 close to lateral margins	25
—	Trichobothria of all abdominal sterna 3 to 7 close to lateral margins (shieldbugs)	26
25.	Opening of dorsal abdominal gland at intersegmental area 5–6 strongly displaced posteriorly (rare, red-and-black insects)	PYRRHOCORIDAE
—	Opening of dorsal abdominal gland at intersegmental area 5–6 not or weakly displaced posteriorly	LYGAEIDAE (part)
26.	Tibiae armed with spines; opening of anterior dorsal abdominal gland a slot, without pores at the ends (Fig. 26)	CYDNIDAE
—	Tibiae at most with stout hairs; all three dorsal abdominal gland openings with a pore at each end	27
27.	Anterior pair of dorsal abdominal gland pores only as far apart as the other two pairs (Fig. 27)	PENTATOMIDAE
—	Anterior pair of dorsal abdominal gland pores more widely spaced than the other two pairs (Fig. 25)	28
28.	Insect almost wholly green or green with a yellowish or reddish suffusion, integument not pigmented with brown or black; usually on bushes or trees	ACANTHOSOMATIDAE
—	Insect with brown or black pigment in integument, never green; usually in soil or among grasses	SCUTELLERIDAE
29.	Rostrum short, stout, curved, 3-segmented (as in Figs 51, 52); prosternum anteriorly with a finely cross-striate stridulatory groove	REDUVIIDAE
—	Rostrum usually more slender, usually straight; prosternum without stridulatory groove	30
30.	Rostrum clearly 4-segmented	31
—	Rostrum apparently 3-segmented; insect often rather flattened	32
31.	Legs long, slender, apex of hind femur reaching to about apex of abdomen (insect often more than 2 mm long)	NABIDAE
—	Legs short, apex of posterior femur reaching to about middle of abdomen; body length 2 mm or less	MICROPHYSIDAE

92 HETEROPTERA

32. Abdominal terga each with several sclerotized plates; all antennal segments short, robust; insects living under bark (Fig. 21) .... ARADIDAE (part)
— Abdominal terga without sclerotized plates 33
33. Antennal segments III and IV filiform, much more slender than segments I and II and more than twice as long as I and II taken together .... CERATOCOMBIDAE and DIPSOCORIDAE
— Antennal segments III and IV often filiform but never twice as long as I and II together 34
34. Most body hairs serrate or apically bifid (Fig. 57); no trace of pads of developing wings or hemelytra .... CIMICIDAE
— Pubescence of normal type; later instars with obvious wing and hemelytra pads .... ANTHOCORIDAE

Aradidae

Even in comparison with countries at similar latitudes, Britain has very few representatives of this medium-sized family, amounting only to five species of *Aradus* (barkbugs) and two of *Aneurus* (flatbugs). The commonest species, and the only one found in Ireland, is *Aradus depressus*.

Most aradids feed on fungi. Their extremely flattened bodies, short legs and short, thick antennae enable them to enter the space between the loose bark and the wood of dead trunks and branches in the first year or two of decay, when the mycelia of the higher fungi are at their most abundant. The aradid rostrum is short but the stylets are very long, being coiled in sacs in the head when not in use. By contracting these sacs, the bug can protrude its stylets to follow the course of fungal hyphae through the decaying wood. Seasonality is not very pronounced and eggs, nymphs and adults of the commoner species have been found in most months of the year. Although the typical habitat is beneath bark, barkbugs are sometimes found on the fruiting bodies of bracket fungi and are occasionally taken by sweeping low vegetation in woods. Massee (1960) reported large numbers of *Aradus depressus* flying around the stumps of recently felled trees that were still exuding sap.

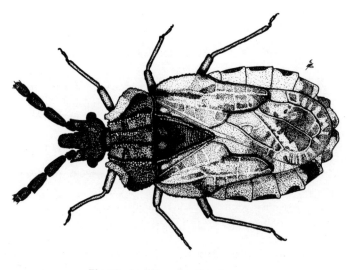

Fig 20. *Aradidae:* Aradus depressus.

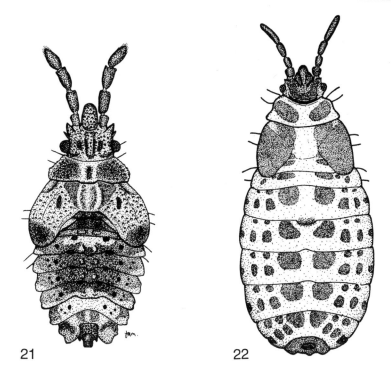

Figs 21–22. *Aradidae nymphs. 21,* Aradus depressus; *22,* Aneurus laevis.

Aradus cinnamomeus is unusual in several ways. It differs from the other British species of the family, which are always macropterous, in displaying wing-polymorphism. Females are brachypterous or macropterous while the males are 'stenopterous': their hemelytra reach almost to the tip of the abdomen but are greatly narrowed in the apical three-quarters of their length. The diet of *A. cinnamomeus* is unique among British aradids in that the insect is a sap-feeder, although its mouthparts are not noticeably different from those of its fungus-feeding relatives. It lives beneath bark flakes on the stems and young branches of scots pine in the heaths of southern England.

Acanthosomatidae

This and the following three families are collectively known as shieldbugs, the name being derived from the large scutellum. The five native acanthosomatids have the typical shieldbug characteristics of five-segmented antennae and a scutellum that reaches at least to the base of the membranes of the hemelytra when the latter are folded at rest. Like all British shieldbugs, they are always macropterous. Their tarsi have only two segments instead of the three that are present in all the other shieldbug families.

All five species have an annual life-cycle, overwintering as adults. *Elasmostethus tristriatus* passes the winter months on the junipers and cypresses on which it feeds and, in consequence, is often brought indoors when branches of its hostplants are cut for winter decorations. Its striking coloration – green and red – makes it a conspicuous insect when the unusual warmth stirs it into activity.

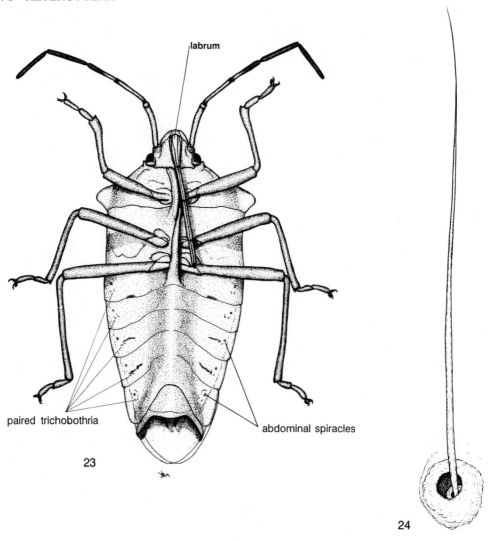

Figs 23–24. *Acanthosomatidae:* Acanthosoma haemorrhoidale. *23, ventral view;* 24 *trichobothrium, greatly enlarged*

Elasmucha grisea, which lives on birch, has long been famous for the phenomenon of parental care. The female lays her eggs, usually between 30 and 40 of them, in a close-packed hexagonal array on the underside of a birch leaf and straddles them until they hatch, shielding them from potential enemies with her body. In the 20 or so days that it takes for the eggs to develop, she never leaves them, feeding only on the mesophyll of the leaf, which becomes peppered with small, white patches of air-filled cells showing where she has taken her frugal meals.

Elasmostethus interstinctus also lives mainly on birch and *Acanthosoma haemorrhoidale* feeds on hawthorn and several related trees and shrubs. These inhabitants of deciduous trees hibernate among grasses and mosses, beneath logs and in similar secluded places.

Cydnidae

The spiny legs of this family of shieldbugs reflect their basic life-style, that of root-sucking burrowers. Only one of the eight British species, *Geotomus punctulatus*, pursues this way of life, in the coastal sand dunes of the extreme south-west. It is unique among native Cydnidae in having a fringe of hairs around the margin of the head. A single specimen of the non-native *Aethus flavicornis* was once taken on the south coast. Its head bears a marginal row of stout pegs, presumably an adaptation for burrowing.

Thyreocoris scarabaeoides differs from the other species in having the hemelytra almost completely hidden beneath a large, rounded scutellum. Its glossy black coloration and spiny legs mark it out as a member of the Cydnidae. It is the smallest of the British shieldbugs with a rounded scutellum, being less than 4 mm long. The food plants are species of *Viola*, especially hairy violet on chalk downs and field pansy in sandy areas.

Four species of *Sehirus* and two of *Legnotus* live in the British Isles. They belong to the subfamily Sehirinae, characterized by the triangular scutellum and the absence of hairs or pegs fringing the head. Most of them have the glossy, black cuticle typical of Cydnidae

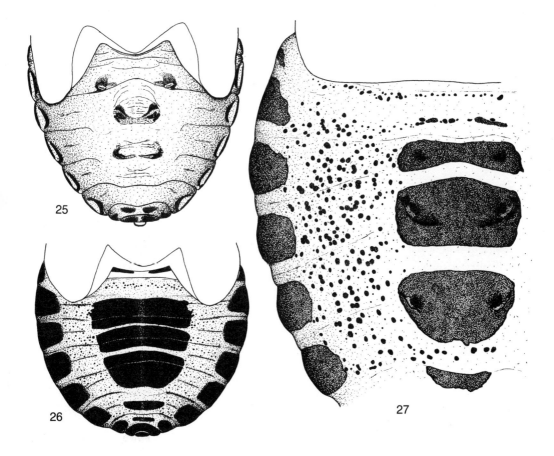

Figs 25–27. *Fifth-instar nymphs, dorsal views of abdomens.* 25, Acanthosomatidae: Elasmostethus tristriatus 26, Cydnidae: Sehirus biguttatus; 27, Pentatomidae: Dolycoris baccarum.

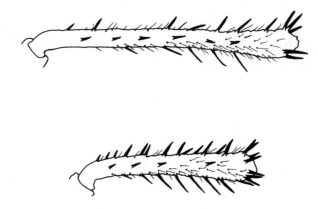

Fig 28. *Cydnidae:* Sehirus bicolor, *mid and hind tibiae, showing stout spines.*

but *Sehirus dubius*, a rare, southern species living on bastard toadflax, is metallic blue or purple and *Sehirus bicolor* is strikingly pied black and white. The biology of this last species is typical of Sehirinae. It feeds on the stems of its host plant – usually white deadnettle – and has a single generation each year, overwintering as an adult. Its burrowing ability is used when selecting a hibernation site and again, by the gravid female, in excavating a shallow depression in the soil in which her ball of about 40 eggs is laid and guarded until hatching. As in all Cydnidae, both sexes have a row of pegs on a vein in the hind wing. When rubbed against a ridged area of the dorsal surface of the body (with the wings and hemelytra held in the normal closed position), these pegs produce a high-pitched vibration which plays a role in courtship (Plate 4, fig. 1).

Scutelleridae

In all four British species of Scutelleridae the hemelytra are almost completely hidden beneath the greatly enlarged, rounded scutellum, with less than a quarter of the membrane area exposed. This degree of development of the scutellum also occurs in *Thyreocoris* (Cydnidae) and *Podops* (Pentatomidae), but the two species of *Odontoscelis* differ from them in being densely hairy and the two native *Eurygaster* species, both over 8 mm long, are much bigger. Adults are very variable in colour, exhibiting a mixture of different shades of brown and, sometimes, black. There is a single generation annually. The overwintering stage is the adult in *Eurygaster* and the half-grown nymph in *Odontoscelis*.

Odontoscelis species live in sandy places, usually near the coast, where both adults and nymphs burrow in loose soil. Collectors have associated both species with storksbill but it is not clear whether this is a food plant or simply an indicator of a suitably friable substrate. A white wax, produced on the underside of the male abdomen, is believed to have an 'aphrodisiac' effect on the female, like the androconial scales of some male Lepidoptera (Carayon, 1984).

Eurygaster species feed on grasses, attacking the grain in the ear. In some parts of Europe and Asia, though not in Britain, they are major pests of cereal crops. The oval eggs are laid in a double row on a grass blade, usually in batches of 14 (showing that only one egg per ovariole matures at a time), with an interval of several days between batches.

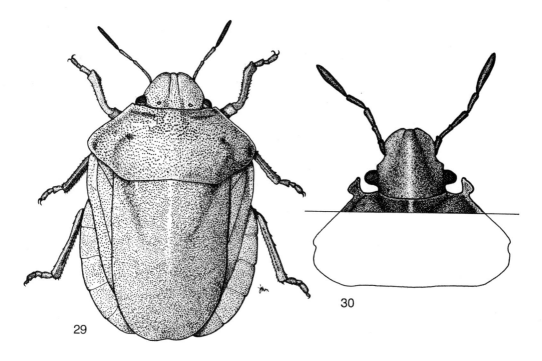

Figs 29–30. *29, Scutelleridae:* Eurygaster maura; *30, Pentatomidae:* Podops inuncta, *head and anterior pronotum.*

Pentatomidae

This is the largest family of shieldbugs, represented in Britain by 17 breeding species, one former native that is probably now extinct (*Pitedia juniperina*, on juniper) and several others, more or less regularly imported, that cannot tolerate the climate. All have tibiae that are at most bristly, rather than spiny, and three-segmented tarsi. In all but one species the scutellum is triangular or, at most, narrowly rounded and does not conceal more that a quarter of the membrane of the hemelytra.

The only species with a broadly rounded scutellum reaching almost to the tip of the abdomen is *Podops inuncta*, which resembles a small (up to 6.2 mm long) scutellerid. This grey-brown insect, which is the only British representative of the subfamily Podopinae, differs from the Scutelleridae (and from the glossy black cydnid *Thyreocoris*, with a similarly enlarged scutellum) in having a small hook on each anterolateral angle of the pronotum, partly embracing the short-stalked eyes posteriorly and laterally. It is always found at ground level, sometimes beneath objects lying on the ground such as stones, sacking, dung or carrion. It is widespread in southern England and probably feeds on grasses.

In four British pentatomids the rostrum is robust, about as thick as the anterior femur. This modification, which is characteristic of the subfamily Asopinae, appears to be an adaptation to their predaceous way of life. Three of these species are bronze-brown with a few orange markings while the fourth, *Zicrona caerulea*, is bright, metallic blue or greenish-blue; the nymphs of *Zicrona* are orange-red with varying amounts of blue-black

98 HETEROPTERA

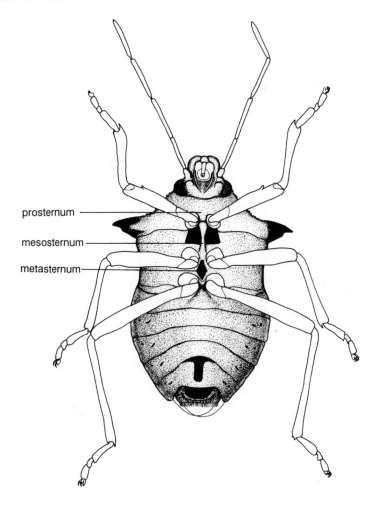

Fig 31. *Pentatomidae:* Picromerus bidens, *ventral view.*

markings. These four Asopinae feed mainly on caterpillars and the larvae of chrysomelid beetles but other slow-moving insects, including adult moths, are also taken. The bugs are 'timid predators', detecting prey by antennal contact and gently inserting the stylets; they make no attempt to subdue active prey or to hold struggling victims with the legs. Vigorous resistance by the prey usually causes the bug itself to flee. *Troilus luridus* lives on trees; the other three species hunt their prey on low plants.

The remaining shieldbugs belong to the subfamily Pentatominae. In all of these the rostrum is more slender than the anterior tibia and the diet is probably strictly vegetarian. Records of predation by *Pentatoma rufipes*, one of the few normally arboreal species, may be due to confusion with one or other of the superficially similar brown Asopinae. *Pentatoma rufipes* is found throughout Britain, as far north as Raasay and Sutherland. Few other representatives of this predominantly southern group occur north of the Trent. Among the more widespread species are the densely hairy, pinkish or yellowish-brown *Dolycoris baccarum* and the green *Palomena prasina* and *Piezodorus lituratus*. The first two

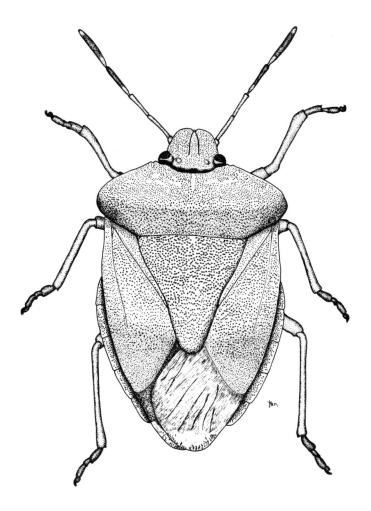

Fig 32. *Pentatomidae:* Palomena prasina.

species are polyphagous, with *P. prasina* often venturing onto trees, especially hazel, while *P. lituratus* is strongly attached to gorse although it sometimes feeds on other plants of the pea family. During the winter, both of these green species are heavily suffused with dark red pigment which is lost at the onset of sexual maturity in the spring.

Pentatomids lay up to about 200 eggs, usually in batches of a dozen or so, either in hexagonal arrays on flat surfaces like leaves or in double rows on twigs and petioles. The eggs are barrel-shaped with a ring of about 20 projections, the aeromicropyles, around the upper rim. There is a circular line of weakness at the upper end of the egg which marks the edge of the operculum, a lid that is forced off when the nymph hatches out. First-instar nymphs do not feed on plants. They stay clustered on the remains of the egg-mass, ingesting the symbiotic bacteria that the female smeared on the eggs as she laid them. There is a single generation annually and in almost all species the adult is the overwintering stage. Exceptions are *Pentatoma rufipes*, which overwinters as third-instar nymphs, and *Picromerus bidens*, which is believed to be able to pass the winter as either eggs or nymphs. In late summer, adults and even the older nymphs of Pentatominae frequently abandon their usual host plants to feed on many different kinds of ripening or fully ripe fruit, which probably enables them to build up fat reserves rapidly. They are sometimes accidentally gathered with blackberries and may ruin a bag full of berries by discharging the nauseous contents of the metathoracic gland over them in response to rough handling.

Coreidae

Ten species of this mainly tropical and subtropical family breed in Britain. None of them is found as far north as Scotland and only one, *Coreus marginatus*, occurs in Ireland. These bugs are always macropterous and the numerous veins in the membrane of the hemelytra distinguish them from all other bugs apart from the related Alydidae, Rhopalidae and

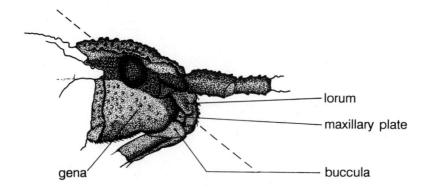

Fig 33. *Coreidae:* Arenocoris fallenii *nymph, lateral view of head. Broken line passes through middle of eye and apex of tylus, below antennal insertion.*

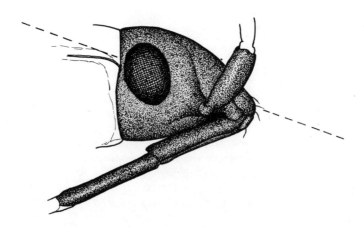

Fig 34. *Lygaeidae:* Rhyparochromus pini *nymph, lateral view of head. Broken line passes through middle of eye and apex of tylus, above antennal insertion.*

Stenocephalidae and some shieldbugs. The only British alydid is a quite unmistakable insect and the distinctive shape of the head of stenocephalids prevents their being mistaken for coreids. The metapleural scent-gland peritremes are well developed in Coreidae, in contrast to the obsolete peritremes of Rhopalidae. All British coreids are brown in colour, have one generation a year, overwinter as adults and feed only on plants. There are two subfamilies, with five species in each.

Pseudophloeinae have smoothly cylindrical tibiae and, usually, conspicuously clubbed

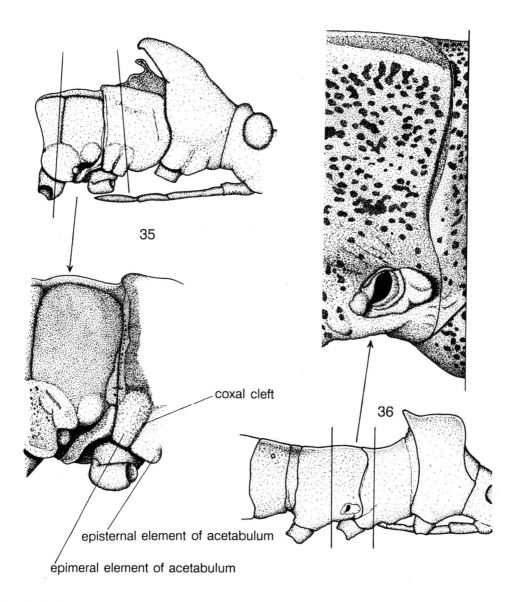

Figs 35–36. *Lateral view of thorax and detail of area around metathoracic gland opening. 35, Rhopalidae:* Rhopalus subrufus; *36, Coreidae:* Coreus marginatus.

posterior femora with a ventral row of spines of various sizes in the distal half. Except for the two *Arenocoris* species, which may be restricted to storksbill, they feed on clovers, trefoils and other members of the pea family. They are more likely to be found beneath their host plants than upon them. The commonest species, *Coriomeris denticulatus*, has a distinctive row of setae arising from prominent, tubercular bases on the lateral margins of the pronotum.

Coreinae have a groove along the dorsal surface of all tibiae and their hind femora are never clubbed or toothed. *Coreus marginatus*, the largest British coreid at 14–15 mm in length, lives on various docks and sorrels or, more rarely, the closely related persicarias, bistorts and knotgrasses and even rhubarb. As in some shieldbugs, the adults and late nymphs may move onto the green or ripe fruits of Bramble to feed in late summer. *Gonocerus acuteangulatus* is the only member of the family that feeds on woody plants in Britain. Here, it breeds only on box and seems to be restricted to a single locality in Surrey, where it has been known for many years.

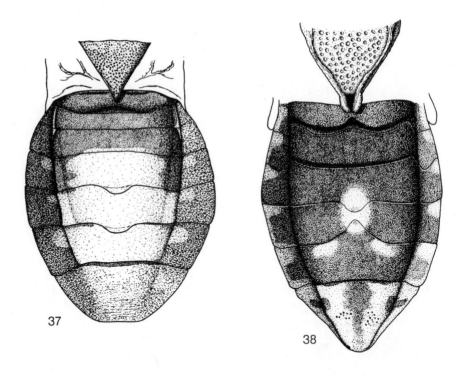

Figs 37–38. *Scutellum and abdominal dorsum. 37, Coreidae:* Coreus marginatus; *38, Rhopalidae:* Rhopalus subrufus.

Alydidae

Alydus calcaratus is the only British alydid. It is widespread in England and Wales, very rare in Ireland and absent from Scotland. It occurs right across Europe, northern Asia and

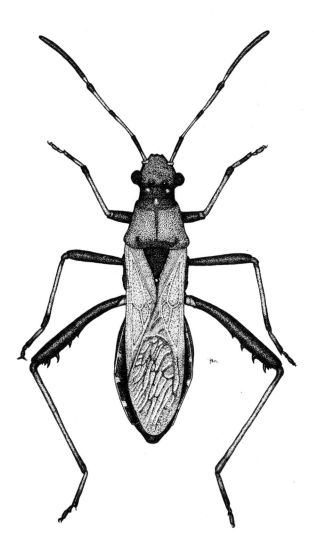

Fig 39. *Alydidae:* Alydus calcaratus.

North America. There are two colour-forms, one grey-brown and the other almost black. In both forms the dorsal surface of the abdomen is bright orange-red. It is a difficult insect to find in dull weather but is very active in sunshine, resembling the similarly coloured pompilid wasps as it flies close to the ground among or just above the vegetation. The clubbed hind femora, with a row of teeth beneath, are reminiscent of those of the pseudophloeine Coreidae but the elongate and slightly curved fourth antennal segment is unique to *Alydus*. The life-cycle is similar to that of Coreidae. It breeds on various plants of the pea family. The dark red-brown nymphs bear a striking resemblance to ants.

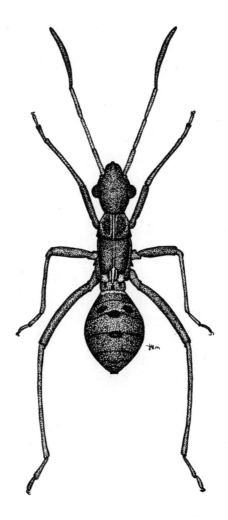

Fig 40. *Alydidae:* Alydus calcaratus, *fourth-instar nymph.*

Rhopalidae

In addition to the diagnostic features of the numerous veins in the membranes of the hemelytra and the greatly reduced metapleural scent-gland peritremes, all Rhopalidae, both nymphs and adults, have the second dorsal abdominal scent-gland aperture displaced forwards into the fifth tergite so that it is close to that of the first gland. This arrangement is unique to the family and universal within it. In all eight British species the egg has a circular operculum and there are two large aeromicropyles, one on the operculum itself and the other on the body of the egg.

Most rhopalids are shortly oblong and macropterous but there are two elongate and usually brachypterous species that feed on the leaves and unripe seeds of grasses in Britain. One of these, *Myrmus miriformis,* is about 8 mm long and is found in many types of grassland throughout England and Wales. Females of this species are usually green, while males may be either green or brown. Macropters occur quite frequently. The related *Chorosoma schillingi* is yellow-brown and about twice as long as *M. miriformis* but scarcely any wider. It shows a marked preference for coarse grasses and is almost, but not completely, confined to coastal dunes and shingle. In both species the eggs, which are the overwintering stage, are laid one or two at a time on vegetation.

The remaining species overwinter as adults and, like their grass-feeding relatives, have only one generation a year. They feed on a range of herbaceous plants, especially those that have sticky hairs. *Rhopalus subrufus* is particularly attached to common St John's wort. Like most Rhopalidae, it has the veins of the corium thickened and the cells between them glassily transparent, like most British Rhopalidae. Only in *Corizus hyosciami* is the corium completely opaque. This striking black and red bug lives in coastal dunes in England, Wales and south-eastern Ireland. *Liorhyssus hyalinus* is perhaps the most widely distributed heteropteran in the World. It is found in almost all of the warmer parts of the globe and occasionally appears in the British Isles, where it is known to have bred more than once. Its preferred hosts are members of the daisy family.

Stenocephalidae

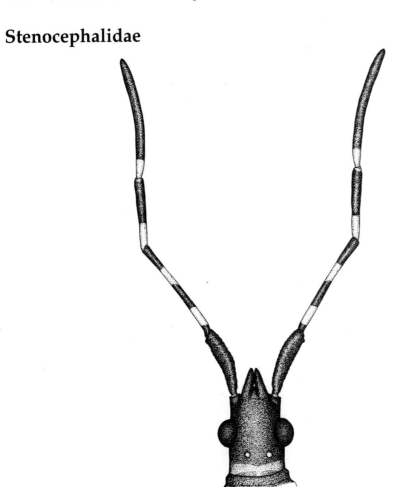

Fig. 41. *Stenocephalidae:* Dicranocephalus agilis, *head and pronotum.*

Two species of these oblong, dark brown insects are British. Their large size, 8–11 mm long in *Dicranocephalus medius* and 12–14 mm in *D. agilis*, their conspicuously light and dark annulated antennae and the bilobed front of the head (Fig. 41) are very characteristic. *Dicranocephalus medius* lives on wood spurge in southern England and *D. agilis* on portland spurge, sea spurge and cypress spurge in England, Wales and southern Ireland. The latter species is found on or close to the sea shore, where its host plants grow. Probably both species have only a single generation annually; adults of both overwinter.

About 50 years ago, it was discovered that *Dicranocephalus* species were vectors of a protozoan disease of spurges. The protozoan is a flagellate of the genus *Phytomonas*, closely related to the organisms that cause Sleeping Sickness and Chagas' Disease in humans and other vertebrates. Although *Phytomonas* (not to be confused with a bacterium of the same name) is so far unreported from Britain, it could be found in any spurges that support its vectors. Under the microscope, the flagellates should be clearly visible in a smear of fresh latex as their thin, colourless, transparent bodies move through the milky medium.

Pyrrhocoridae

Pyrrhocoris apterus is the only British species of its family. The pattern of its black and red markings makes it impossible to confuse with any other indigenous bug. In native-

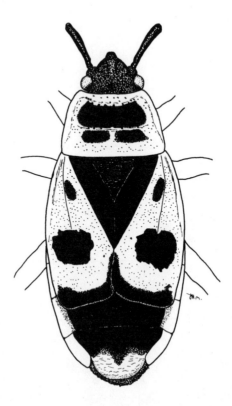

Fig 42. *Pyrrhocoridae:* Pyrrhocoris apterus, *red-and-black colour pattern.*

bred specimens the corium is almost fully developed but the hemelytral membrane is almost completely lacking. Full macropters that have been found in Britain may be immigrants. On the continent of Europe it feeds on lime (*Tilia*) and various plants of the mallow family, such as hollyhock, but in Britain it seems that its only host plant is tree mallow. It has been found on the coast in several counties of England and Wales. There is a well known colony on a rocky islet called the Oar Stone in Torbay.

Dysdercus species are found in many parts of the world on plants related to mallow (lime, baobab etc). Some of them are notorious pests of cotton ('Cotton Stainers') and, being easy to rear on cotton seeds, are sometimes cultured in laboratories.

Lygaeidae

About 80 species of Lygaeidae occur regularly in Britain, making this the second largest family of Heteroptera in the country. Most small, hard-bodied bugs taken from the ground surface or among leaf litter will be found to belong here. In macropters and usually also in submacropters and brachypters there are four longitudinal veins in the membrane of the hemelytron. Ocelli are always present but in those species in which the head is coarsely sculptured they may be difficult to see. Anthocoridae might be mistaken for Lygaeidae but they have an apparently three-segmented rostrum, in contrast to the four-segmented rostrum of Lygaeidae. Macropterous anthocorids always have a cuneal fracture in the costal margin of the hemelytron, which is never the case in lygaeids, and lygaeid antennae never have the last two segments more slender than the second, as is the case in most brachypters and some macropters of Anthocoridae. There is usually a single generation each year and the overwintering stage is the adult, though a few late nymphs may survive the winter to mature in the spring. The ovipositor is elongate and the eggs are inserted into crevices among moss or leaf litter or in the soil.

The largest subfamily, Rhyparochrominae, contains about 56 British species. The shape of the suture between the fourth and fifth abdominal sternites is diagnostic of the group. This suture curves forwards at its outer ends and does not reach the lateral margins of the body. There are three pairs of trichobothria on the head, as in saldids and pondskaters; these specialized sensory hairs are not present in the other subfamilies. It has been known for a long time that many Rhyparochrominae are associated with particular plants but it was only quite recently that they were shown to be seed-feeders. Sweet (1960) reared many species on ripe sunflower seeds despite the close associations of some of them with other plants in the field. Perhaps the season of availability of ripe seeds is more important to the bugs than their nutritional qualities. The anterior femora are thicker than the other two pairs and in many species they bear a row of spines of different sizes underneath, near the apex. The three *Stygnocoris* species lack these spines. These are small (2.5–4.2 mm), dark brown bugs, found in many open habitats. *Peritrechus* species are somewhat larger (4–6 mm) bugs, often found together with *Stygnocoris* species, and are mostly also dark brown. They have spines on the anterior femora and a rather indistinct pale spot in the middle of the lateral margin of the pronotum on each side. *P. lundi*, an inhabitant of areas with light, sandy soils, is rather less drab than its congeners, with a variegated pattern on the corium and a pale spot on the membrane. *Drymus* species, too, are brown, with toothed anterior femora; they are mostly intermediate in size between *Stygnocoris* and *Peritrechus* and lack the pale pronotal spots of the latter. *Scolopostethus* species are superficially similar to *Drymus* but more brightly patterned with black, off-white and many shades of brown. Species of both genera occur commonly in the leaf litter that accumulates beneath clumps of stinging nettle. *Scolopostethus decoratus* is often abundant beneath heather. The only tree-dwelling Rhyparochrominae

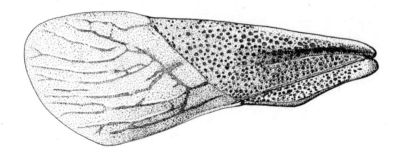

Fig 43. *Coreidae:* Ceraleptus lividus, *hemelytron.*

in Britain are the two species of *Gastrodes*. They are flattened, oval, brown bugs about 6–7 mm long, living much of the time in the cones of Pine and Spruce but feeding, at least partly, on the needles.

The four species of Cyminae, all in the genus *Cymus*, differ from all other native Lygaeidae in having a very short scutellum, about half as long as the claval commissure. In the other subfamilies the scutellum is about as long as the commissure. For this reason, some authorities have placed the Cyminae among the Berytidae. They feed on the seeds of rushes and sedges but not those of grasses.

Two species of *Henestaris* represent the subfamily Henestarinae in southern Britain. They are distinguished by their stalked eyes, which are especially prominent in the commoner species, *H. laticeps*. Both species live in coastal habitats.

Reedmace (*Typha*) is the host plant of the sole British species of Artheneinae, *Chilacis typhae*. The bugs gains access to the dense seed-heads of the plants where these have been damaged by caterpillars. This yellowish-grey and black insect bears a pale V on its dark scutellum, like some Rhyparochrominae, but lacks the characteristic abdominal suture of that subfamily. It has recently become established in the eastern USA.

Sunny nettle beds in England and Wales often support the gregarious *Heterogaster urticae*. It is a large lygaeid, 6–7 mm long, and its boldly black and white annulated antennae and tibiae are very distinctive. In the autumn the adults move to overwintering sites and hundreds of them may congregate in dwellings at this time of year, to the understandable but unnecessary alarm of the human occupants. There is a distinct cell bounded by the veins at the base of the hemelytral membrane; this is characteristic of the subfamily Heterogastrinae which, in Britain, has only two species, both in the same genus.

In the subfamily Orsillinae, the hemelytra at rest extend widely beyond the margins of the abdomen and have a slightly translucent appearance, particularly in the rufous-brown *Kleidocerys* species. Bugs of this genus are often numerous on heather, birch and rhododendrons. Several greyish-brown species of the worldwide genus *Nysius* are British. The commonest of these, *N. ericae* (which has been confused in the past with the mainly coastal *N. thymi*), feeds on many plants, especially weeds of the daisy family. In long, hot summers habitats with much bare ground, such as stubble fields and cinder tips, may be found teeming with countless thousands of the bugs. *Orsillus depressus*, which lives on cypress, has been spreading northwards in Europe from its original Mediterranean home and has recently reached Britain. It remains to be seen whether or not it will become a permanent member of the fauna.

Lygaeinae are not native to Britain but specimens of *Lygaeus equestris*, a black and red insect about a centimetre long, have very occasionally been found in the wild. They must

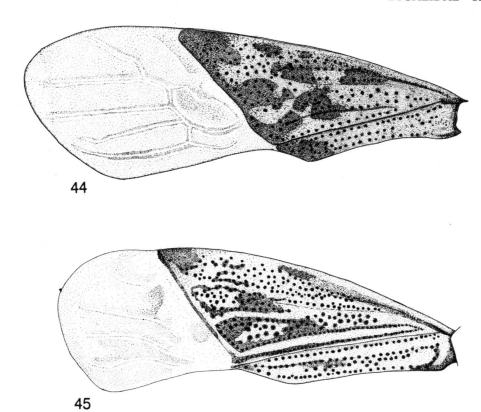

Figs 44–45. *Lygaeidae: hemelytra.* 44, Heterogaster urticae; 45, Peritrechus nubilus.

be casual immigrants since the only known host plant of the nymphs is *Vincetoxicum*, which does not grow here.

Macroplax preyssleri feeds on rockrose on the Mendip hills and *Metopoplax ditomoides* has been found a couple of times in man-made habitats but is probably not permanently established there. These two species belong to the subfamily Oxycareninae. Their anterior femora are thicker than the other two pairs and spined beneath, as in most Rhyparochrominae, but all the ventral abdominal sutures reach the lateral margins of the body. Both species are clothed in peculiar blunt bristles which are unknown outside the subfamily.

The subfamily Blissinae is closely associated with grasses and some of its members are important pests of pastures and cereal crops. Fortunately, the two British species, both belonging to the genus *Ischnodemus*, are harmless, One of them is extremely localized but *I. sabuleti* is often abundant in southern Britain. It seems to be a twentieth century addition to the fauna and is probably still extending its range northwards. The elongate, parallel-sided, dorsoventrally flattened and often brachypterous adults, which are patterned in black, brown and white, and the red larvae can be found in all months of the year on or among reeds and grasses in waterside habitats, marshes and damp meadows. Unlike most Lygaeidae, Blissinae feed not on ripe seeds but on the sap of their host plants.

Berytidae

Nine species of stiltbugs are found in the British Isles. Their elongate bodies and slender legs, with thickened 'knees' at the apices of the femora, are very distinctive. They overwinter as adults and have a single generation annually. All of them live on or beneath low-growing plants, generally in well drained, sunny places, but *Metatropis rufescens*, a brown insect about a centimetre long and reminiscent of a cranefly in appearance, lives on Enchanter's Nightshade in damp woods in England and Ireland. Its feeding results in perforating and tattering of the leaves, very like the feeding damage of some Miridae, but no mirid feeds habitually on this plant. *Gampsocoris punctipes* is about half as long as *M. rufescens* and feeds on restharrow, especially on sandy soils. In Ireland and Scotland it is apparently restricted to coastal dunes. It has been observed repeatedly sucking out the liquid contents of the glandular tips of the hairs of its host plant. In *Neides tipularius* and the six species of *Berytinus* the front of the head is produced into a flat, triangular process

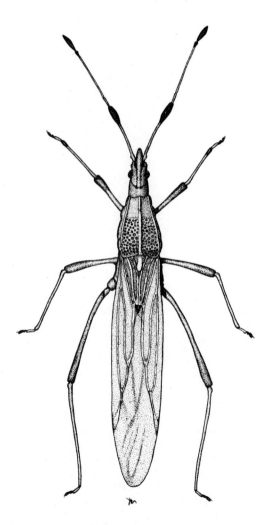

Fig 46. *Berytidae*: Berytinus minor.

that looks like a spine when seen from above. This process is lacking in *Metatropis* and *Gampsocoris*. Like these two bugs, *B. montivagus* is always macropterous but the other species all occur in both macropterous and bachypterous forms. The fore wings of brachypters are very little shorter than those of macropters but the membrane is noticeably more narrowly rounded at the apex and the convexity of the pronotum is much less. *Neides tipularius* is about as big as *M. rufescens* and *B. hirticornis* is not much smaller. The remaining five species of *Berytinus* are 4–8.5 mm long. Some Berytidae, including *B. minor*, are partially predaceous but all British species could probably develop to maturity and reproduce on vegetable food alone. *Berytinus hirticornis*, which used to be very rare but is now expanding its range rapidly in southern England, has been reared on cocksfoot grass. The other species feed mainly or entirely on dicotyledonous herbs, including mouse-ear chickweed (*B. crassipes*) and various low-growing members of the pea family (other *Berytinus* species). *Neides tipularius* is very polyphagous, feeding on both dicotyledonous plants and grasses.

The European Berytidae were monographed by Péricart (1984). This work contains much biological information and has descriptions and figures of the immature stages as well as the adults.

Piesmidae

The little bugs belonging to this family have the same general appearance as the lacebugs (Tingidae). The scutellum, however, is not covered by the pronotum as it is in tingids and ocelli are present though difficult to see. Both British species feed on the sap of various species of goosefoots and oraches. *Piesma quadratum*, with three longitudinal keels on the pronotum, is largely confined to salt marshes where it also feeds on sea purslane. *Piesma maculatum*, with only two pronotal keels, is also found inland and may include beet and spinach in its diet. Woodroffe (1966a) described a new species of the same genus, *P.*

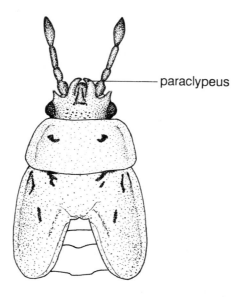

Fig 47. *Piesmatidae:* Piesma quadratum *nymph, fore body.*

spergulariae, from the Scilly Islands, where it was associated with cliff sand-spurrey. This plant belongs to the family Caryophyllaceae while the other plants mentioned are all members of the Chenopodiaceae. Despite this difference in host plants, Heiss & Péricart (1975) reduced *P. spergulariae* to the status of a subspecies of *P. quadratum*, which it resembles in having three pronotal keels. All three forms probably have two generations, at least in favourable years, and overwinter as adults.

Tingidae

The lacebugs, with their reticulated hemelytra, and often reticulate or at least punctate pronotum as well, have a very characteristic appearance and cannot easily be confused with any other bugs except the Piesmidae. Both families have two-segmented tarsi but in Tingidae the scutellum and clavus are concealed by a triangular extension of the posterior lobe of the pronotum; in Piesmidae these structures are exposed. Piesmids also have ocelli, which tingids lack. There are two dozen British species of lacebugs, all feeding exclusively on plants and probably all with only a single generation each year. They are small insects and few of them attain a length of even 4 mm. The host plants of several species are still unknown. In those species whose method of feeding has been studied, the main food is the contents of palisade and other mesophyll cells. Damaged leaves have a pale-speckled appearance like those attacked by typhlocybine leafhoppers. Dense populations of lacebugs may completely kill the leaves. Some foreign species cause galls to form but none of the British forms is believed to do this; at most, their oviposition stimulates a slight proliferation of cells around the point where the eggs are inserted into the host plant.

The genus *Acalypta*, with five British species, is intimately associated with mosses, among which the bugs live and on whose capsules and vegetative parts they feed. Adults hibernate among the host plants and lay their eggs in the leaf tissues in the spring. Brachyptery is frequent in this genus. Several of the species are found throughout the British Isles.

Dictyonota strichnocera and *D. fuliginosa* both feed on broom and the former also on gorse. Like *Kalama tricornis*, whose host plant is unknown (it may be polyphagous) and *Stephanitis rhododendri*, which feeds on rhododendron leaves, they overwinter as eggs. In all other British lacebugs the adult is the overwintering stage. *Stephanitis rhododendri* is a twentieth-century introduction and at first spread rapidly and became sufficiently numerous to cause alarm to nurserymen; but its numbers have since declined and it may even have died out.

Apart from *Stephanitis* and *Dictyonota*, only *Physatocheila* species live on woody plants in Britain. They prefer mature trees with a growth of foliose lichens to afford them some shelter. All four British species are confined to southern England where one species lives on maple and sycamore, one on alder and birch and two on hawthorn, sloe and fruit trees such as apple and cherry. An american lacebug, *Corythucha ciliata*, has recently appeared in southern Europe and has spread rapidly, severely damaging the leaves of plane trees (which, in its native New England, are known as sycamores). If it reaches Britain and can withstand the cool summers it could become a serious pest here, too.

The most frequently encountered lacebugs are probably *Tingis ampliata* and *T. cardui*. Both live on various kinds of thistles but they are most abundant on creeping thistle and spear thistle respectively. Adults of *T. cardui* move with agility among the spines clothing the flower-heads of their favoured host plants, where they are often found in company with their spiny, flattened nymphs.

Agramma laetum differs from all other British tingids both in its choice of host plants –

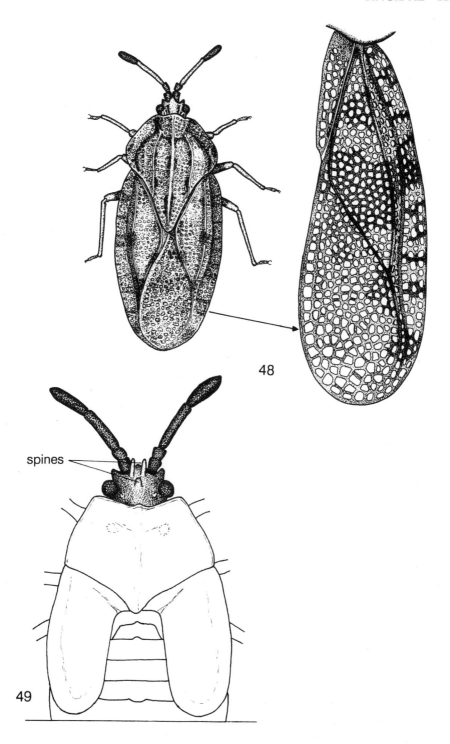

Figs 48–49. *Tingidae:* Tingis cardui. *48, adult and detail of hemelytron; 49, nymph, fore body.*

sedges and rushes – and in the form of its pronotum, which lacks lateral flanges. Reticulate flanges, sometimes reflexed and adpressed to the surface of the pronotum, are present in all the other species, which often have longitudinal pronotal keels and sometimes a hood overhanging the head as well.

Péricart (1983) published a monographic treatment of the European species of the family, with figures of the adults and also the nymphs of many species.

Reduviidae

The great predaceous family of assassin bugs is very largely a tropical one. It is represented in Britain by a mere half-dozen species and in Ireland by only two. The very short, curved rostrum is unique to the group and the transverse constriction of the head between the eyes is encountered elsewhere only in Berytidae. All reduviids (apart from a handful of species) are able to stridulate by rubbing the apex of the rostrum in a transversely striated groove at the front of the prosternum. The sounds made by the larger species are readily audible to the human ear and may be produced in response to disturbance. The larger reduviids can also inflict a painful bite if handled incautiously. Most reduviids are raptorial predators, pouncing on their victims and holding them with their front legs, which are variously modified for the purpose. Although few in number, the six native species, all of them predators, show considerable diversity of form and habitat.

All species of the subfamily Triatominae live by sucking the blood of vertebrates. One of them, *Rhodnius prolixus*, is a well known laboratory insect, much used in physiological research. Most triatomine species are confined to the warmer parts of the New World, where several of them are vectors of protozoan diseases of Man and other animals. The subfamily has a few species in the warmer parts of the Old World, but none of them occurs in Europe.

The genus *Empicoris*, which has three British and two Irish species, belongs to the subfamily Emesinae. These are delicate, gnat-like insects, 4.5–7 mm long. They walk on the hind two pairs of legs, which are long and extremely slender, and hold the raptorial anterior pair close to the body as mantids do. Their prey consists of small, soft-bodied insects like booklice, which they hunt by slowly walking about on walls, tree-trunks and branches. Adults hibernate in dry leaf-litter, woodpiles, haystacks and similar situations and also on evergreen trees.

The nymphs of *Reduvius personatus* are more often noticed than the adults. They cover their bodies with detritus after the manner of certain lacewing larvae. Since they live in houses, particularly around sash windows, and in outbuildings, church roofs and other dry, frost-free places frequented by people, their strange appearance readily attracts human attention. The covering of detritus has earned the bug both the Latin epithet *personatus* and its splendidly melodramatic English name of masked assassin bug. Its prey consists of the insects and spiders that inhabit or stray into the places where it lives. It was once valued as a predator of the bedbug. Adult *R. personatus* are oblong, blackish-brown insects about 16 or 17 mm long. In the summer they occasionally fly at night through open windows into lighted rooms. Both adults and nymphs can survive long periods without food. Nymphs overwinter for two successive years. The species is clearly on the edge of its range in Britain, where it is the sole representative of the subfamily Reduviinae. It is probably unable to survive in the open and, although it is transported widely around the World among dry cargoes in ships, it is not found even indoors in northern Britain or Ireland.

The most widespread species of the family is *Coranus subapterus*, which belongs to the

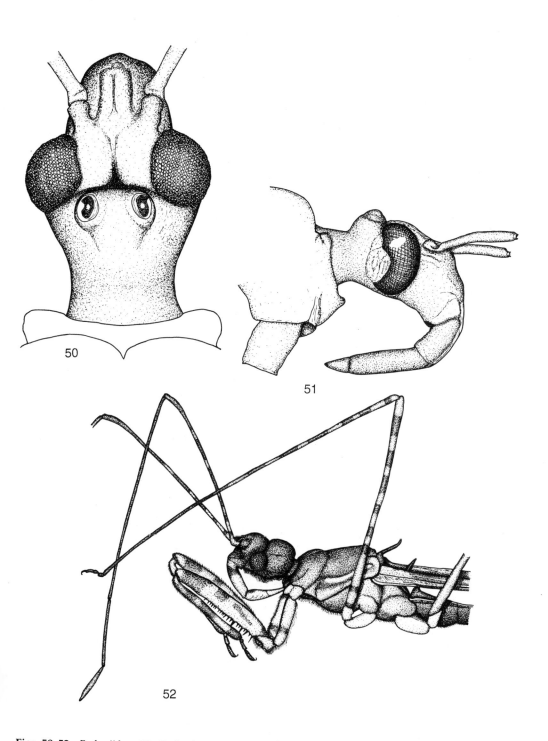

Figs 50–52. *Reduviidae. 50*, Reduvius personatus, *head in dorsal view; 51, same in lateral view; 52,* Empicoris culiciformis, *fore body in lateral view.*

subfamily Harpactorinae. This bug is frequently strongly brachypterous but the fore wings, less than half the length of the abdomen, have clearly defined apical membranes, so that they look like slightly distorted, miniature versions of the fore wings of macropters. The insect occupies two distinct habitats: heaths and dunes; it is not found in grasslands. There is some evidence that macropters are more common in coastal populations, which also differ slightly from those on heathlands in the shape of the head, the length of the antennae and the parameres. These differences were described by Woodroffe (1959) and are not confined to British populations. Some authorities believe the two forms represent distinct species. Both forms are about a centimetre long and mottled grey with some black markings. They overwinter as eggs and there is only one generation a year.

The rarest British assassin bug, perhaps now locally extinct, is *Pygolampis bidentata*. It is about as long as *Reduvius personatus* but paler brown and thinner, with an elongate, parallel-sided head. Only three specimens have been taken, all in southern England. The last British capture was in 1920. It is the only member of the subfamily Stenopodainae ever to have been found in the British Isles.

Nabidae

In contrast to the hemipteran fauna of warmer countries, that of Britain has more Nabidae than Reduviidae. Like the reduviids, the nabids are aggressive predators, using the front legs to hold the prey, but the legs are less strongly modified for this purpose in the present family. The slender, curved rostrum is four-segmented and reaches well beyond the prosternum, unlike the short, stout, three-segmented reduviid rostrum. Nabids are sometimes confused with Miridae but the hemelytra of macropters do not have a cuneus, marked off by a costal fracture, and the trochanters are undivided. All 12 native British species have slender anterior tibiae of roughly equal diameter throughout, a character that places them in the subfamily Nabinae (damselbugs). They are generalist predators, feeding on eggs, immature stages and adults of a variety of other insects and spiders. One species of the subfamily Prostemminae, with the anterior tibiae strongly broadened towards their apices, formerly occurred in Britain but is now extinct. This bug, *Prostemma guttula*, is still found in the Channel Islands and in northern France. Like most prostemmines, it preys on other Heteroptera.

Flightlessness is of frequent occurrence in the family, perhaps because the distances covered on foot during the day-to-day business of hunting are sufficient to ensure adequate dispersal. *Anaptus major* and *Nabis ferus*, however, are always macropterous and *N. pseudoferus* is more often macropterous than submacropterous. Of the other *Nabis* species, *brevis* is always sub-brachypterous while *ericetorum* and *rugosus* are usually sub-brachypterous, with very occasional submacropters and macropters respectively. Submacropters are difficult to distinguish from macropters without careful examination. Their wings are slightly shorter than the hemelytra and the hemelytral membranes are very slightly reduced. Sub-brachypters have the wings strongly reduced and their hemelytra are scarcely as long as the abdomen. *Nabicula flavomarginata*, *Stalia boops*, *Himacerus apterus* and *Aptus mirmicoides* are all most comonly brachypterous, with the fore wings reaching the middle of the abdomen, and all occasionally occur as macropters, with some intermediate forms as well. The most extreme reduction is encountered in *Nabicula limbata* and *N. lineata*, which are almost always staphyliniform brachypters with short, truncate, membraneless hemelytra; macropters of these species are extremely rare. In the genera *Nabicula* and *Himacerus* there are only four nymphal instars, at least in brachypters.

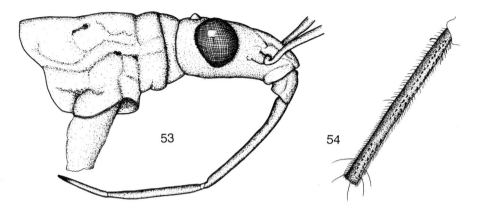

Figs 53–54. *Nabidae. 53,* Nabis rugosus, *head and prothorax in lateral view; 54,* N. rugosus, *anterior tibia in ventral view, showing long and short hairs and very short spines.*

The first four nymphal instars of *Aptus mirmicoides* are mimics of ants. They are shiny brown in colour with the basal tergites of the abdomen coloured white laterally and brown medially, giving the impression of a stalked gaster. The fifth instar is altogether less ant-like, with the pale basal areas of the abdomen less distinct.

Specialization of hunting in the different species is manifested chiefly in choice of habitat rather than in choice of prey. Thus, *Nabis rugosus* is most at home in gardens and waste ground, *N. ferus* in dry grassland, *N. ericetorum* on sandy heaths and *Nabicula limbata* among lush vegetation in marshes. Some species, like *A. major*, rarely leave the ground while others, like *N. ferus*, are easily taken by sweeping. There are also differences in timing of the life-cycles. All species have a single generation annually but the *Nabis* species and *A. mirmicoides* hibernate as adults while the others overwinter as eggs. Nabids typically lay a few hundred eggs in the course of a lifetime. They are inserted, usually in rows, into plant stems in batches of five to ten at a time, though batches of several dozen have been reported.

Only *H. apterus* lives on trees and bushes. Some authorities claim that it oviposits in the young twigs of these plants but it seems that the eggs are usually laid in the stems of herbaceous plants, like those of other nabids. The first two instars live on low plants but the third-instar nymphs ascend woody plants to hunt.

Kerzhner (1981) published a monograph, in Russian, on the Nabidae of Europe and temperate Asia and Péricart (1987) made much of the same information available in French. The latter work contains keys, descriptions of all stages, biological information and distribution maps.

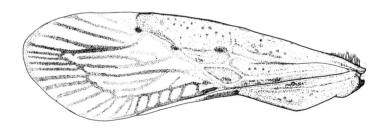

Fig 55. *Nabidae:* Nabis ferus, *hemelytron.*

Anthocoridae

Flowerbugs are small insects, mostly less than 4 mm long, that prey on other small invertebrates. They are 'timid' predators, selecting defenceless and often completely immobile prey. They do not attempt to hold the prey with their legs and back away if serious resistance is offered. Some species are able to detect and feed on leaf-mining caterpillars, inserting the stylets into the mines. The family is sometimes merged with the bedbug family, Cimicidae. Bedbugs are almost certainly only a specialized branch of Anthocoridae that have taken the comparatively short step from preying on invertebrates to sucking the blood of vertebrates. About 30 species of Anthocoridae have been recorded in Britain. They have a distinctive appearance, being rather strongly flattened and generally oval or oblong in outline, with the front of the head produced into a short, parallel-sided 'snout'. Ocelli are present and macropters have a costal fracture in the hemelytron, marking off a distinct cuneus like that of Miridae. A longitudinal suture parallel with the costal margin demarcates a narrow costal strip of the hemelytron called the embolium, which extends from the base of the wing to the cuneal fracture. There are no cells in the hemelytral membrane, though up to four longitudinal veins may be visible. The tarsi are three-segmented and the rostrum also appears to have the same number of segments. The male genitalia are strongly asymmetrical and so highly modified that it is difficult to correlate the structures in them with those visible in more normal Hemiptera. There is only one paramere in all British Anthocoridae except *Lyctocoris*, which has two of unequal sizes. Insemination is always traumatic, involving the introduction of sperm through the female's body wall. In most species the aedeagus pierces the intersegmental membrane between abdominal segments VII and VIII where, in Anthocorinae, there is a special copulatory device, the ectospermalege, to receive the sperm. In Lyctocorinae there is usually no ectospermalege but some *Xylocoris* species have one between segments VII and VIII and *X. galactinus* has one in a right-dorsal position, between tergites II and III. A well developed, piercing ovipositor is present in most Anthocoridae but this organ is greatly reduced in *Xylocoris flavipes* and lacking in the lyctocorine tribe Cardiastethini (*Dufouriellus, Cardiastethus, Xylocoridea* and *Brachysteles*).

Members of the subfamily Lyctocorinae, as both adults and nymphs, are recognized by the long hairs on antennal segments III and IV. These hairs are at least twice as long as the width of the segments on which they are inserted and the segments themselves are usually distinctly thinner than segments I and II. *Xylocoridea brevipennis* and *Dufouriellus ater* occur beneath bark flakes on the trunks of deciduous trees. Trees attacked by scolytid beetles often support large populations of *D. ater*. Scolytids are a major food source for this species, although it preys upon many other kinds of insect. *Brachysteles parvicornis*, apparently, feeds only on oribatid mites among moss and herbage and on shrubs, usually in sandy places. The other British lyctocorines usually live among various kinds of vegetable debris, where they prey on small arthropods such as booklice and the detritus-feeding larvae of various moths and beetles. *Xylocoris formicetorum* seems to be strictly associated with ants of the genus *Formica* and *Lasius*, living deep within their nests. *Xylocoris galactinus* has been found in ants' nests and sometimes infests flour mills, where other lyctocorines also may occur, including the frequently imported species *Xylocoris flavipes*. *Lyctocoris campestris* is found in a variety of habitats including the nests of birds and mammals, animal houses and industrial and domestic premises. In such situations it is a facultative feeder on the blood of birds and mammals. Several species of lyctocorines living in factories and warehouses have been known to cause annoyance by biting the people working there.

The larger of the two British subfamilies of flowerbugs is the nominate subfamily, Anthocorinae. Its members never have the last two antennal segments noticeably thinner

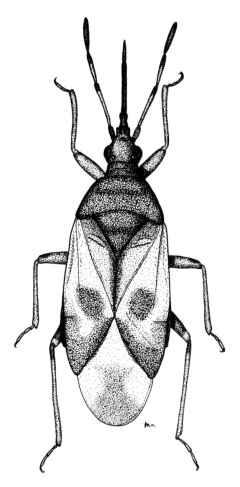

Fig 56. *Anthocoridae:* Anthocoris nemorum.

than the first two, nor do the antennae bear hairs as long as twice the diameter of the segments. *Elatophilus nigricornis*, the two species of *Acompocoris* and *Tetraphleps bicuspis* all live on conifers. *E. nigricornis* feeds on scale insects and the others prey on Aphidoidea, especially Lachnidae. There are 12 British species of *Anthocoris*. Three of them are common and highly polyphagous predators of insects inhabiting trees and shrubs. *Anthocoris nemoralis* and *A. confusus* are largely restricted to woody plants but *A. nemorum* commonly lives on herbaceous plants as well. Some species of this genus are highly specialized predators. For example, *A. butleri* and *A. visci* are closely associated with Psyllidae on box and mistletoe respectively, while the immature stages of *A. gallarumulmi* and *A. minki* live in pemphigid galls on the leaves of elm and the petioles of poplar respectively. Adults of these species are more catholic in their choice of prey. Half a dozen species of *Orius* live in the British Isles, on both herbaceous and woody plants, preying on thrips, aphids, mites, whiteflies and insect eggs, among other things. Some foreign species feed only on pollen and the native *O. vicinus* can develop to adulthood on this food although it needs to feed on animal prey before it can reproduce. *Orius majusculus* may cause damage to chrysanthemums and asters by feeding on the leaves. Both this species and *Anthocoris nemorum* may cause a nuisance by biting people working out of doors.

Most Anthocoridae are fully winged but in *Temnostethus gracilis* and *Xylocoridea brevipennis* the brachypterous morph is much more common than the macropter. *Xylocoris flavipes* and *X. cursitans* occur commonly in both brachypterous and macropterous morphs. All species occurring out of doors overwinter as adults. In some, at least, the number of generations varies from one to three per year, depending on the local climate and the genetically determined diapause mechanisms of local populations. Péricart (1972) monographed the European members of the family.

Cimicidae

Bedbugs and their close relatives are plain yellowish or reddish-brown, micropterous insects with no ocelli. Their bodies are broadly oval and very flat unless they have fed recently, in which case their abdomens become plump, purplish and nearly cylindrical. Their only food is the blood of birds or mammals. They are active only in darkness. Having no adaptations for clinging to fur or feathers, they are only very rarely found on the bodies of their hosts. All the active stages, especially the adults, can go for long periods without food but Mayakovsky's (1929) suggestion of a bedbug surviving for 30 years in a frozen cellar is surely fanciful. The eggs are elongate, like those of Anthocoridae and Miridae, and are cemented to the walls of the dwellings, lairs or nest-cavities of their hosts. The manner of copulation is bizarre and is probably inherited from an anthocorid ancestor. In the confined spaces where the bugs often live, normal copulation is difficult. The females have a secondary copulatory aperture, the paragenital sinus or Ribaga's organ, on the fourth abdominal sternum. Spermatozoa, injected here, migrate through the haemocoel to the ovaries, where fertilization takes place. The normal female genital aperture, situated as in other bugs between the seventh and eighth abdominal sterna, is used only for egg-laying.

Five species, all but one of them belonging to the genus *Cimex*, are known to have bred in the British Isles but one of these may be extinct and another is an occasional introduction from the tropics and has not succeeded in establishing itself as a permanent resident.

The common bedbug, *Cimex lectularius*, now mercifully much less comon than formerly, has been recorded in almost every county in the British Isles. The adult bug is about 5-6 mm long and has a broad pronotal flange that embraces the posterior part of the head. Its life cycle, in heated buildings, is independent of the seasons. Its only requirement, apart from warmth and the blood of humans or domestic animals on which to feed, is somewhere to hide during the day. The narrow spaces behind skirting boards, the architraves round doors and wooden panelling are favourite haunts. The bugs often betray their presence by leaving spots of their excrement on walls at the points where they habitually emerge from concealment. Eggs are laid in cracks and crevices in furniture, behind wallpaper and similar situations, usually on a rough surface if one is available. The whole period of development from egg to adult, via five nymphal instars, takes about 10 weeks at normal indoor temperatures. In a warm house, each female bug, on average, feeds about once a week and lays eggs at the rate of two or three a day. Her total production of eggs over a lifetime amounts to 150 or so. Adults may live for several years. Feeding and development cease if the temperature falls below about 13C and recommence when it rises above this figure again. There is an association in the public mind between bugs and dirt but there is no stage in the life cycle that requires detritus for

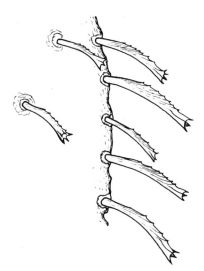

Fig 57. *Cimicidae:* Cimex lectularius *nymph, lateral body setae, greatly enlarged.*

food or for any other reason. Infested premises have a distinctive smell, redolent of the herb coriander.

Some fortunate people never experience discomfort when bitten by bedbugs but most react as they would to mosquito (gnat) bites. A small, itching bump arises at the site of each bite and persists for several days. Loss of sleep from constant irritation is the main danger to health. Bedbugs have never been convicted of transmitting any human disease, despite repeated attempts to incriminate them. Nevertheless, Environmental Health Officers make quick and effective arrangements to eliminate any infestation reported to them. New infestations probably originate with stray insects brought home in holiday luggage or in furniture carried in removal vans.

With the increasing popularity of foreign travel, the importation of the tropical bedbug, *Cimex rotundatus*, has become a distinct possibility and at least two established populations of this species have been exterminated in recent years. The chief difference between the two species is the much narrower pronotal flange of *rotundatus*.

Cimex columbarius, the pigeon bug, is very like *lectularius* in appearance. It may be extinct in Britain, though it still occurs in parts of Europe in the nests of various birds in tree-holes or nest-boxes. Wooden dove-cotes are a favourite habitat. Most reports of *Cimex* species from poultry houses, however, turn out to be infestations by *lectularius*.

Lansbury (1961) showed that there was only one species of batbug in Britain, *Cimex pipistrelli*. It seems to be confined to bat roosts and shows no preference for any particular kind of bat. It has been reported from several counties of England and Ireland. It differs from *lectularius* and *columbarius* in the narrower pronotal flanges, which make it more like *rotundatus*.

Apart from the *Cimex* species, there is only one other cimicid in the British Isles. It is the martin bug, *Oeciacus hirundinis*. While in adults of *Cimex* the anterior margin of the pronotum is curved forwards at the sides to embrace the head to some extent, in *Oeciacus* it runs straight across. The martin bug is also smaller than its relatives (3.0–3.5 mm long, compared with 3.6–6.0 mm for *Cimex* species). It is quite common in southern England and may occur in Ireland. The usual host is the house martin. It has been found occasionally in the nests of other birds nesting on or in buildings and in hollow trees.

There is a single generation a year and most of the blood consumed by the bugs must be that of nestlings. Adults overwinter, occasionally entering houses at this time of year and even biting the occupants, but normally they remain in and around the nests without feeding, waiting for the birds to return in the spring.

The Cimicidae of Europe were monographed by Péricart (1972) and those of the world by Usinger (1966).

Microphysidae

There are only six British species of these tiny bugs, which resemble small Anthocoridae in general appearance. They differ from anthocorids in having only two tarsal segments. Males are macropterous. Their hemelytra, which have a costal fracture demarcating the cuneus, amply cover the abdomen and extend beyond it on all sides. Females lack the hemelytral membrane and the cuneal fracture; in *Myrmedobia coleoptrata* the convex, coriaceous hemelytra completely cover the abdomen, giving the insect a beetle-like appearance, while in the other species the hemelytra are short and truncate like the elytra of rove beetles. The most extreme hemelytral reduction occurs in females of *Loricula elegantula*, in which they are reduced to small flaps that cover only the metathorax, leaving the whole of the disc-shaped abdomen exposed.

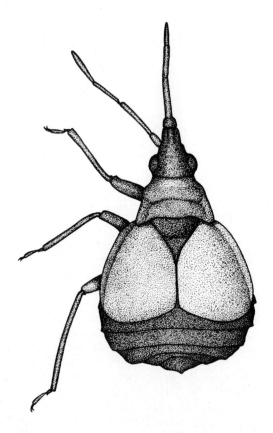

Fig 58. *Microphysidae:* Loricula pselaphiformis, *female.*

There is probably only one generation a year in all British species, with the eggs overwintering. The bugs are predaceous on small arthropods such as booklice, aphids, springtails and psyllids. *Loricula* species live among lichens growing on tree bark and, in the case of *L. elegantula*, also on rocks and walls. *Myrmedobia* species live among lichens, mosses and fallen conifer needles as well as under flakes of bark on tree trunks. The elongate eggs of Microphysidae are laid among mosses or lichens or in chinks in bark, with the micropylar end projecting. The micropylar cap is coarsely reticulated, with spiky processes radiating from its rim, quite unlike the egg of any other bug. The European members of the family were monographed by Péricart (1972).

Miridae

More than 200 species of this family are British. Several of them are of interest to farmers and horticulturists as either pests or predators of pests. Works on economic entomology often refer to them as capsid bugs (from the former name of the family, Capsidae). They are generally rather delicate, soft-bodied insects, usually between 3 and 10 millimetres in length, with a four-segmented rostrum and three-segmented tarsi. The last two antennal segments are often thinner and more flexible than the first two. In macropters, the apical part of the corium is marked off by a costal fracture as a separate area called the cuneus, as in Anthocoridae and male Microphysidae. Anthocorids, however, have a three-segmented rostrum and microphysids have two-segmented tarsi. All British mirids lack ocelli. The membrane of the hemelytra has one, moderately large cell with no veins arising from it and, usually, a much smaller cell between the larger one and the apical margin of the cuneus. The apparently two-segmented trochanter of both adults and nymphs is universal within the family and unique to it. The legs can be shed by muscular action, the break occurring between the two halves of the trochanter. Legs may be shed if they are seized by a predator or trapped in a film of moisture and are often lost if the live insect is dropped into preserving fluid.

Much of the success of mirids as a family may be attributable to the ease with which they can walk on leaves. Contact with varying kinds of leaf surfaces seems to have been a mainspring of adaptive radiation in the family as the major groups within it are largely defined by the various modifications of the structures at the tips of the tarsi that are concerned with adhesion to the substrate. Nymphs, when disturbed, may evert the rectum, which adheres to the substrate and provides an additional hold. When the disturbance subsides, the bug retracts the rectum and appears to suffer no harm from the temporary prolapse.

Typically there is one generation a year and the overwintering stage is the egg. A lifetime's egg production usually amounts to one or two hundred eggs, which may be laid a few at a time or all at once. Mirid eggs are elongate and slightly curved, with a yolk plug at one end. They are inserted into plant stems, beneath the bark of twigs and in similar situations, depending on the species, with the yolk plug projecting or just inside the plant tissue, facing outwards.

There are three small subfamilies and three large ones in the British fauna. The smallest is the Bryocorinae, with two species. These small, brown bugs, between 2 and 4 millimetres long, are the only native Heteroptera that feed on ferns. Macropters have only one cell in the membrane of the hemelytron. The little cell that is present between the large one and the cuneus in all other British mirids is lacking in this subfamily. A pronotal collar is present. The third tarsal segment is distinctly thicker than the other two, another feature unique to the subfamily. The parempodia are bristle-like, pulvilli are present and the claws are not toothed. *Monalocoris filicis*, feeding on bracken and other ferns, is always macropterous. It overwinters as an adult and often lays its eggs in the sporangia of its host plants. *Bryocoris pteridis*, which feeds on several different ferns but not usually on bracken, overwinters as eggs in the midribs of the fronds or those of the leaflets. Adults of this species may be macropterous or brachypterous.

There are six species of Deraeocorinae in Britain. The smallest are the brachypterous females of *Bothynotus pilosus*, 3.5–4.0 mm long (males and macropterous females are larger) and the always macropterous *Deraeocoris lutescens*, 3.8–4.6 mm long. The largest is *Deraeocoris olivaceus*, which is 8.5–10.5 mm long. All Deraeocorinae are predaceous. *Alloeotomus gothicus* lives only on scots pine and *D. olivaceus* on hawthorn. Both are recent introductions to the south-east. *Deraeocoris lutescens*, which lives on many deciduous

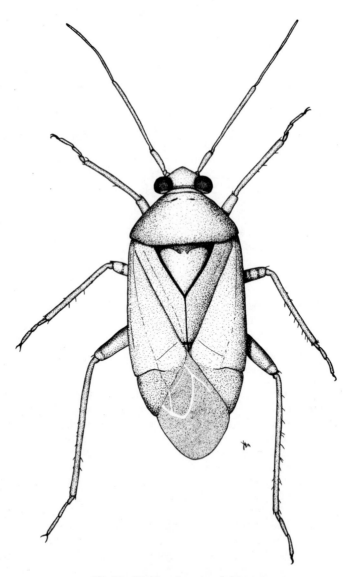

Fig 59. *Miridae:* Lygus maritimus.

trees in southern Britain, is the only British deraeocorine that overwinters in the adult state. A relative, *D. ruber*, is often abundant in England and Wales on rank herbage such as nettle. It varies from orange-red to nearly black in colour, the males being darker than the females. The rare *D. scutellaris*, whose habitat requirements are unknown, is dimorphic: it may be entirely black or the scutellum may be bright red. A rounded pronotal collar is present in this subfamily; the parempodia are bristle-like and the third tarsal segment is not thickened. In nymphs the claws bear membranous pulvilli at the base but these become sclerotized in the adult to give the impression of a tooth at the base of the claw. This form of adult claw is unique to Deraeocorinae. Apart from *B. pilosus*, all species are macropterous.

Dicyphinae are more slender and elongate than the robust, oval bugs of the preceding two subfamilies. Like them, they have bristle-like parempodia and a rounded pronotal collar. They have neither the incrassate third tarsal segment of Bryocorinae nor the toothed adult claws of Deraeocorinae. The right paramere is atrophied and probably non-functional. There are 11 British species, most of them living on hairy and often stickily hairy plants. Several *Dicyphus* species are dimorphic, occurring in both brachypterous and macropterous forms. Four species of this genus overwinter as adults. *Campyloneura virgula* is probably the only parthenogenetic heteropteran in Britain; males are very rare indeed. It lives on various broadleaved trees, preying on other insects. The other members of the subfamily live on herbaceous plants with the exception of one restricted to rhododendrons and another on brambles. Some of the species on herbaceous plants are host-specific and probably vegetarian while others, in particular the common *Dicyphus errans*, are at least partially predaceous.

Members of the large subfamily Phylinae, like those of the three small subfamilies mentioned above, have bristle-like parempodia. There is no pronotal collar except in three species of ground-dwelling ant-mimics. Phylines live on both herbaceous and woody plants, including conifers, and most of them are highly host-specific. They are generally small and delicate, even for mirids, and most of them are between three and five millimetres long. The largest species are *Oncotylus viridiflavus*, a greenish-yellow bug about 7 mm long living on hardheads (*Centaurea*), and *Harpocera thoracica*, coloured in various shades of grey and brown, 6–7 mm long and living on oak. This last species feeds on the flowers of its host tree and is one of the earliest oak-dwelling mirids to appear in the spring. It spends about 11 months of the year in the egg. Some phylines are wholly predaceous, like *Tytthus pygmaeus*, which feeds on eggs and nymphs of planthoppers. Others, like *Psallus betuleti* on birch, have a mixed diet of plant and animal food. Others again are wholly phytophagous. The small, dark-coloured *Chlamydatus* species are often brachypterous and live close to or on the ground. Their hind legs are modified for jumping and they can easily be confused with *Halticus* and its allies (Orthotylinae) with similarly enlarged hind femora. *Chlamydatus evanescens*, which feeds on stonecrop, is the only British phyline that overwinters as an adult. It has two generations a year like the common and widespread *Ch. pullus*. Brachyptery is found elsewhere in this subfamily in the black, red-headed females of *Orthonotus rufifrons*, which feed on nettle, and in the three ant-mimics. The two ant-mimicking *Hallodapus* species are almost invariably brachypterous in both sexes; these reddish bugs have a 'petiole' indicated by white, wedge-shaped marks on the somewhat shortened hemelytra, giving them a superficial resemblance to ants of the genus *Myrmica*. The female of *Systellonotus triguttatus* is altogether more convincing as a mimic of *Lasius niger*, with strongly abbreviated hemelytra and a globose abdomen. Males of *Systellonotus* are macropterous and their resemblance to *L. niger* is due, as in *Hallodapus*, to the presence of strategically placed pale wedges on the hemelytra.

Orthotylinae have ribbon-like parempodia that are parallel or convergent towards their

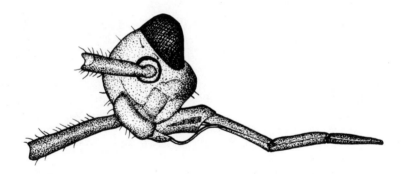

Fig 60. *Miridae:* Calocoris quadripunctatus *nymph, lateral view of head.*

apices. Usually there is no pronotal collar; if one is present, it is clearly depressed below the level of the rest of the pronotum. On average, orthotylines are somewhat larger than phylines and more of them are green rather than brown. The four *Pilophorus* species are ant-mimics. The divisions between the different regions of the ant's body are paralleled by transverse bands of silvery pubescence across the hemelytra. These bugs feed on tree-dwelling aphids. Presumably, they gain protection from other predators by their resemblance to the ants that attend the aphids. They evade the ants themselves by their own alertness and agility. Unlike the ant-mimics in other subfamilies, which live on the ground, these tree-dwelling bugs are always fully winged. *Halticus apterus* is usually brachypterous. It is a black bug with the hind femora enlarged for jumping and feeds on various low-growing plants, especially clovers and bedstraws. In the related *Pachytomella parallela*, only the female is brachypterous, while in *Orthocephalus*, another closely related genus of black, jumping bugs, one species is usually brachypterous in both sexes and the other is brachypterous only in the female. Sexual dimorphism, with macropterous males and brachypterous females, is also the norm in the non-jumping *Mecomma* and *Globiceps*. Both of these genera are polyphagous and *Globiceps* species, at least in captivity, are strongly inclined to carnivory. Some orthotylines, like the black-kneed capsid, *Blepharidopterus angulatus*, are wholly predaceous, feeding on insects and mites. Even closely related species may differ in their feeding habits. Waloff (1968) reported that three of the four orthotylines and one phyline living on broom were generalist predators but the fifth orthotyline, *Orthotylus virescens*, fed mainly on phloem sap. All orthotylines overwinter as eggs.

The apically divergent, ribbon-like parempodia of the subfamily Mirinae are very distinctive. The majority of species are 5–10 mm long. Most mirines have a rounded pronotal collar like that of the three small subfamilies but the collar is lacking in two tribes, Stenodemini and Pithanini, in which the first segment of each tarsus is the longest, in contrast to Mirini and all Orthotylinae, the only other mirids with ribbon-like (but not divergent) parempodia. Stenodemini are entirely vegetarian, feeding on the leaf tissues or flower and seed-heads of grasses, sedges and rushes. *Stenodema* species overwinter as sexually immature adults, *Notostira* as mated females and the other members of the tribe as eggs. Various degrees of wing-reduction are encountered in the subfamily. In the nominate tribe, Mirini, with over 50 species, females of two *Capsodes* species are short-winged, otherwise macroptery is the rule. In Stenodemini, females of *Leptopterna* and both sexes of *Teratocoris* species are usually brachypterous and females of *Notostira elongata* are submacropterous; females of *Acetropis* vary from brachypterous to macropterous. The two British species of Pithanini are both micropterous. *Pithanus maerkeli* is one of the commonest and most widespread of all British bugs. Grasses and rushes are the usual food plants and the bug can be abundant in damp meadows. Macropterous specimens have been found occasionally. The related *Myrmecoris gracilis* is a rare insect of sandy heaths in southern England. It is mainly predaceous. This species is the only native mirine that mimics ants; the resemblance is very striking, particularly in the living bug, and is due to modifications of body shape rather than to a deceptive colour pattern. Predators, among them the agile, mottled and mainly arboreal *Phytocoris* species, are also found in the tribe Mirini, which includes mixed feeders, for example *Calocoris* and *Adelphocoris* species, and largely or wholly phytophagous ones like *Lygus* and *Lygocoris*. Adults of these last two genera hibernate, as do those of *Agnocoris*, *Charagochilus* and some *Orthops* species. Several Mirini are minor pests, for example *Calocoris norvegicus* on potato, *C. fulvomaculatus* on hop, *Lygocoris pabulinus* on cane fruits and *Plesiocoris rugicollis* on apple. *Lygus rugulipennis* feeds on a wide variety of low-growing plants, including some crops, and has two generations a year. *Capsus ater*, a robust, black mirine that feeds on grasses, has a common variety in which the black colour of the head and pronotum is replaced by orange-brown; this coloration is most pronounced in females.

Wagner (1952) and Wagner & Weber (1964) dealt with the Miridae of Germany and France, respectively, providing illustrated identification keys and biological and distributional information.

Dipsocoridae

These tiny bugs might be mistaken for small anthocorids. The first two antennal segments, however, are much shorter than the last two segments, which are delicate, slender and clothed with long, fine hairs. The ocelli are so close to the eyes that they are easily overlooked. There is a well marked costal fracture in the hemelytron, extending about half-way across it. The hemelytron is not differentiated into corium and membrane. The rostrum is very short and does not reach beyond the prosternum. It is supposed that the two British species are predatory. *Pachycoleus waltli*, which is 1.1–1.6 mm long, is usually brachypterous, with the hemelytra reaching beyond the abdominal apex but not overlapping. Macropters are rare. The bug lives among wet mosses, including *Sphagnum*, by running water. *Cryptostemma alienum* is distinctly larger, 1.8–2.4 mm long, and always macropterous. It is found among gravel and beneath stones beside rapidly flowing upland streams throughout the British Isles. Adults of both species have been found in the winter. The rather elongate eggs have a peculiar seam or narrow sole along one side.

Ceratocombidae

Ceratocombus coleoptratus is the only British member of this family. It differs from the two dipsocorids in its longer rostrum, which reaches nearly to the apices of the mid coxae, and its very short costal fracture, extending no more than one-seventh of the way across the hemelytron. The hemelytra extend beyond the apex of the abdomen but do not overlap except in the rare macropterous form. The bug, which is 1.5–2.0 mm long, is believed to be predaceous and lives in damp places, particularly in tufts of the moss *Polytrichum*. There is probably one generation a year, overwintering as adults. The egg is elongate, with a yolk plug at one end, like that of Miridae and Anthocoridae.

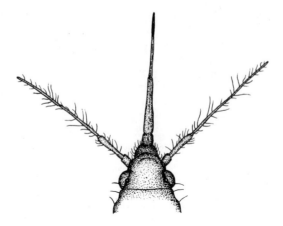

Fig 61. *Ceratocombidae:* Ceratocombus coleoptratus, *head, rostrum and antennae.*

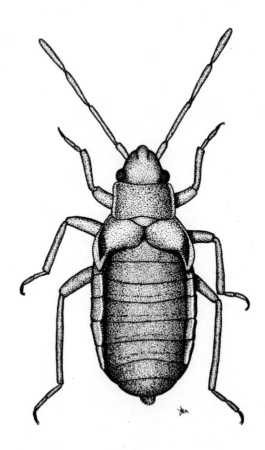

Fig 62. *Aepophilidae:* Aepophilus bonnairei.

Aepophilidae

The only species of this family, *Aepophilus bonnairei*, lives in the intertidal region of rocky shores on the Atlantic coast of Europe, including those of Ireland, the Isle of Man, Wales and south-west England. It is quite impossible to mistake the adults, which are reddish-brown and about 3 mm long, for any other species because of the strongly concave margins of the greatly shortened hemelytra. Macropterous specimens have never been found and dispersal is probably effected by the action of the sea. Adults and nymphs usually live in crevices in rocks or beneath the edges of boulders embedded in sand but they have been seen walking actively about on the sides of small rock pools beneath the water. *Aepophilus* is believed to be a predator but much remains to be discovered about this and other aspects of its biology. It has a number of structural features that show it is related to Saldidae but its extreme wing-reduction, minute ocelli and small eyes show that it has long departed from a typically saldid way of living.

Saldidae

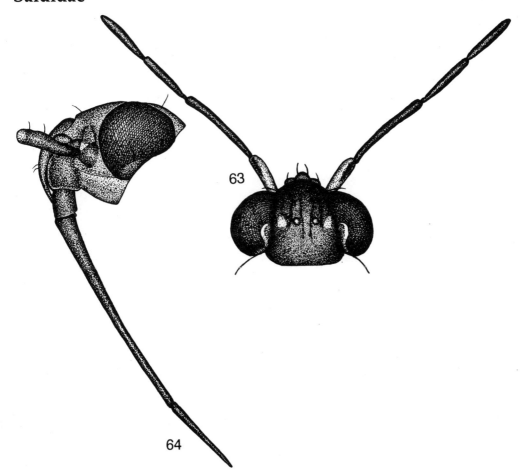

Figs 63–64. *Saldidae. 63*, Saldula saltatoria, *head; 64,* Salda littoralis *nymph, lateral view of head.*

Twenty-one species of these big-eyed predators are native to the British Isles. Wing-polymorphism is frequent in the family. Subbrachypters, in which the membranes of the hemelytra scarcely overlap and in which the wings are useless for flight, are the predominant form in *Salda* and *Halosalda*. Macropters are known to occur in all British species of these genera but those of the upland species *S. morio* and *S. muelleri* have yet to be found here; the subbrachypters of these two species have the membrane area sclerotized and darkened almost as much as the corium. In all the other British Saldidae, both macropters and submacropters occur. In submacropters, the hemelytra and wings are about as long as the abdomen and can probably be used for short 'flits', adding to the agility of these active hunters. Macropters predominate in *Chartoscirta cincta*, *Saldula scotica*, *S. saltatoria*, *S. pilosella* and *S. pallipes* and are rarely found in the other species. Although saldids are active predators their fore legs, unlike those of Reduviidae, Nabidae and the waterbugs, are not modified for seizing or manipulating the prey.

Bare ground is the favourite habitat of Saldidae; they need to be unencumbered by vegetation when hunting. Only *Saldula orthochila* occurs away from water; it may be found on heaths, dunes, acid grassland and tracks and pathways on sandy soils throughout the British Isles. The other species hunt on the bare sand, mud or peat at the margins of moorland pools (the two *Salda* species mentioned above), lakes and streams (for example, *Saldula saltatoria* and *S. pallipes*) or tidal flats and saltmarshes (e.g. *Saldula pilosella* and *Halosalda lateralis*). When pursued they run, leap and flit with great agility, except for *Salda littoralis*, which hides among vegetation. This bug is also unusual in having two alternative year-cycles, with either eggs or adults overwintering. Normally, in this family, there is a single generation each year, overwintering as adults. The eggs, which are superficially rather like those of Miridae, are laid amongst moss or, in *Saldula* species, inserted into the tissues of vascular plants.

Woodroffe (1966b) gives notes on the taxonomic differences between two closely related species, *Saldula pallipes* and *S. palustris*. Péricart's (1990) monograph covers the Saldidae of western Europe and north-western Africa.

Mesoveliidae

The pondweed bug, *Mesovelia furcata*, is the only British member of this family. It is the only green bug associated with the water surface and the only surface-dwelling aquatic bug to overwinter as an egg. It is also the only one with a well-developed ovipositor. The eggs are inserted in plant stems which die off in the winter and sink to the bottom of the pond, lake or canal. On hatching, the nymphs rise to the surface of the water and break through the surface film without difficulty. There may be two generations annually, at least in favourable years and localities. The bug is as much at ease running over the floating leaves of aquatic plants as it is on the water surface. It will feed on moribund insects floating on the water if they are not moving and will spear live crustaceans and insects through the surface film, especially, perhaps, those trapped in shallow pools or bays on the surface of broad, floating leaves. Although *M. furcata* is usually totally apterous, macropters are not at all rare. They exhibit the phenomenon of autotomy in which, probably following a dispersal flight, the hemelytra and wings are deliberately broken off by grooming movements of the legs. The flight muscles are then broken down metabolically, releasing nutrients that are channelled into reproduction.

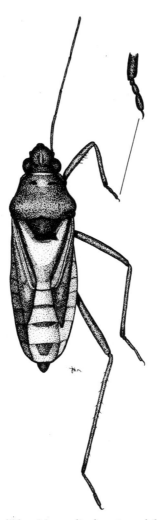

Fig 65. *Mesoveliidae:* Mesovelia furcata *and detail of tarsus.*

Hebridae

Two species of 'sphagnum bugs', both belonging to the genus *Hebrus*, live in Britain. These tiny bugs, 1–2 mm long, live among wet moss (not always *Sphagnum*) at the margins of ponds, lakes, streams and ditches. Unlike all other bugs except the shieldbugs, they have five-segmented antennae. The nymphs have four antennal segments and the apical one divides into two at the final moult. The division is incomplete in *H. pusillus*. Adults overwinter and there is only one generation a year. The large eggs mature a few at a time and are laid amongst mosses. *Hebrus pusillus*, a southern species, is always macropterous but *H. ruficeps* is usually micropterous, only rarely producing macropters. The bugs are believed to be predaceous; reports of their feeding on mosses have been discounted as probably arising from observations of the bugs probing for prey amongst the plants.

Hebrus, like *Mesovelia, Microvelia* and *Hydrometra*, has a remarkable technique for ascending the meniscus where the surface film is in contact with an emergent object such as a plant stem or the side of an aquarium. The bugs are so small that they cannot reach up the slope of the meniscus to seize the emergent object; nor can they run up the curved slope. Instead, they lift some of the feet, with the wettable pretarsal structures still in contact with the surface, so that the part of the meniscus in contact with the upraised feet is raised above the surrounding surface instead of depressed as it is when bearing the weight of the bug. Surface forces then cause the bug to glide up the curved meniscus until it makes contact with the solid object.

Hydrometridae

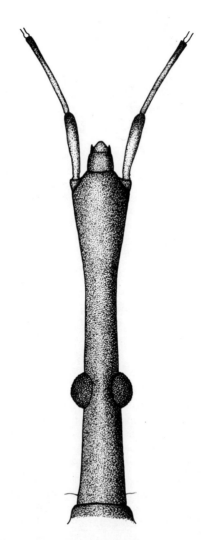

Fig 66. *Hydrometridae:* Hydrometra stagnorum, *head.*

The extraordinarily elongated heads of the two water-measurers, *Hydrometra stagnorum* and *H. gracilenta*, are unique among British Hemiptera. These slender bugs, 7–12 mm long when adult, walk slowly over floating vegetation and the surface of still waters in search of small aquatic insects and crustaceans, which they seem to be able to detect beneath the surface, and spear them with the stylets. They also feed on drowned insects but will not take large prey that is still struggling. All British specimens of *H. gracilenta*, which is restricted to Hampshire and Norfolk, have been found to be micropterous. The widespread *H. stagnorum* is usually micropterous, with a few brachypters; fully winged individuals are rare but occasionally a population is found that consists entirely of macropters. Eggs are attached to vegetation or stones at about water-level. There is probably one generation a year, and adults overwinter.

Veliidae

Like Gerridae, Veliidae have the tarsal claws inserted subapically but, unlike them, they have the middle pair of legs inserted equidistantly from the other two pairs. Two of the five British species belong to the genus *Velia*, in the subfamily Veliinae. They are 6–8 mm long and have three-segmented tarsi when adult. The remaining three species belong to the genus *Microvelia*, in the subfamily Microveliinae. They are tiny insects, 1.4–2.0 mm long; their anterior tarsi are single-segmented and the other two pairs are two-segmented.

Microvelia species are so small that they can walk about on the water surface as easily as they can on land or on the floating leaves of water plants. The early instars of *Velia* and

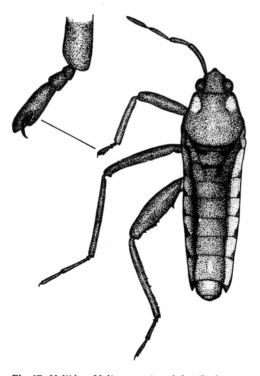

Fig 67. *Veliidae:* Velia caprai *and detail of tarsus.*

other surface-dwelling bugs of similar size, whether adult or immature, can similarly make use of the surface tension of water in walking. Adult *Velia* can both walk slowly and row rapidly with simultaneous strokes of the middle pair of legs, like pondskaters. Both of the veliid genera can glide forward, as if by magic, without moving the limbs at all. They do this by reducing the surface tension of the water behind them by exuding a detergent fluid, possibly saliva, from the tip of the rostrum.

Apters are the usual form encountered but macropters are found occasionally in some species and quite frequently in others. Adults overwinter and the eggs are laid in moss. The number of generations in a year probably depends on the local climate. The largely northern *Velia saulii* is said to be univoltine, while *Microvelia reticulata* may have as many as three generations in southern Britain. *Microvelia* species, being so small, are quite at home in reed beds and among floating water plants, using the free water between the stems and leaves as well as walking on the plants themselves. The bigger *Velia* species prefer weed-free pools, slow streams and the backwaters of rivers. Food consists of drowning insects, springtails, small crustacea and the floating eggs of mosquitos.

Gerridae

The pondskaters are more highly modified for life on the surface film than any other bugs. They row themselves along with the elongated middle pair of legs, which thrust simultaneously instead of alternating as in walking. This gives them a speed and agility which, coupled with their acute vision, makes them difficult to catch. The pre-apical insertion of the tarsal claws separates nymphs and adults from all other families except Veliidae, from which they differ in having the middle pair of legs inserted much closer to the hind pair than to the front pair, instead of equidistantly from both. All 10 British species belong to the subfamily Gerrinae.

Like most surface-dwelling bugs, pondskaters are densely clothed with short, virtually unwettable hairs, which are velvety black or brown above and silvery beneath. The legs, including the tarsi, are also unwettable except for the pretarsal structures, which penetrate the surface film. The weight of the bug is supported by the weight of water displaced from the six little dimples in the surface film where the feet are resting. Forward motion, if unimpeded by vegetation or detritus, is very efficient because of the low frictional and viscous forces involved but the bugs can also make energetic leaps both on land and water. This less economical leaping behaviour is most evident when danger threatens: the bug skips rapidly over the surface at a remarkable speed. Each leap, when the insect completely leaves the surface of the water, is followed by a longer phase of gliding along the surface. In normal, unhurried locomotion, progress is entirely by gliding, with the feet always in contact with the surface film.

Because of their specialized method of locomotion, pondskaters require habitats that include considerable expanses of open water. Ponds, canals, wide ditches and lake shores are among the most favoured places. The large (13–17 mm long) and generally flightless *Gerris (Aquarius) najas*, however, prefers slow-flowing rivers. Apterous females of this species, when full of eggs, are very bulky insects with the abdomen noticeably distended. The slightly larger *Limnoporus rufoscutellatus* almost certainly used to breed in southern Ireland and may still do so. The few, scattered records from Britain probably all refer to migrants from continental Europe that have failed to breed here. Between them, the British species cover most of the available aquatic habitats. *Gerris thoracicus* even thrives on salty pools by the sea, though it is also found inland. Other species of this genus, notably *G. lateralis* and *G. costai*, have a preference for acid, peaty pools in the highlands and islands of western Scotland. Several genera of water-surface bugs (not all of them gerrids) live permanently on the surface of the sea, attaching their eggs to mangrove roots or to floating debris. Most are found close to the shore or in estuaries but five species of the gerrid genus *Halobates* live on the surface of the tropical oceans hundreds or even thousands of miles from land.

The food of gerrids consists of dead or dying insects trapped in the surface film. The bugs can detect the ripples made by struggling insects from a distance of several centimetres and home in on them. At close quarters, sight also plays a part in the detection of prey. Generally, there are two generations a year but the northern or upland species have only one. Wing polymorphism is frequent. In the common *Gerris lacustris* all sorts of intermediate forms occur between total aptery and full macroptery. The genetic and environmental factors that govern wing-length in this species are complex.

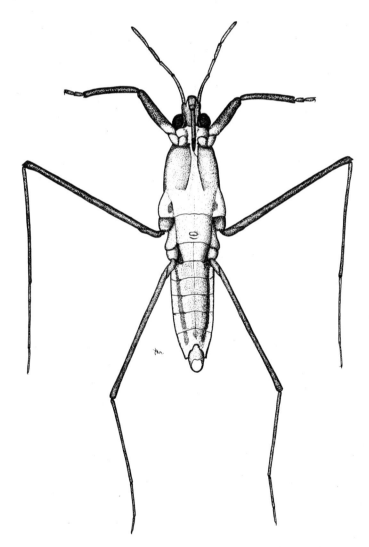

Fig 68. *Gerridae:* Gerris gibbifer, *in ventral view, showing insertions of legs and position of metathoracic scent gland opening.*

Macropterous gerrids migrate freely from place to place during the summer and, in consequence, the ratio of fully-winged to flightless individuals in a population can vary considerably from day to day. In the winter, macropters may fly some distance from water in search of hibernation sites. Flightless morphs, of course, spend the winter close to the water. Sometimes they may be found in a cataleptic state among pond-side vegetation at this time of year. Seasonal flooding and freezing can kill a high proportion of the flightless individuals, so that the spring population consists largely of macropters after a severe winter. Eggs are laid among vegetation or attached to floating debris. Vepsäläinen & Krajewski (1986) provide a key to the nymphs of the northern European species.

Nepidae

The two British water-scorpions are unmistakeable. *Nepa cinerea* is a dark brown, leaf-shaped bug, about 2 cm long, with raptorial forelegs and a slender, bristle-like 'tail' about half as long as the body. *Ranatra linearis* is long-legged, stick-like, pale brown and about 3 cm long. Its fore legs, too, are raptorial and its 'tail' is proportionately longer than that of *Nepa*. The 'tails' are rigid, hollow respiratory siphons, derived from paired abdominal outgrowths.

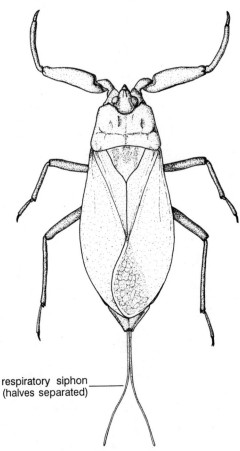

Fig 69. *Nepidae:* Nepa cinerea.

Nepa is found throughout Britain, and in parts of Ireland, in shallow waters at the margins of ponds. The bug lies motionless in the water with its respiratory siphon just breaking the surface, waiting for suitable prey to pass by. It is said to be able to snatch flies running on the surface film and also to extract caddis larvae from their cases. *Ranatra* is restricted to southern Britain. It lives in deeper water than *Nepa*, usually clinging to the stems of reeds, though young nymphs can float suspended by their respiratory siphons from the surface film. The food, especially that of the nymphs, consists mainly of water-fleas (Cladocera). In general, *Ranatra* takes smaller prey than *Nepa*, though it can tackle small tadpoles.

In both species there is only a single generation annually, with adults overwintering. Eggs are inserted into the tissues of aquatic plants, in batches of a few dozen. Each *Ranatra* egg has two slender micropylar respiratory processes projecting from the oviposition slit; those of *Nepa* have about seven such processes. Because of their large size and their projecting processes the eggs are quite easy to find by examining likely oviposition sites.

Aphelocheiridae

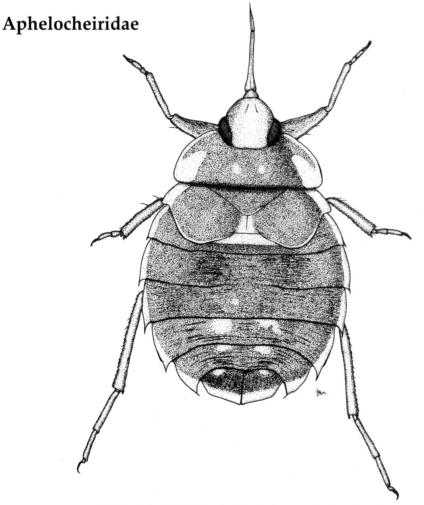

Fig 70. *Aphelocheiridae:* Aphelocheirus aestivalis *micropter.*

Only *Aphelocheirus aestivalis* represents this family in Britain, where it occurs from Sussex to the Scottish borders, and in Ireland. It lives in the middle reaches of well oxygenated, reasonably fast-flowing rivers and sometimes in quite small streams. It is a flattened, oval bug, about a centimetre long, and is usually strongly brachypterous with the hemelytra reduced to short pads, leaving most of the abdomen exposed. Macropters do occur occasionally. Most British populations belong to the dark brown form *'montandoni'* rather than the typical yellow-brown european form. The bug is largely covered with a coating of fine hairs that trap a layer of air round its body. The exchange of oxygen and carbon dioxide between this air-layer and the surrounding water is so efficient that the animal never needs to visit the surface to replenish its air supply. *Aphelocheirus* preys on invertebrates typical of its river-bottom habitat, principally the larvae of mayflies and midges. Unlike the naucorid *Ilyocoris*, its front legs are not raptorial: they are three-segmented with paired claws, like the other two pairs. Its rostrum is longer than that of *Ilyocoris*, reaching the apex of the mesosternum. Only the hind tibiae and tarsi have fringes of swimming hairs. There is also a fringe of long hairs on the dorsal surface of the anterior femur. The ventral surfaces of the femora and tibiae of the first two pairs of legs have pads of dense, short bristles throughout their length that probably serve either to grip the prey or to cling to the vegetation among which the bug lives.

Naucoridae

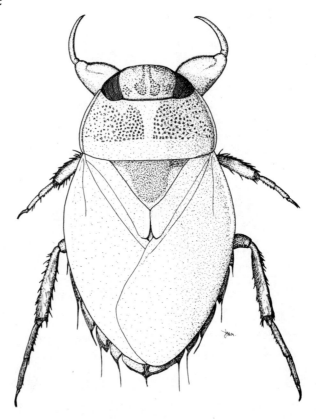

Fig 71. *Naucoridae:* Ilyocoris cimicoides.

The saucer bug, *Ilyocoris cimicoides*, a dark brown, oval insect about 14 mm long, is the sole British member of its family. It lives in still waters in southern Britain, preying on crustaceans and insects. The front legs are strongly modified for seizing prey, with robust femora and with the tibia and single-segmented tarsus of each fused into a curved blade that closes against the femur. The tibiae and tarsi of the other legs bear fringes of swimming hairs on their outer surfaces. The rostrum is very short, scarcely reaching beyond the posterior margin of the head. There is a single generation each year. Adults overwinter and the eggs are laid in the spring in rows in slits cut in the stems of water plants.

Notonectidae

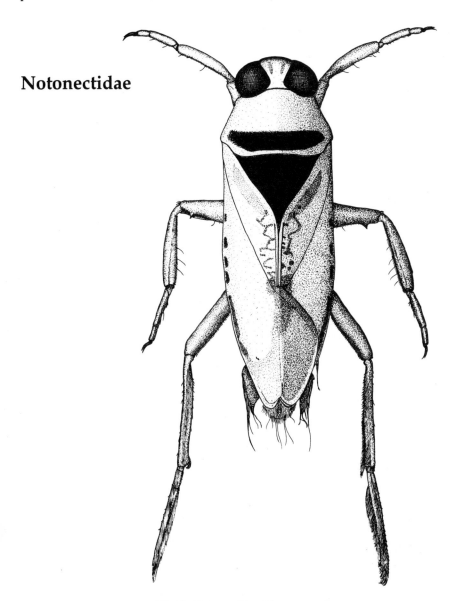

Fig 72. *Notonectidae:* Notonecta glauca.

Four species of water-boatmen, all in the genus *Notonecta*, live in Britain. All are about 15 mm long and swim with the ventral side uppermost, in contrast to Corixidae, which swim dorsal side up. Another point of difference between the two families is that, in Notonectidae, the first two pairs of legs are similar in structure while Corixidae have each pair of legs specialized for a different function. The underside of the oblong body bears a bubble of air trapped by rows of hydrofuge hairs. In adults the bubble extends also to fill the space between the hemelytra and the abdominal dorsum. The air needs to be changed frequently, particularly in warm weather when the bugs are active and the dissolved oxygen content of the water is low, requiring repeated visits to the surface, where the abdominal apex is briefly exposed to the air.

The biology of the common and widespread *Notonecta glauca* is typical of the genus in Britain. There is only one generation a year, overwintering as adults. About 60 eggs are laid, in batches of eight or so, in the leaves or stems of water plants in late winter or spring. The adults of the new generation appear in late summer. *Notonecta maculata*, which replaces *glauca* in such weed-free and apparently unproductive water bodies as spring-fed ponds, water troughs and concrete reservoirs, differs from the other three species in that only a minority of individuals follow the *glauca* pattern: most adults die off at the onset of winter, leaving the eggs, which are attached to the substrate, to overwinter.

Prey consists of almost any aquatic invertebrate or vertebrate that is not too much bigger than the bug. *Notonecta maculata* must feed mainly on insects that fall into the barren water bodies that it inhabits. The prey is grasped by the front two pairs of legs, which are also used for clinging to inanimate objects. The long, hair-fringed hind legs are used only for swimming. In still waters, when danger does not threaten, adults and especially nymphs may be seen hanging head-down beneath the surface film with the front two pairs of tarsi resting against the underside of the surface film and the rosette of hairs at the apex of the abdomen spread out on the surface, putting the ventral air-bubble in direct communication with the atmosphere.

Pleidae

Plea minutissima is the only British species of this small family. It resembles a compact, miniature notonectid, only 2–3 mm long, swimming with the ventral side uppermost. *Plea* lives in clean, weedy, still or slow-flowing waters throughout Britain except in the north. Water-fleas (Cladocera) are the principal prey. Eggs are laid in slits in the leaves of aquatic plants in midsummer and the new generation is adult by early autumn. There is only a single generation each year.

Corixidae

This is the largest family of water bugs in the British Isles, with 34 native species. The other five wholly aquatic families can muster only nine British species and there are only 20 surface-dwelling bugs. Corixids, like notonectids, are often called water-boatmen because they row their oblong bodies through the water with oar-like strokes of the hair-fringed hind legs. They differ from notonectids in swimming with the dorsal side uppermost. The middle pair of legs are long and slender, adapted for clinging to vegetation or other inert objects. The anterior pair are strongly modified for food-gathering. The fore tibia is very short and there is a single tarsal segment, termed the pala, with rows of bristles extending throughout its length. The single claw is often scarcely distinguishable from the bristles. Female Micronectinae have the fore tibia and tarsus fused. The rostrum of corixids is very short and triangular, forming a ventral continuation of the face, and its apical aperture is larger in proportion to the size of the bug than in most other Hemiptera.

Many Corixidae are known to stridulate in one or both sexes. Males of *Micronecta* make a surprisingly loud noise by rotating the genital capsule so that the right paramere is rubbed against a field of ridges on the eighth abdominal segment. Both sexes of other Corixidae have a field of pegs on the inner surface of the anterior femur which can be rubbed against the sharp keels at the sides of the head. Some corixine males also have pegs on the femora of the middle pair of legs. When a male of these species mounts a female, he rubs his femora against the sides of her hemelytra.

Two other structures in Corixidae have been wrongly supposed to be stridulatory. A small, ridged structure on the male abdominal dorsum has been named the 'strigil' for this reason. Its true function is not well understood but a recent, plausible suggestion is that it forms a conduit between the air-bubbles of the copulating male and female. It is

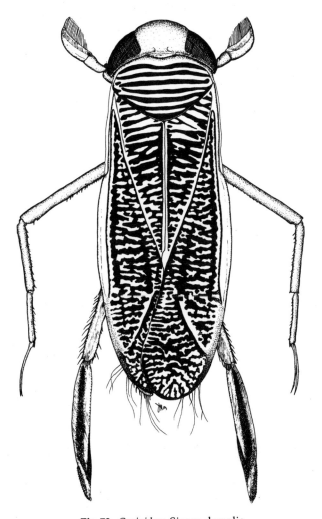

Fig 73. *Corixidae:* Sigara dorsalis.

difficult to see how otherwise the female, held beneath the male, could replenish her air supply at this time. A row of pegs on the pala (anterior tarsus) of male Corixinae is almost species-specific in its design. The pegs are not rubbed against the head, as was once thought (the femoral pegs are the ones involved in stridulation). Instead, their function is to aid in gripping the sides of the female's hemelytra.

Three subfamilies occur in Britain. The little Micronectinae, less than 3 mm long, are represented by three species, all in the genus *Micronecta*. The scutellum is visible in this genus; in other British corixids it is concealed by a posterior extension of the pronotum. Cymatiainae, with two species of *Cymatia*, lack transverse furrows on the rostrum and alternating transverse light and dark lines on the pronotum. Both of these features are present in Corixinae.

Flight polymorphism is widespread in this family. In Micronectinae and Cymatiainae, macropters are extremely rare and distinctly larger than the usual brachypterous forms. Brachypters have the hemelytral membranes reduced and not completely overlapping and their wings are reduced to rudiments. The flight muscles are atrophied and the shape of the pronotum consequently differs between brachypters and macropters. Polymorphism in Corixinae, where it occurs, involves only the degree of development of the flight muscles and the consequent differences in the shape of the thorax.

Jansson (1986) considered Corixidae to be mainly predaceous and suggested that the plant material found in the gut of some species was originally present in the gut of the prey and was not deliberately ingested by the corixids. Some genera, notably *Cymatia* and *Glaenocorisa*, are active predators, pursuing individual prey which they detect by sight. Lansbury (1983) noted that *Cymatia coleoptrata* and *C. bonsdorffi* fed on other corixids, water-fleas and the larvae of mayflies and chironomid midges. Most Corixinae, however, use their bristle-fringed palae to sift food items from the sediment that accumulates at the bottom of the water-bodies they frequent. Recently dead animals, as well as living ones, are consumed. Some corixines have been seen to prey on the eggs of their own or other species. The method of feeding and type of food in *Micronecta* are not yet known.

Micronecta species overwinter as third or fourth-instar nymphs. All other European Corixidae overwinter as adults. Eggs are laid in the spring or early summer attached, usually in short rows, to water plants or stones. In southern Britain most corixids probably have two generations a year and perhaps even three. It is unlikely that more than one generation occurs in the north, although many species are present in the Scottish highlands.

The type of habitat preferred varies with the species, though running water is generally avoided by all of them. In the absence of predaceous fish, open water is preferred but the presence of fish restricts the bugs to weedy areas where the chances of escape are greater. Some corixids have pronounced requirements, or at least special tolerances, for water bodies of a particular kind. *Sigara selecta* and especially *S. stagnalis* can tolerate water so salty that it would be lethal to most corixids. *Hesperocorixa castanea* is confined to bog pools, while the related *H. moesta* prefers shady ponds in woods with many dead leaves on the bottom.

Since the publication of the identification guides by Southwood & Leston (1959) and Macan (1965), *Corixa iberica* has been found to occur in Britain. It is restricted to the north-western coasts of Ireland and Scotland, including the Shetland Islands, and southern Spain and Portugal. Perhaps it has been replaced elsewhere by the common and widespread *C. punctata*, as the latter species is not known to occur together with it. *Corixa iberica* is included in Savage's (1989) edition of Macan's keys. Jansson's (1986) synopsis includes keys and distribution maps for all European Corixidae. Cobben (1960) produced a preliminary key to the nymphs of the Dutch corixids.

15

AUCHENORRHYNCHA

The position of the rostrum is diagnostic of this group. It arises at the back of the head, in contact with the anterior margin of the prosternum. It is not displaced between the anterior coxae, as in Sternorrhyncha, nor is there a bridge of cuticle behind it, closing the head capsule posteriorly, as in Heteroptera. There are always three tarsal segments in the adult, in contrast to the maximum of two in adult Sternorrhyncha. The only Heteroptera likely to be confused with Auchenorrhyncha are Corixidae which, when dead and pinned, can look surprisingly like Cicadellidae. Their hind legs are fringed with long hairs for swimming and have only a single tarsal claw.

Most Auchenorrhyncha lay their eggs in slits in plant tissues, often in rows. Exceptions are Tettigometridae, Issidae and Cixiidae. In the first two of these families the ovipositor is plate-like and the eggs are deposited on the substrate. Cixiidae have an elongate ovipositor, which is used to insert the eggs into the soil. Extensive investigations into the ovipositor of Auchenorrhyncha were made by Müller (1942). There are five nymphal instars in most families but more in Cicadidae. A few species are obligately gregarious, at least in the earlier instars, but the great majority can develop to adulthood in solitude.

All Auchenorrhyncha feed on vascular plants. There are no predaceous, parasitic or aquatic forms. Stems, leaves and underground parts are attacked but reproductive organs and buds are usually spared. Phloem sap is the usual food but Cicadidae, Cercopidae and some Cicadellidae feed on xylem sap instead and most typhlocybine cicadellids feed on the contents of mesophyll cells. A few species have been shown to take more than one of these types of food. The kind of faeces voided varies with the diet and the normal method of feeding soon becomes apparent if the insects are confined in glass tubes with fresh material of a suitable host plant. Phloem-feeders produce small amounts of colourless, sugary honeydew; xylem-feeders void large quantities of colourless, watery faeces and mesophyll-feeders produce brown drops of excreta. Many Cicadellidae also produce from the anus, in addition to the normal faeces, occasional drops of opaque, white or yellow fluid containing brochosomes, which they distribute over the body surface and wings.

Male cicadas are well known to produce sounds audible to the human ear, by vibrating the tymbals, a pair of plates at the base of the abdomen. Tymbals occur in males of all families of Auchenorrhyncha but the sounds produced in families other than Cicadidae are inaudible to us under normal conditions. Their vibrations seem to be transmitted through the substrate and they play an important part in mate-selection by the females. Songs of related species often differ markedly and it is thought that the different species-specific songs of the males and the responses to them by the females act as barriers to hybridization between species.

Much recent work on Auchenorrhyncha has involved the analysis of the faunas of particular geographical areas or types of habitat. British studies of community structure in various grassland habitats were compared by Waloff & Solomon (1973) and many European studies are cited in the extensive list of references accompanying Günthart's (1987) paper.

The families of Auchenorrhyncha are grouped into two series or infraorders, each with four British families. Fulgoromorpha, comprising the families Tettigometridae, Issidae, Cixiidae and Delphacidae, are characterized by the possession of tegulae, a pair of small flaps on the mesonotum, covering the articulation of the fore wings. The two veins of the clavus in the fulgoromorph families are united into the shape of a Y and there is a suture separating the postclypeus from the frons. The antennae are inserted on the sides of the head, below the eyes. In all of these families except the Tettigometridae there are conspicuous longitudinal keels on the face, vertex, pronotum and scutellum of the adults and the nymphs bear conspicuous, circular, sensory pits on various parts of their bodies. The middle coxae of Fulgoromorpha, again with the exception of the tettigometrids, are elongate and their insertions are rather far apart. The cicadomorph families, Cicadidae, Membracidae, Cercopidae and Cicadellidae, all lack tegulae, do not have the claval veins united into a Y and do not have the postclypeus separated from the frons by a suture. Their antennae are inserted low down on the face. The middle coxae are short and inserted close together and they lack the characteristic longitudinal keels and nymphal sensory pits of the Fulgoromorpha. *Megophthalmus*, in Cicadellidae has an x-shaped arrangement of keels on the face but no longitudinal keels on the head or thorax.

The major works dealing with the identification of British Auchenorrhyncha are those of Edwards (1894–1896), LeQuesne (1960, 1965b, 1969) and LeQuesne & Payne (1981). The last includes a checklist. Ossiannilsson's (1978, 1981, 1983) account of the Scandinavia Auchenorrhyncha is also useful for identification and contains more biological information than the others. Nast (1972) catalogued the Palaearctic fauna.

Key to families of British Auchenorrhyncha adults

1. Posterior tibia with large, mobile spur at apex (fig. 79) .... DELPHACIDAE
— Posterior tibia without such a spur 2

2. Bases of fore wings covered by tegulae (Fig. 74); clavus of fore wing with two veins uniting posteriorly to form a Y (Figs 75, 77; indistinct in Tettigometridae) 3
— Tegulae absent; clavus of fore wing with two parallel veins (Figs 86, 87), these rarely uniting and then separating again 5

3. Fore wing with cells glassy, pronotum deeply incised posteriorly in middle and with three or five longitudinal keels (Fig. 77) .... CIXIIDAE
— Fore wing with cells opaque, leathery; pronotum not or weakly indented posteriorly and without longitudinal keels 4

4. Head dorsally, pronotum, tegulae, scutellum and fore wings pitted with punctures (Fig. 74) .... TETTIGOMETRIDAE
— Dorsal surface not punctate anywhere (Fig. 75) .... ISSIDAE

5. Posterior tibiae keeled longitudinally (Figs 84, 87) 6
— Posterior tibiae cylindrical (Figs 81, 86) 7

6. Pronotum produced posteriorly in a spine extending beyond apex of scutellum (Fig. 84) .... MEMBRACIDAE
— Pronotum not produced posteriorly (Fig. 87) .... CICADELLIDAE

7. Fore wing between veins leathery; top of head with two ocelli (Fig. 86) .... CERCOPIDAE
— Fore wing between veins glassy (Fig. 81); top of head with three ocelli (Fig. 82) .... CICADIDAE

Key to families of British Auchenorrhyncha nymphs

1. Postclypeus separate from frons. Middle pair of coxae long, their bases widely separated (Fig. 78). (Fulgoromorpha) 2
— Postclypeus not separate from frons. Middle pair of coxae short, their bases set close together. (Body without sensory pits) (Cicadomorpha) 5

2. Body without sensory pits. (One species, uncommon) .... TETTIGOMETRIDAE
— Large, circular sensory pits present on body (Fig. 76) 3

3. Hind tibia with a movable spur at the inner side of apex .... DELPHACIDAE
— Hind tibia without apical spur 4

4. Second segment of hind tarsus with row of teeth at apex. Living at the ground surface or subterranean .... CIXIIDAE
— Second segment of hind tarsus with teeth only at sides. Living on trees, shrubs and ivy .... ISSIDAE

5. Fore legs modified into digging organs, rather like the pincers of a crab (Fig. 83). Subterranean. (One species, rare) .... CICADIDAE
— Fore legs not modified for digging 6

6. Last segment of abdomen about as long as rest of abdomen; pronotum elevated and swollen (Fig. 85) .... MEMBRACIDAE
— Last segment of abdomen not much longer than the one preceding it 7

7. Abdominal sterna flanked by membranous lateral lobes derived from terga and pleura of segments 3–9; in 'cuckoo-spit' .... CERCOPIDAE
— Abdominal sterna not flanked by membranous lobes; insects not surrounded by 'cuckoo-spit' .... CICADELLIDAE

Tettigometridae

Tettigometra impressopunctata is the only British species of this family. A rather small, brown hopper, 4.0–4.5 mm long, with opaque fore wings, it might be mistaken for a cicadellid or even, because of the numerous impressed punctures on the head, thorax and fore wings, a cercopid. The presence of tegulae, the Y-vein in the clavus and the suture separating the postclypeus from the frons, however, all show that it is more closely related to the fulgoroid families Delphacidae, Cixiidae and Issidae. Adults of Tettigometridae differ from those of the other three families in having opaque, punctate fore wings with only weakly differentiated veins. An equally striking difference is the complete absence, in tettigometrids, of keels on the face, vertex, pronotum or scutellum. The broad, dorsoventrally flattened nymphs lack the characteristic circular sensory pits of the other fulgoroid families. In both nymphs and adults the middle coxae are short and their insertions are close together, more like those of Cicadellidae than their closer relatives.

Little is known of the biology of *T. impressopunctata*. In Britain it is a scarce insect, confined to the southern half of the country. The usual habitat is the surface of the ground in well drained, chalky or sandy soils, including downs and dunes. Adults have been found in most of the winter months and there is probably only a single generation each year. The eggs of *Tettigometra* species are laid on or under stones, not inserted into plant tissues or into the soil and not mantled with detritus like those of issids. Some members of the genus are attended by ants (Lesne, 1905), a fact that implies that they are phloem-feeders.

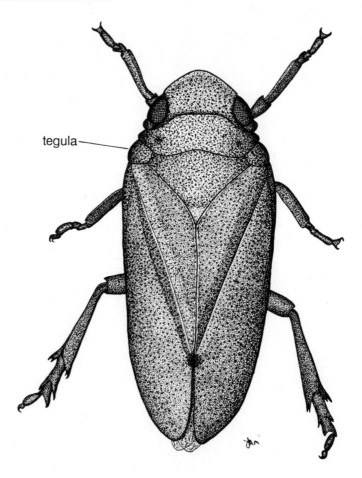

Fig 74. *Tettigometridae:* Tettigometra impressopunctata.

Issidae

Two species of *Issus* are the only British representatives of this family. The genus is easily recognized as belonging to the Fulgoromorpha by its keeled head and scutellum. Adults are 5.5–7.0 mm long and broadly rhomboid in shape. The tibiae lack the movable spurs characteristic of Delphacidae. There is a network of cross-veins between the major veins of the fore wings, which are opaque and rather horny in appearance; in Delphacidae, Cixiidae and Tettigometridae the cross-veins are few and are restricted to the apical half of the fore wing. Nymphs have the circular, sensory pits characteristic of most Fulgoroidea. They are distinguished from those of Cixiidae by the disposition of teeth on the second segment of the hind tarsus: only at the sides in Issidae but in a complete apical row in Cixiidae.

Issus coleoptratus is widespread in southern England and South Wales and is also found in Ireland. It lives on trees and shrubs, often oak, ivy and whitebeam. Males are grey with a slight greenish tinge and, usually, a fuscous discal spot on the fore wing. Females are pale brown with a similar discal spot and a tendency to develop two transverse, fuscous

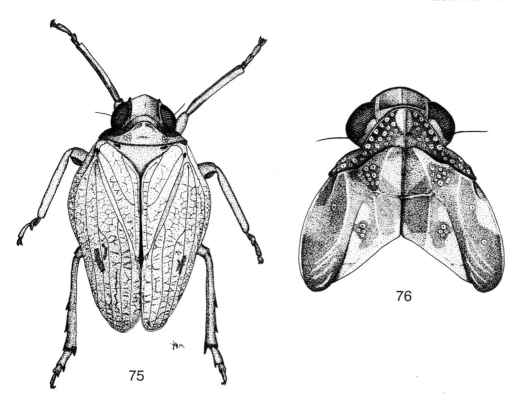

Figs 75–76. *Issidae:* Issus coleoptratus. 75, adult; 76, nymph, fore body, showing circular sensory pits.

or even piceous bands on the fore wings. These bands may, rarely, be joined together by a longitudinal band near the commissure; they are almost obliterated in the palest females, while some males may show traces of banding. The second (more distal) of the two bands, when present, includes the discal spot.

Eggs are laid on the ground and covered with a mantle of soil particles by means of the specially modified ovipositor. Young nymphs can be found throughout the winter on ivy and on the bare twigs of some deciduous trees. Adults can be found as early as May and, very rarely, as late as November but they are most frequent in August, which is the latest month in which fifth-instar nymphs have been seen.

The presence of a second species of *Issus* in Britain was not suspected until Payne's (1979) publication of a record of *I. muscaeformis* on oak at Arneside, Lancashire. Alexander (1981) beat another specimen from yew at Silverdale, two miles away from Payne's locality. Both these records refer to adult females taken in August. W. E. China collected a female and two fourth-instar nymphs of the same species from ivy on the walls of Wray Castle, Cumbria in October, 1954, but the record has never been published. The capture of fourth instars at this time of year suggests that the life-cycle of *I. muscaeformis* must differ from that of *I. coleoptratus*. Old records of *coleoptratus* from the southern Pennines and Strathclyde may well refer to *muscaeformis*. Males of the latter species have never been taken in Britain or, it seems, anywhere in northern Europe, yet only *muscaeformis* is reported from Scandinavia. The females are undoubtedly distinct from *coleoptratus* but until males are examined there must remain some doubt as to the identity of this northern species, since *muscaeformis* was originally described from the Balkans.

The two British species are most easily separated by the arrangement of cross-veins between the longitudinal veins of the fore wing. At least half of the cross-veins in the anterior half of the wing of *muscaeformis* run straight between the main veins so that there is mostly a single row of roughly quadrangular cells between the main veins. In *coleoptratus* the cross-veins mostly branch and meet each other to form two or more rows of mainly pentagonal or hexagonal cells between adjacent longitudinal veins. This reticulation is particularly rich in the female fore wing, in which the apical parts of the first two long veins become lost and obliterated in the network of small cells.

Cixiidae

Eleven species of Cixiidae, distributed among four generally recognized genera, are found in the British Isles. Adults can be found on trees and bushes, amongst grasses and sometimes under stones. They somewhat resemble macropterous Delphacidae but most of them are larger than the largest delphacids, ranging in length from 4.1 mm to 8.0 mm but mostly in the range of 5–7 mm. They lack the tibial spurs of Delphacidae and the numerous crossveins of Issidae. The nymphs have a row of spines across the apex of the second segment of each hind tarsus. Females secrete much white, woolly wax, which is carried as a tuft at the hind end of the body.

Eggs are laid into the ground with the awl-like ovipositor. Probably all the species have a similar life-cycle, with nymphs overwintering and a single generation each year. The nymphs are generally subterranean except for those of the widespread and common *Tachycixius pilosus*, which can be found throughout the winter in the litter layer among grasses in dry, sunny places. In this species, they may ascend shrubs or trees before undergoing the final moult. Adult Cixiidae occur from May to October.

Little is known about host-plant requirements but there are definite habitat preferences. *Oliarus leporinus* and *Cixius remotus* favour salt marshes and so, to a lesser

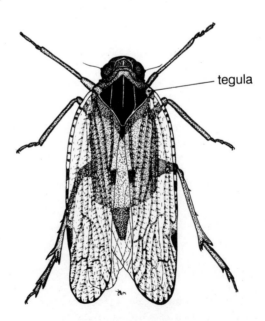

Fig 77. *Cixiidae:* Cixius nervosus.

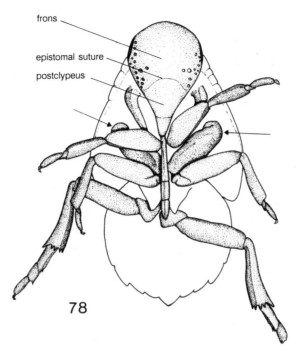

Fig 78. *Cixiidae:* Tachycixius pilosus *nymph in ventral view. Arrows indicate basal articulations of mid coxae.*

extent, does *C. simplex*, which is also found in woods, the preferred habitat of *C. nervosus* and *C. distinguendus*. *Cixius similis* prefers wet heaths and *Oliarus panzeri* frequents fields that become dry and cracked in the summer. Most species are restricted to southern Britain, or at least are more common there than in the northern counties, but *Cixius caledonicus* is known only from Scotland and *C. cambricus* is confined, in Britain, to grassy mountain slopes in Scotland and Wales. Woodroffe (1962) described the nymphal habitat of *cambricus* on the island of Rhum: beneath stones on wet, sandy terraces at an altitude of 2000 feet (600 metres), in small, wax-lined cells surrounding sedge roots. The wax was secreted as tufts of threads from glands on the last three visible abdominal tergites.

Since the publication of LeQuesne's (1960) handbook, the european species *Trigonocranus emmeae* has been recorded from Ashford in Kent, Leatherhead in Surrey and the edge of a meadow bordering woodland in coastal Lancashire. It is about 4 mm long, with three pronotal keels as in *Cixius* and *Tachycixius* but with a longer head, like that of *Oliarus leporinus*. It was figured and briefly described by LeQuesne (1965a).

Delphacidae

More than 70 delphacid species are native to Britain, where they are the second largest family of Auchenorrhyncha, after Cicadellidae. Both adults and nymphs can be distinguished from all other auchenorrhynchan families by the presence of an articulated spur at the apex of each hind tibia. In the youngest nymphs this spur is very small and might be overlooked but the circular sensory pits on the body are sufficient to distinguish delphacid nymphs from those of all other families except Issidae and Cixiidae.

Many Delphacidae are brachypterous, sometimes strongly so, in one or both sexes.

150 AUCHENORRHYNCHA

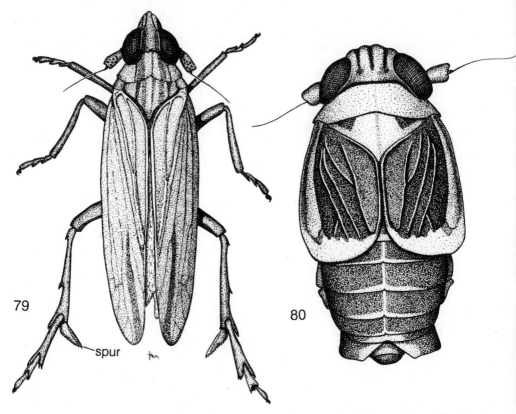

Figs 79–80. *Delphacidae. 79,* Stenocranus minutus; *80,* Criomorphus albomarginatus *brachypter.*

Usually at least a few macropterous females occur in normally brachypterous species. Rothschild (1964) found that brachypters of *Conomelus anceps* outnumbered macropters by about ten to one in both sexes but Yin Yin May (1975) found that macropters and submacropters of *Stenocranus minutus* were produced in about equal numbers, the proportion of macropters declining through emigration. Both of these authors found that flightless females were more fecund than macropters but Mochida (1973) could find little difference in total egg-production in the two forms of *Javesella pellucida*. Lifetime egg-production varies from fewer than 50 in *C. anceps* to several hundred in some larger-bodied species.

All species live on or close to the ground at all stages of the life cycle. Most feed on grasses or sedges and none of them lives on trees or shrubs. Host preferences are often marked. *Stenocranus minutus,* for example, seems only to feed on cocksfoot (*Dactylis glomerata*) and all five *Chloriona* species are found only on common reed (*Phragmites australis*). *Kelisia, Anakelisia, Megamelus* and some others feed exclusively on sedges, *K. vittipennis* apparently only on cottongrass (*Eriophorum* species). Rothschild (1964) made a detailed study of *Conomelus anceps,* which feeds only on rushes (*Juncus* species) under natural conditions.

Only three British Delphacidae are known to feed on plants other than grasses, sedges and rushes. *Ditropis pteridis* has long been known to feed on bracken. Drosopoulos (1982) found *Megamelodes quadrimaculatus* in Greece on stems and even roots of dicotyledonous

Plate 5

Heteroptera. **1**, *Heterogaster urticae* (Lygaeidae) mating on nettle. **2**, *Anthocoris nemorum* (Anthocoridae) preying on a hoverfly. 3–6, Miridae. **3**, Ant-mimicking nymph of *Pilophorus* sp. **4**, *Deraeocoris ruber* preying on a ladybird. The first two segments of the bug's labium (sheath of rostrum) are elbowed back to enable the stylets (guided across the resulting gap by the labrum) to penetrate deep into the prey. **5**, Brachypterous female of *Leptopterna dolabrata* feeding on seedheads of a grass, whose coloration it resembles. **6**, *Megacoelum infusum* nymph grooming its long antennae.

Plate 6

Auchenorrhyncha. **1**, Pair of *Centrotus cornutus* (Membracidae). **2**, Pair of *Cercopis vulnerata* (Cercopidae). **3**, Nymph of *Philaenus spumarius* (Cercopidae) in its protective froth ('cuckoo spit'). **4**, *Cicadella viridis* (Cicadellidae).

plants growing on and near water and reported rearing it through several generations on water mint. Asche & Remane (1982) confirmed that it fed on dicotyledonous plants but found it feeding mainly on monocotyledonous ones in Germany. They also reared *Asiraca clavicornis* for more than one generation on daisy. Probably all Delphacidae suck phloem sap. *Dicranotropis hamata* and four *Javesella* species are known to transmit virus diseases of cereals.

There are one or two generations a year, depending on the species and the local climate. In the majority, including all five species of *Javesella*, it is the half-grown nymph that overwinters. Adults of *Asiraca*, at least two of the four *Stenocranus* species (including *S. minutus*), *Megamelus* and *Delphacodes* overwinter, and eggs overwinter in *Conomelus*, *Muellerianella* and some, if not all, of the species of *Kelisia* and *Anakelisia*. Some females of *Anakelisia fasciata* may sometimes overwinter, as well as the eggs.

Booij (1982) discovered a peculiar triploid form in the grass-feeding genus *Muellerianella*. It occurred naturally in Britain and continental Europe and could be synthesized in the laboratory by hybridization of *M. fairmairei* and *M. brevipennis*. All triploids were female and reproduced asexually but their eggs matured only in response to otherwise superfluous matings with males of *M. fairmairei*. Booij also established that *M. extrusa* is a valid species and not a synonym of *fairmairei* as it appears in LeQuesne & Payne's (1981) checklist.

Adult Delphacidae can be identified with LeQuesne's (1960) handbook, though the nomenclature is now very much out of date. Some nymphs may be identified by using Vilbaste's (1968) work. Host plant and other biological data were summarized by Ossiannilsson (1978). More information on these topics was given by Waloff & Solomon (1973), Drosopoulos (1982) and Asche & Remane (1982).

Cicadidae

Cicadetta montana is the only British cicada and is quite unmistakeable. Its transparent wings are 19–25 mm long and its black body is 15–22 mm long. The stridulation of the male is too high-pitched for most people to hear, unlike the penetrating calls of many species that live in warmer climates. The insect is known only from Hampshire and Surrey and is apparently on the verge of extinction even there, although it is still widespread in Europe. Little is known about its biology. Adults are about in May, June and July, usually on trees. The eggs are laid in large numbers in slits in plant stems, often in the fronds of bracken. They hatch in a few weeks and the young nymphs immediately burrow into the soil, where they suck xylem sap from roots. There is some doubt as to the number of nymphal instars (5 or 6) and the duration of the nymphal stage (a few months or several years). The emergence of the nymphs from the soil is synchronized, with the majority of them in a particular area appearing on the same day. The empty moult-skins, still attached to the stems of low plants, are conspicuous and may alert the fortunate observer to the possibility of finding adults in the trees above. Adults live for about a month, feeding on the xylem sap of young twigs.

The life-history of *Cicadetta* is typical of the family as a whole. In some species the number of years spent underground is precisely fixed. The best-known example of this phenomenon is the North American species known as the seventeen-year 'locust'. As its name implies, the entire cycle of development occupies 17 years. The most striking consequence is that, in each locality where the species occurs, there may be some years when no cicadas emerge from the ground at all and others when they are very abundant. There are 17 year-classes, which do not interbreed because adult life occupies only a few weeks. Peaks and troughs of abundance therefore recur at 17 year intervals. A great

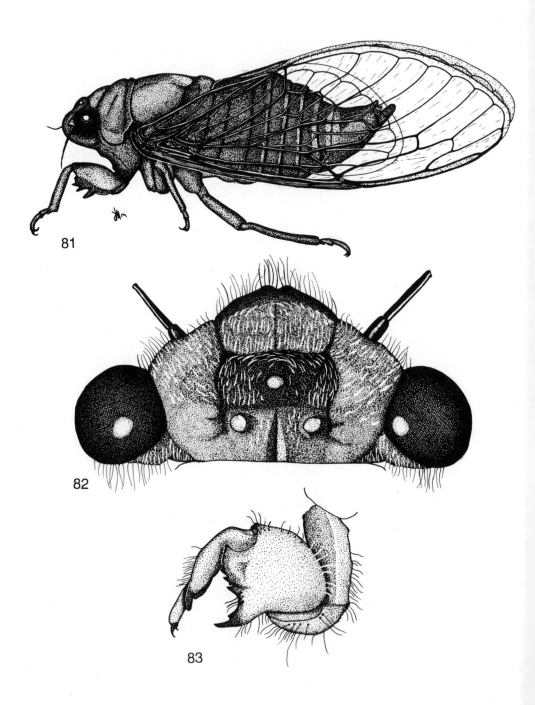

Figs 81–83. *Cicadidae:* Cicadetta montana. *81, lateral view; 82, head, showing the three ocelli; 83, nymph, fore leg.*

variety of vertebrate predators (including birds and cats) attack the cicadas when they emerge but, because of the close synchronization of emergence, the predators can destroy only a small proportion of them before they have mated and the eggs of the next generation have been laid. If natural fluctuations in the fortunes of the underground nymphs have reduced a year-class to very low numbers, the predators may wipe it out completely at emergence. That year-class then becomes locally extinct. On the other hand, an abundant year-class will have a high rate of breeding success and will appear again in abundance 17 years later. Investigation of the biology of the seventeen-year 'Locust' has shown that the picture is more complicated than this simple outline suggests. There are three quite distinct species and all of them have a tendency to precocious emergence after 13 years (but never fewer and very rarely 14, 15 or 16 years).

Cicada nymphs moisten the walls of their burrows with copious, watery excreta, cementing the soil particles into a smooth, durable lining. When it is ready to metamorphose into the adult form, the mature nymph constructs a special chamber, near the surface of the soil, where the final moult occurs. Mature nymphs of some species, including those of the seventeen year 'locust' in certain circumstances, may actually push mud up above the ground surface to form structures variously called chimneys, cones, huts and turrets. These are typically 10–15 cm in height and about half that in width, with a chamber in the middle. An exit-hole is made at the base of the turret before the moult, which takes place in the protection of the central chamber.

Membracidae

The treehoppers are mainly tropical insects and only two species of the family occur in Britain. Neither of them lives on trees. The posterior prolongation of the pronotum of the adults and the long anal tube of the nymphs distinguish them from all other Auchenorrhyncha. Both species are dark brown with short, golden pubescence and transparent wings.

Like most of the closely related leafhoppers, treehoppers feed on phloem sap. *Gargara genistae*, about 5 mm long when adult, lives on broom, sainfoin and species of *Medicago* and *Genista*. It is found only in southern England. Its life-cycle occupies a single year. Eggs overwinter and hatch in June. Nymphal development takes about a month. Both nymphs and adults are sluggish in their movements and are attended by ants. *Centrotus cornutus*, about 9 mm long, is widespread in Britain and is polyphagous on dicotyledonous herbaceous plants, brambles and saplings of woody plants. The eggs are laid in rows in plant stems or petioles. Nymphs are found much less often than adults because they live at the very base of plant stems, often half buried in the soil or covered with dead leaves and detritus. Müller (1984a,b) investigated its life-cycle in Germany and illustrated all five nymphal instars. He found that most individuals lived for two years, overwintering in the first year in the third instar and in the second year in the final instar. A few individuals completed their development in one year and yet others took three years to become adult, so the 'odd-year' and 'even-year' broods were not completely isolated genetically.

Cercopidae

Adult froghoppers could be mistaken for leafhoppers (Cicadellidae) but their tibiae are cylindrical, with numerous, fine, evenly distributed setae and a few stout spurs, not longitudinally keeled, with rows of robust setae restricted to the keels as in the great

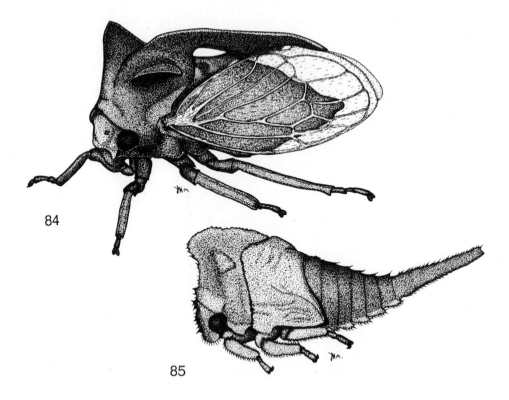

Figs 84–85. *Membracidae:* Centrotus cornutus. *84, adult; 85, nymph.*

majority of cicadellids. The lack of tegulae separates them from the fulgoromorphan families. Most are coloured in various shades of brown and grey but one is black and red.

The familiar, frothy masses of 'cuckoo-spit' are produced by nymphs of all ten British species of froghoppers. Being xylem feeders, they produce copious amounts of watery excreta, which are mixed with air to make a foam stabilized with albumen-like proteins. The fully-fed fifth-instar nymph blows an extra large bubble which bursts, allowing the final moult to take place exposed to the air. The continuous shower of surplus fluid falling from the branches of trees heavily infested with cercopid nymphs has earned trees so afflicted the name of rain trees.

The most common species, found on all manner of shrubs, herbs and grasses in gardens and elsewhere, is *Philaenus spumarius*. Its eggs are laid in the autumn in small groups at ground level or a few centimetres above in various natural chinks and fissures such as the spaces between the culms and leaf-sheaths of grasses. Each group of eggs is surrounded by a small quantity of a whitish secretion that hardens off and affords them protection through the winter. In the spring the small, dark-coloured nymphs seek suitable places at which to start feeding, often ascending shrubs and perennial herbs to a height of a metre or more. Each nymph secretes its own blob of froth, in which it grows rapidly, turning pale yellow or greenish in later instars and becoming adult in early summer. The adults, which are about 6 mm long, do not become sexually mature for several months. They exhibit 11 distinct colour-patterns which result from the 28 possible pairings of 7 alleles, whose expression is affected by some non-allelic modifier genes and whose dominance hierarchy depends on the sex of the individual. Details were given by Halkka *et al.* (1973,

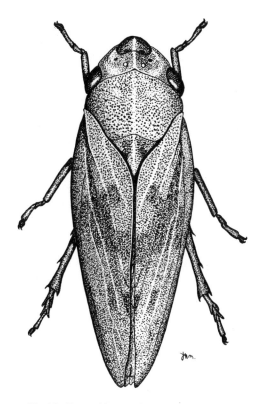

Fig 86. *Cercopidae:* Aphrophora maculata.

1975). The patterns vary from uniformly straw-coloured to uniform pitch-brown via several bold patterns, made up of these two colours, and also a speckled form with two pale patches on each fore wing. The polymorphism is probably maintained by predation during the long adult life, with predators concentrating on hunting the commonest colour-variety until it has become difficult to find and then switching their attentions to the next commonest variety and so on. Lees & Dent (1983) studied populations of *P. spumarius* around a source of intense industrial pollution and found that, in the worst-polluted areas, the darker forms were much more common than the paler ones, whose frequency in the population increased with increasing distance from the source of pollution. This froghopper may cause damage to crops, pasturelands and garden plants both by the direct effect of feeding and by carrying virus diseases.

All the froghoppers found in Britain have only a single generation each year, and in all but one of them the timing of the life-history is broadly similar to that outlined above. Colour polymorphism is usually lacking and is never as elaborate in the other species as it is in *P. spumarius*, probably because the adults do not live as long.

Aphrophora species are the largest of the British cercopids, up to 12 mm long. Nymphs of *Aphrophora salicina* and *A. alpina* live gregariously in big, dripping masses of froth on the young twigs of sallow and willow. The eggs are laid in the wood of dead twigs. *Aphrophora major* lives only on bog myrtle. The solitary, red or reddish grey nymphs of the common and widespread *Aphrophora alni* live very close to ground-level on wild strawberry, hogweed and many other dicotyledonous plants but the adults can often be beaten from trees and shrubs. Four species of *Neophilaenus*, which are about the same size as *P. spumarius* or smaller (*N. exclamationis* is only 3.7–4.6 mm long), live on grasses in meadows and moorlands. Their nymphs, like those of *A. alni*, are solitary. Whittaker (1965) studied the biology of two of them.

The adult of *Cercopis vulnerata* is a striking insect, about 10 mm long and black with six bright red spots. It resembles the common burnet moths in size and appearance. China's (1925) notes suggest that the life-cycle of this froghopper in Britain is similar to that detailed for it in the southern Alps by Mauri (1982). Adults are found, mainly in May and June, in clearings in woods, roadside verges and other grassy places, usually near trees. Females enter cracks in the soil to lay their eggs. These hatch in the late summer of the same year and the pallid nymphs live gregariously in balls of sticky froth formed on the roots of grasses at a depth of 15–20 cm. The moult from fourth to fifth instar occurs in January and the final moult in late April or early May. Adults emerge from the soil as soon as they have hardened off. When they are very numerous they may cause noticeable damage to fruit trees, hops and various weeds such as Dock. In some populations there occur a few individuals in which the red coloration is replaced by grey-brown, pink or fawn (Gibson, 1976, Measday, 1979).

Adult froghoppers can be identified from LeQuesne's (1965*b*) key. As regards nymphs, the genera *Aphrophora*, *Neophilaenus* and *Philaenus* and some of their species are separable with the works of Vilbaste (1982) and Adenuga (1971).

Cicadellidae

With well over 200 species, distributed among 13 subfamilies, the leafhoppers constitute one of the largest hemipteran families in Britain. Cicadellid adults differ from those of the fulgoromorph families in lacking tegulae and from adult Cicadidae, Cercopidae and Psyllidae in having longitudinally keeled tibiae; the tibiae are cylindrical in these other groups. The tibiae of Membracidae are keeled like those of Cicadellidae but the posterior prolongation of the pronotum of membracids never occurs in cicadellids. The nymphs of ground-dwelling cicadellids are most likely to be confused with those of delphacids but they never have either circular sensory pits on the body or a movable spur at the apex of the hind tibia. The absence of a suture between the postclypeus and frons sets them apart not only from delphacids but also from nymphs of the other three fulgoromorph families (Tettigometridae, Issidae and Cixiidae). They lack the digging adaptations of the fore legs of cicadid nymphs and the long anal tube of membracids and are never surrounded by froth like immature cercopids.

Cicadellidae occur in almost every kind of vegetation. Most leafhoppers of the subfamilies Typhlocybinae and Macropsinae and all British Idiocerinae, Ledrinae and Jassinae live on trees and shrubs (rockrose, host of one British jassine, is technically a shrub), as do a very few Deltocephalinae. The others live on and among low plants, where the deltocephalines are the dominant group. Arboreal forms are always macropterous but brachyptery is frequent among those groups that live on or very close to the ground surface.

Cicadellids feed on phloem or xylem sap or cell contents and may damage their host plants either directly, through the debilitating effects of the removal of sap and protoplasm, the toxicity of their saliva and the obstruction of the vascular system, or indirectly, by introducing viruses or mycoplasmas that cause diseases. The economic impact of cicadellids on crops has stimulated a great deal of research in recent years. The volume edited by Nault & Rodriguez (1985) provides a useful outline of the state of knowledge about their systematics, morphology and biology, including their interactions with plants and disease organisms. Knight (1966) listed the species occurring on

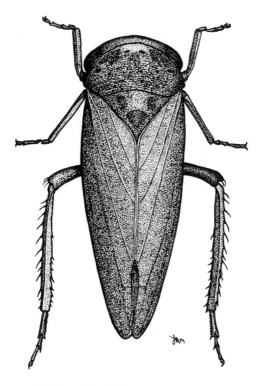

Fig 87. *Cicadellidae:* Jassus lanio.

economically important plants in Britain and Delong (1971) published a valuable introduction to the biological literature on the family.

In addition to the identification guides by Ossiannilsson, Edwards, LeQuesne and LeQuesne & Payne, cited among the general works on Auchenorrhyncha, Ribaut's (1936, 1952) volumes on French Cicadellidae are very useful, particularly for their excellent illustrations of genitalia. Giustina's (1989) supplement brought Ribaut's work up to date. Vilbaste (1982) dealt with the identification of many genera and species of North European cicadellid nymphs. Knight (1965) and Prior (1965) described techniques for preparing material for identification. It should be borne in mind that the genitalia, like other characters, are apt to vary geographically (LeQuesne & Woodroffe, 1976) and, in one instance, seasonal variation in the genitalia has been demonstrated (Müller, 1954).

Just over 100 species of Deltocephalinae (often called Euscelinae in the literature) have been found in Britain, making this the largest subfamily of leafhoppers in the country. The ocelli are situated at the junction of the face and the vertex, a position unique to the group. Deltocephalinae are the dominant Hemiptera of grasslands and there can be few grassy places that lack them altogether. Because of the mixture of plant species growing together in such habitats it is difficult to establish their host plant preferences merely by collecting in the field. Many of them are known to feed on grasses or sedges and some have been definitely shown to develop on herbaceous Dicotyledones. A few will accept both kinds of plant. A rather small number of species are dependent on trees and shrubs. Among these are *Grypotes puncticollis* on pine and two introduced species, *Opsius stactogalus* and *Placotettix taeniatifrons*, living on tamarisk and rhododendron respectively. *Thamnotettix* and *Speudotettix* species are found on trees as well as on low plants and the species of *Allygus*, whose nymphs feed on grasses, occur regularly on deciduous and coniferous trees as adults. Most deltocephalines are believed to feed only on the phloem sap of their hosts but Port (quoted by Prestidge, 1982) found that the polyphagous *Euscelis incisus* is mainly a xylem feeder, taking about 70 per cent of xylem sap and 30 per cent of phloem sap. Waloff & Solomon (1973) established the outlines of the life histories of 21 species in southern England. Of these, only their three *Euscelis* species overwintered as nymphs; they were all bivoltine. *Mocydiopsis parvicauda* and *Balclutha punctata* were univoltine and hibernated as adults. The remaining 16 species overwintered as eggs; 5 were univoltine and 11 bivoltine. The number of generations per year is known to vary with the geographical latitude of the populations, with more generations in the south. Some species, *Doratura stylata* for example, are habitually brachypterous, with macropters occurring only rarely. There is pronounced variation in colour and pattern in some species. In *Mocydia crocea* nymphs there are six quite distinct colour morphs which have been studied extensively by Müller (1974, 1979, 1982, 1987). The same author (Müller, 1954) demonstrated seasonal polymorphism in size, coloration and the form of the male genitalia in *Euscelis incisus*. Since LeQuesne's (1969) handbook was published, an additional species of *Cicadula* has been found in Britain (LeQuesne, 1983). Walter (1975, 1978) gave detailed descriptions and a key to the nymphs of numerous European Deltocephalinae, supplementing the information given by Vilbaste (1982).

About a dozen species of Aphrodinae occur in Britain. Opinions differ as to the status of some of the named forms and comparison of the two recent treatments of the group is complicated by the fact that Ossiannilsson (1981) recognized four genera where LeQuesne (1965b) recognized only two. The ocelli are situated on the vertex, just behind its anterior margin, as in no other British Cicadellidae. Adults appear in June or July and survive until the autumn. Both adults and nymphs of all species are found at or near the ground surface. Some species favour damp habitats while others favour dry ones. There is also a difference in the degree of shade or exposure to sunshine in the places where they are usually found. Such habitat preferences suggest at least a measure of host-plant specifity. *Aphrodes bicinctus* is polyphagous on dicotlyedonous plants, including clover and plantain, while *A. bifasciatus*, *A. albifrons* and *A. albostriatus* feed on various grasses. Probably all species have one generation a year; this is certainly the case with at least one species of each of Ossiannilsson's genera. *Stroggylocephalus agrestis* has been seen to oviposit in the sedge *Carex riparia*. Different authors have claimed that it overwinters either in the egg or in the adult state. Adults of *S. livens* are also reported to hibernate. The other species probably overwinter as eggs.

The subfamily Cicadellinae, whose members, unlike most cicadellids, are believed to be xylem-feeders, is represented by only three species. They are unique among British Cicadellidae in having the ocelli placed so far back on the vertex of the head that they are nearer to its posterior margin than to its anterior one. *Cicadella viridis* is a narrow-bodied species coloured in various shades of green, brown and yellow. The fore wings are green in most females and purplish-black in most males but intermediately coloured specimens occur in both sexes. Males are distinctly smaller than females (5.7–7.0 mm and 7.5–9.0 mm respectively). The eggs of this widely distributed species are laid in the stems and leaves of rushes but the usual food-plants of the nymphs are grasses, especially species of *Holcus*. Adults feed on a wide variety of herbs and even shrubs and trees. In Britain there is a single generation per year but in the warmer climate of southern Europe there may be up to four. A second species of this genus, *C. lasiocarpae*, has recently been reported from Anglesey and several Irish localities (Le Quesne, 1987). Both males and females may be up to a millimetre longer than *viridis*. It appears to be capable of breeding only on the sedge *Carex lasiocarpa*. *Graphocephala fennahi* is widespread on rhododendrons. This striking insect, 8.4–9.4 mm long, is a twentieth century introduction from North America. It was first noticed in 1935, in Surrey. Adults are yellow and green with black markings on the head; the fore wings are green with two longitudinal red bands; the abdominal dorsum and scutellum are also orange-red; the hind wings are dark grey. It has a single generation annually, with adults present from June to October. The eggs, which overwinter, are laid into the flower buds, which may be made more susceptible to 'bud-blast' disease by the damage caused during oviposition.

Both British species of the subfamily Evacanthinae belong to the genus *Evacanthus*. They are easily recognized by the presence on the dorsum of the head of a raised keel running just behind the front margin from eye to eye. About half way between each eye and the apex of the head this keel is kinked sharply inwards; the ocelli lie close behind these kinks. These are rather large (5–7 mm long) leafhoppers with long legs, living in grassy places. They may be xylem-feeders, like the Cicadellinae. Adults of both species are found from June to October. Eggs of *E. interruptus* are known to overwinter. This species is polyphagous on dicotyledonous herbs and shrubs.

Six species of Agalliinae are reported from the British Isles. All are small (2.3–4.0 mm long), stoutly built, straw-coloured insects with bold, black spots on the head and pronotum. The ocelli are situated on the face, as in Typhlocybinae and the mainly tree-dwelling Idiocerinae, Jassinae and Macropsinae. The ridges above the antennae are weakly defined, not reaching dorsally as far as the eye. All species live on or close to the ground where they feed on dicotyledonous herbs. Collecting dates indicate that in most species adults overwinter and that there is probably only a single generation each year. *Agallia ribauti*, according to Günthart (1987), breeds on a wide range of low plants whereas *A. venosa* is restricted to horseshoe vetch. *Agallia brachyptera* is the only species that frequents marshy places and also the only one in which brachyptery is known. Its fore wings are usually truncate, exposing more than half the abdomen.

Eupelix cuspidata has a broad, triangular or rounded head whose lamellate lateral margins embrace the eyes anteriorly. It is the only British representative of the subfamily Dorycephalinae (or Eupelicinae), found throughout the British Isles on the ground in dry places among the various grasses on which it feeds. The usual overwintering stage is the nymph, though adults have been found in almost every month of the year. Males are 5.3 to 5.8 mm long and greyish-brown or greyish-yellow; females, 5.6 to 7.2 mm long, are predominantly straw-coloured. This colour difference between the sexes is already apparent in the nymphs. In males of the typical form the outline of the head is rounded, almost semicircular, but in males of the variety *depressa* and in females its sides are straighter, giving it a generally triangular shape.

The subfamilies Ulopinae and Megophthalminae are represented in Britain by two species each of *Ulopa* and *Megophthalmus*. These are small (2.5–4.1 mm long), chunky insects living at or close to the surface of the ground. *Ulopa* differs from all other British cicadellids in having no spines at the tip of the hind femur and *Megophthalmus* is instantly recognizable by the x-shaped arrangement of prominent keels on the upper part of the face, with the two, large ocelli in the angles between the keels. *Ulopa reticulata* is found under heather throughout the British Isles. It is almost invariably brachypterous but the thick, convex fore wings completely cover the body. The predominant colour of the insect is a slightly pinkish brown with two incomplete, whitish bands across the fore wings and some darker markings on the head. There is a general resemblance to the litter that accumulates beneath the hostplant. Waloff (1981) found that in southern England there were two or three generations a year and that the species overwintered mainly as immature adults with large fat reserves, though a small part of the overwintering population consisted of nymphs, mainly in the fourth instar. *Ulopa trivia* is confined to the southern counties of England, where it occurs in sandy, maritime habitats as well as on chalk soils. Le Quesne (1965b) suggested an association with viper's bugloss but Morris (1972) denied this and suggested instead ribwort plantain and lady's bedstraw. He found that only females overwintered. Both sexes are straw-coloured and males are strongly marked with black on the head and pronotum and along the veins of the fore wings. Like *U. reticulata*, this species is habitually brachypterous. Macropters are extremely rare. The two *Megophthalmus* species are straw-coloured, with extensive dark brown or black markings on the head, pronotum and scutellum of the male. They are found in grassy places. In Europe, *M. scanicus* is univoltine, hibernating in the egg, but adults have been found in January in Britain. Both species have the hind wings somewhat reduced in length. Full macropters, if they occur at all, are rare.

More than 90 species of Typhlocybinae have been found in Britain. These delicate little hoppers, none of which is more than 5.5 mm long, belong to the group of subfamilies with the ocelli situated clearly on the face. They are always macropterous and the veins of the fore wing run almost straight, without forks or cross-veins, through at least its basal two-thirds. Almost all typhlocybines feed not by sucking sap, like most other cicadellids, but by emptying leaf mesophyll cells of their contents. This activity results in the white stippling of air-filled cells often seen on the leaves of rose, plum, mint and many other plants. Although the hoppers live on the lower surfaces of the leaves they feed mainly on the upper (palisade) layer of green mesophyll cells, so that the white stippling is visible only from above. Similar stippling is caused by some thrips (Thysanoptera), spidermites, lacebugs and Malcid bugs, which feed in the same way. Günthardt & Wanner (1981) showed that *Empoasca decipiens* is unusual in being able to live on the mesophyll cells of the stems as well as of the leaves of broad bean. On the leaves, it feeds on both the upper and the lower layers of mesophyll. The related *E. vitis* is very unusual in being a phloem-feeder (Claridge & Wilson, 1981). The majority of British Typhlocybinae live on broadleafed trees; Claridge & Wilson (1976, 1981) give the results of detailed studies of their hostplant preferences. Exceptions to the tree-dwelling habit are the large genus *Eupteryx* on dicotyledonous herbs, especially Labiatae, and (*E. filicum*) ferns (Stewart, 1986c, 1988), the seven species of the tribe Dikraneurini on herbs and grasses (Vidano, 1965), *Hauptidia maroccana*, *Arboridia parvula* and some *Empoasca* and *Zygina* (or *Flammigeroidea*) species on various low plants, *Zyginidia scutellaris* on grasses and *Aguriahana* (or *Wagneripteryx*) *germari* on pine. The species living on herbaceous plants lay their eggs mainly into leaf tissues while those living on trees insert them just beneath the bark of young shoots. In nearly all species it is the egg that overwinters. A few hibernate as adults: *Zyginidia scutellaris*, *Dikraneura variata*, *Arboridia parvula* and the species of *Zygina*, *Linnavuoriana* and *Empoasca*. Overwintering adults of the species that breed on trees shelter either on evergreen trees and shrubs or among grasses. Some species of Typhlocybinae typically have only one generation a year; others have two. The different life histories were summarized by Jervis (1980c). According to Waloff & Solomon (1973), *Z. scutellaris* has many generations in the course of the year. Some of the bivoltine species show a rudimentary form of host plant alternation. Claridge & Wilson (1978b) reported that *Lindbergina aurovittata* lays its overwintering eggs in the evergreen leaves of bramble and holm oak. The first generation must necessarily feed on these plants as nymphs but they migrate when adult to a variety of deciduous trees and shrubs where their progeny develop, returning to the evergreens when they, in turn, become adult. The same authors reported that the second generation of *Edwardsiana rosae* has a wide range of hosts in addition to rose, the only host plant of the first generation. Stiling (1980) found that *Eupteryx aurata* has its first generation on nettle and its second on various low plants; two other species of the same genus, although also bivoltine, were restricted to nettle throughout the year. Seasonal dimorphism is almost unknown in this subfamily, although in some bivoltine species of *Zygina* (subgenus *Flammigeroidea*) the adults of the two generations are differently coloured. Distinct colour forms occur contemporaneously in adults of *Fagocyba cruenta*, *Typhlocyba bifasciata* and *Alebra albostriella*. Nymphal colour-polymorphism occurs in two species of *Eupteryx* (Stewart, 1986a,b). The recent handbook by LeQuesne & Payne (1981) provides illustrated keys to adults of this subfamily. The only reliable diagnostic characters in many instances, especially in the large genus *Edwardsiana*, are furnished by the male genitalia. Wilson (1978) published a key to the British genera of tree-dwelling typhlocybine nymphs and Stewart (1986c) described and keyed the nymphs of *Eupteryx*.

The 18 British Idiocerinae range in length from 4.4 to 7.0 mm. They belong to a group of subfamilies in which the ocelli are situated on the face. Within this group they are unique in having the antenna and the ocellus on each side almost joined by a suture. The head is wider than the pronotum and is broadly convex anteriorly and broadly concave posteriorly, with the two curvatures approximately parallel. Adults of *Idiocerus vittifrons* have been found on their host tree, field maple, from August to December. All the other species feed on the twigs of willows, sallows, poplars and aspen with varying degrees of specificity. One species, *Idiocerus vitreus*, occurs on both black poplar and sallow according to Le Quesne (1965b) but the others are restricted to a single host species or a group of related species within a genus. Most of them are univoltine and overwinter as eggs laid in the twigs or buds but in *Idiocerus poecilus* adults are known to hibernate. *Rhytidodus decimusquartus* is bivoltine on black poplar (including its Lombardy variety). Eggs and some mated females of this species overwinter. Its young nymphs feed on the leaves but later stages move to the twigs. The young nymphs of *I. distinguendus* feed on the upper surfaces of the unfolding leaves of white poplar, moving later to the undersides.

Three species of robust leafhoppers belonging to the subfamily Jassinae occur in Britain. The ocelli are situated on the face and there is a well developed, almost horizontal ridge above each antenna extending from the frons to the inner margin of the eye. *Batracomorphus irroratus* (4–5 mm long) is wholly green with the fore wings and sometimes the whole visible dorsal surface peppered with minute black dots. Adults and the densely bristly nymphs are found in summer on rockrose over much of England. *Jassus lanio* is univoltine on oak, overwintering in the egg. It is considerably larger than *irroratus* (7.0–8.5 mm long). Newly moulted adults of the typical form are green but the head, pronotum and scutellum eventually turn brown, sometimes with a more or less pronounced reddish tinge. There is also a distinct form in which the fore wings are brown. The nymphs also occur in green and brown forms. The species is found in oakwoods throughout the British Isles. *Jassus scutellaris* is similar in size to *J. lanio* and has the same colour pattern as its green morph. It is sometimes abundant on elms in southern England but was only recognized as a British species by Wilson in 1981.

Another subfamily in which the ocelli are situated on the face is Macropsinae. Like Jassinae, these have a well developed ridge above each of the antennae but here it is strongly oblique, due to the greater length of the face. The head is not or only slightly broader than the pronotum and is often sharply narrowed in the midline to accommodate the angulate forward thrust of the pronotum. Adults of all 21 British species occur in the field from June or July until September or October. All species whose biology has been investigated have been found to be univoltine and to overwinter as eggs. *Hephathus nanus*, which is known to have this type of life cycle, is the smallest of the British Macropsinae, being only 2.8–3.5 mm long; the other species are between 3.6 and 5.9 mm in length. It inhabits the short grasslands of southern England; its host plants are unknown. *Macropsis scutellata* feeds on nettle and two other species of the same genus feed on bramble. With these four exceptions, all native macropsines live on the twigs of trees and shrubs, laying their eggs into the dormant buds or into the young twigs themselves (Claridge & Reynolds, 1972). Each species is confined to a single host genus and most show specificity within the genus, or at least a preference for laying eggs on one species rather than another. Two of the tree-inhabiting *Macropsis* species feed on elm and the other eight on various willows and poplars, including sallow, white and dwarf willows and aspen. *Pediopsis tiliae* feeds on lime. *Oncopsis* has one species each on alder, hazel and hornbeam and a complex of about four species or races on birch, differing in their courtship songs (Claridge & Reynolds, 1973; Claridge & Nixon, 1986) and chromosomes (John & Claridge, 1974) as well as in details of the male genitalia and their relative preferences for silver birch and downy birch (Claridge, Reynolds & Wilson, 1977). *Oncopsis flavicollis* displays a pronounced colour polymorphism in both nymphs and adults (Claridge & Nixon, 1981).

The largest cicadellid found in the British Isles, *Ledra aurita* (13 to 18 mm long), is the sole representative here of the subfamily Ledrinae. Its broad, shovel-shaped head and the pair of erect, semicircular lobes on the posterior angles of the pronotum are quite unlike anything encountered in the other British Cicadellidae. The pronotum, in its bizarre shape, recalls that of Membracidae but it is not extended backwards over the scutellum. Despite its bulk, *Ledra* is an inconspicuous insect when resting with its body and flat head pressed against the branches of the deciduous trees on which it lives. Its greyish, mottled coloration and irregular profile provide an excellent camouflage. Hazel, alder and especially oak are known hostplants. The insect is confined to the deciduous woods of southern England. Adults can be found throughout the summer months. Both small and large nymphs have been taken in midsummer and it is thought likely that the life-cycle occupies two years, the first winter being passed in the egg and the second as a half-grown nymph. In Italy it damages the twigs of hazel by its oviposition. The eggs are laid, 4–5 at a time, in each of two parallel slits under the bark (Viggiani, 1971).

16

STERNORRHYNCHA

The insects grouped in this suborder are characterized by the position of the rostrum, which appears to arise from the prosternum, between the fore coxae. Sternorrhyncha never have more than two tarsal segments, unlike adult Auchenorrhyncha and most adult Heteroptera, which have three. They are small insects and almost all of them feed on the phloem sap of vascular plants.

The most highly modified Hemiptera belong to this suborder. Aleyrodidae and Coccoidea approach the kind of complete metamorphosis, with a pupal stage, that is characteristic of the holometabolous insects, rather than the gradual kind of development shown by other bugs.

Most adult Psylloidea can be identified with the aid of a hand lens or a low-powered dissecting microscope and are traditionally preserved dry on micropins, card-pointed or even gummed down flat on cards. It is often impossible to identify other Sternorrhyncha to species without mounting them on microscope slides for examination under high magnifications with transmitted light. Nevertheless, the general appearance and host-plant associations are often sufficient to allow the experienced entomologist to make reasonably reliable identifications of many of them without resource to microscopy at all.

General entomologists have sometimes regarded the Psylloidea as a sort of honorary Auchenorrhyncha, because of the similar methods of study required, but have ignored the other superfamilies, while specialists have often elected to work only on Coccoidea (sometimes with the addition of Aleyrodoidea, though both groups have their exclusive devotees) or Aphidoidea and Adelgoidea. In temperate regions, the majority of economically damaging Hemiptera are Aphidoidea, but there are many serious tropical pests among the other Sternorrhyncha.

Key to superfamilies of adult Sternorrhyncha

1. Wings of equal size, white and opaque (Fig. 97) .... ALEYRODOIDEA
— Hind wings smaller than fore wings, usually transparent (Figs 134, 140), or absent 2

2. Tarsi each with a single claw (Figs 152–155, 158) or legs absent .... COCCOIDEA
— Tarsi each with a pair of claws (Figs 90, 106) (tarsi very rarely absent in Aphidoidea) 3

3. Legs robust (Fig. 90); antennae with ten segments, last segment without a processus terminalis (Figs 88–91) .... PSYLLOIDEA
— Legs slender (Fig. 4); antennae with three to six segments, last segment bearing a processus terminalis (Figs 103, 107, 115, 129) 4

4. Fore wing with three oblique veins (Figs 104, 108); antennae of apterae with three segments; cauda broadly rounded; cornicles absent; head plus thorax of apterae greater in volume than abdomen (Figs 101, 102, 105, 106); eyes of apterae with only three ommatidia; abdominal spiracles present on segment I only or I–IV or I–V. All females oviparous .... ADELGOIDEA
— Fore wing with four oblique veins (Fig. 134); antennae of apterae with four to six segments; cornicles present, prominent (Figs 123, 126, 127), represented by pores (Fig. 136) or (rarely) absent; head plus thorax of apterae not greater in volume than abdomen; eyes of apterae often with more than three ommatidia; abdominal spiracles present on segments I–VII or II–V; parthenogenetic females viviparous .... APHIDOIDEA

Key to superfamilies of immature Sternorrhyncha

1. Last segment of antenna with a processus terminalis (as in Figs 103, 107, 115, 129) 2
— Last segment of antenna without processus terminalis 3
2. Cornicles absent; eyes with three ommatidia; abdominal spiracles present on segment I only, or I-IV or I-V; head plus thorax generally greater in volume than abdomen .... ADELGOIDEA
— Cornicles usually present; eyes often with more than three ommatidia; abdominal spiracles present on segments I-VIII or II-V; head plus thorax generally smaller in volume than abdomen .... APHIDOIDEA
3. Anus situated dorsally, on apex of a two-segmented tube recessed into the vasiform orifice (Fig. 98) .... ALEYRODOIDEA
— Anus usually ventral or terminal, if visible dorsally then surrounded by a ring (Fig. 154) or at the anterior end of a cleft extending forwards from posterior margin (Fig. 157) or, if at the end of a tube, the tube of only one segment 4
4. All legs terminating in paired claws .... PSYLLOIDEA
— Legs atrophied or each terminating in a single claw .... COCCOIDEA

Psylloidea

Adult psylloids are always winged; both their richer venation and the presence of a clavus in the fore wing set them apart from other Sternorrhyncha, as do the robust hind legs modified for jumping. There are only two tarsal segments in contrast to the three segments present in Auchenorrhyncha. The external genitalia are conspicuous in both sexes. Some Psocoptera superficially resemble Psylloidea but they have biting mouthparts equipped with obvious palps instead of the simple rostrum of Hemiptera. Psylloid nymphs have the rostrum arising between the anterior coxae and cannot, therefore, be confused with the immature stages of Heteroptera or Auchenorrhyncha. They differ from Aphidoidea and Adelgoidea in lacking a primary rhinarium and processus terminalis on the last antennal segment, from Aleyrodoidea in lacking the characteristic vasiform orifice, and from Coccoidea in having the tarsal claws paired, not single.

All Psylloidea feed on phloem sap, so far as is known. The eggs are shallowly embedded in plant tissue and are covered with a protective wax coating. Each female typically lays 200–500 eggs, up to as many as 1000 in a few cases. There are five nymphal instars. A comprehensive review of the biology of the group worldwide was given by Hodkinson (1974). Hodkinson & White (1979) summarized the biological information available for the British species (much of it derived from continental European studies) and gave bibliographic references to published work on each species. Adults often occur away from the plants on which they breed and are often found on conifers in winter, but nymphs are very closely restricted to their food-plants which, consequently, are a useful guide to their identity. The effects of feeding activity on the host plants are usually slight but several species cause obvious distortion of the leaves and a few provoke the development of simple galls. Nymphs of all species secrete a powdery wax from a circum-anal ring of pores and many have additional glands that secrete tufts of wax filaments. The wax from the circum-anal ring coats the droplets of liquid excreta and is produced most copiously by the gall-forming and leaf-rolling species such as *Trioza alacris*, *Trichochermes walkeri*, *Psyllopsis fraxini* and *Psylla buxi*, in which the danger of contamination from the honeydew is greatest. Adults flick the drops of honeydew away with their fore wings. Ants are known to be attracted to the honeydew produced by the nymphs of some British *Psylla*. The nymphs of a few kinds of Psylloidea, especially some genera of the Australian family Spondyliaspididae, retain drops of honeydew at the abdominal apex until most of the water content has evaporated, sticking the resulting sugary pellets together into tent-like coverings called lerps. The lerp varies in shape according to the species that constructs it, but generally resembles a single cockle or mussel shell. Because of their high sugar content, lerps are important in the diet of some Australian birds. Humans, too, gather this 'leaf-manna' as food.

There is little sexual dimorphism among adult Psylloidea apart from size – males are slightly smaller than females – and the obvious differences in the genitalia. In many species of Psyllidae and Triozidae with a long adult life the adults change colour dramatically, from green to red and brown or black, as they mature. This is particularly true of species that overwinter as adults, such as *Psylla melanoneura*, but also occurs in *Psylla mali*, which has a single generation of long-lived adults in the summer and passes the winter in the egg. Seasonal dimorphism is very unusual (it is found in *Psylla pyricola* and *Trioza chenopodii*) and parthenogenesis is unknown in British species.

Adults of both sexes can be determined with the aid of the keys of Hodkinson & White (1979) supplemented by Hodkinson & Hollis (1980). Final-instar nymphs are keyed by White & Hodkinson (1982). In the account below, the classification used by Hodkinson & White is followed except for the separation of *Calophya* into a family of its own, the separation of Homotomidae from the non-British Carsidaridae and the reduction of Aphalaridae to a subfamily of Psyllidae. In Burckhardt's (1987) classification, Spondyliaspididae and Aphalaridae are treated as synonyms of Psyllidae.

Key to families of adult British Psylloidea

1. Antennae flattened, with long, dense pubescence (Fig. 95). (One species, breeding on fig trees.) .... HOMOTOMIDAE
— Antennae cylindrical, with short pubescence (Figs 88–89). (Not breeding on fig trees.) 2

2. Fore wing with Cu, R and M diverging from a single point (Fig. 93) .... TRIOZIDAE
— Fore wing with Cu and M arising together from R, separating further along wing (Figs 90, 96) 3

3. Head with genal cones (Fig. 91) 4
— Head without genal cones (Fig. 89) 6

4. Fore wing with cell cu1 more than three times as long as high (Fig. 96). (One small species, fore wing less than 1.9 mm long, confined to *Eucalyptus*.) .... SPONDYLIASPIDIDAE
— Fore wing with cell cu1 usually much less than three times as long as high (Fig. 90); never more than three times as long as high. (Breeding on various trees and shrubs but not on *Eucalyptus*) 5

5. Basal segment of hind tarsus without spines; antennae shorter than width of head. (One small species, fore wing less than 1.6 mm long, breeding on *Cotinus* and *Rhus*) .... CALOPHYIDAE
— Basal segment of hind tarsus with at least one stout, black spine; antennae longer than width of head. (Not breeding on *Cotinus* or *Rhus*) .... PSYLLIDAE (Psyllinae)

6. Antennae with second segment wider than first and longer than third (Fig. 88). (Breeding on rushes and sedges.) .... LIVIIDAE
— Antennae with second segment narrower than first and shorter than third (Fig. 89). (Breeding on dicotyledonous herbs and trees.) .... PSYLLIDAE (Aphalarinae)

Key to families of final-instar nymphal British Psylloidea

1. On fig trees .... HOMOTOMIDAE
— On *Eucalyptus* .... SPONDYLIASPIDIDAE
— On rushes or sedges .... LIVIIDAE
— On *Cotinus* or *Rhus* .... CALOPHYIDAE
— Not on Fig, *Eucalyptus*, rushes, sedges, *Cotinus* or *Rhus* 2

2. Wing-pads fused with non-fragmented terga; pads of fore wings embracing head (Fig. 94). (Eighteen species, several of them common.) .... TRIOZIDAE
— Wing-pads separate from terga, which are fragmented; pads of fore wings not embracing head (Fig. 92) 3

3. Antennae usually more than four-fifths as long as pads of fore wings. (On various herbaceous plants in families Polygonaceae and Compositae, rosebay willowherb, poplars, maples, heaths and heather.) .... PSYLLIDAE (Aphalarinae)
— Antennae usually less than four-fifths as long as pads of fore wings. (On various trees and shrubs, including dyer's greenweed (*Genista*); never on herbaceous plants, heaths, heather, maple or poplar.) .... PSYLLIDAE (Psyllinae)

Liviidae

The two British members of this small family belong to its sole genus, *Livia*. They have thick, opaque yellowish forewings and distinctively shaped heads (Fig. 88). Adults of the widespread *L. juncorum* have been found in every month except April and October; evidently this is the overwintering stage. *L. juncorum* induces the formation of 'tassel-galls' of clustered leaves on its hostplants, which include most of the commoner species of rushes (genus *Juncus*). The number of generations annually is still unknown. Adults of *L. crefeldensis* have been found in September on sedges (*Carex* species). Neither the overwintering stage nor the number of generations per year is known for certain in this species.

Psyllidae

The sixteen British species of the subfamily Aphalarinae lack genal cones and feed on dicotyledonous plants. *Strophingia ericae*, which is common and widespread on ling, may take one or two years to complete its life cycle. All the other species are believed to have one generation per year except for the rare *Aphorma lichenoides* (formerly *A. bagnalli*), whose host plants and biology are unknown.

The four *Aphalara* species feed on various species of dock, bistort and knotgrass, but adults are often encountered sheltering on evergreens from late summer through to spring. *Camarotoscena speciosa*, an extremely rare species on Poplar, also overwinters in the adult state. All other British aphalarids overwinter as nymphs on their hostplants. *Craspedolepta* is our largest genus, with five species on various Compositae (*C. nervosa* is common on yarrow, at least in the south) and two on rosebay willowherb. One species, *C. subpunctata*, is unusual in that its early nymphal stages are subterranean, galling the roots of the willowherb host. The fourth-instar nymph of this species overwinters in the galls and ascends the plants early in the summer, moulting to the fifth instar which, like the adult, lives wholly on the leaves and stems. The eggs are laid here and the first-instar nymphs of the new generation descend to the roots on hatching (Lauterer & Baudys, 1968). *Strophingia cinereae*, on heathers (*Erica* species), is much more local than *S. ericae*,

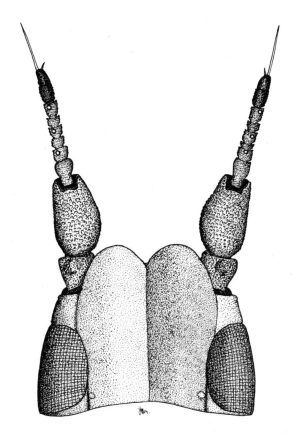

Fig 88. *Liviidae:* Livia juncorum, *head.*

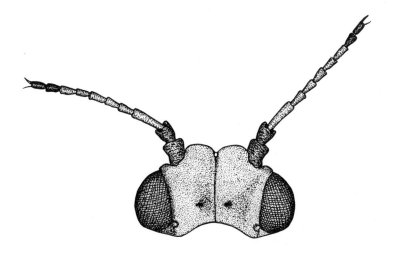

Fig 89. *Psyllidae:* Aphalara polygoni, *head.*

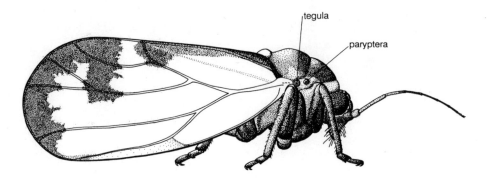

Fig 90. *Psyllidae:* Psyllopsis fraxini.

which lives on ling (*Calluna vulgaris*). *Rhinocola aceris* is locally common on field maple and some other maples, though not on sycamore. Most aphalarines do not cause gall-formation although their feeding activities may result in slight distortion of their hostplants.

Loginova (1979) showed that the name *Aphalara exilis*, as used by Hodkinson & White (1979), included both the true *exilis* and *A. pauli* Loginova; it is the latter that occurs in Britain.

The forty British species of Psyllinae feed on trees and shrubs. Adults have genal cones and, in the fore wings, the veins Cu and M are united for a short distance after separating from R. Thirty of the species belong to the genus *Psylla*, including two pests of fruit trees: *P. mali*, the 'apple sucker', and *P. pyricola*, the 'pear sucker'. *Psylla mali* overwinters in the egg state and passes through a single generation in the spring, crumpling the young leaves and damaging both leaf and flower buds of apple. *Psylla pyricola*, unusually for this genus, has three generations a year, and overwinters in the adult state. Pear leaves attacked by this species suffer from crumpling and rolling of the margins towards the base. Two other species have been taken on pear in Britain but both are very rare. Whitebeam, rowan, damson and sloe are all attacked by at least one *Psylla* species and hawthorn supports four. Five species of this genus are associated with sallows and willows. One of these, *P. ambigua*, overwinters as a nymph, as does *P. visci*, which lives on mistletoe. All the other Psyllidae pass the winter either as eggs or adults, depending on the species. Some *Psylla* species hibernating as adults spend the winter on evergreen trees and shrubs, returning to their true food-plants in the breeding season. *Psylla buxi*, on box, overwinters in the egg, so any species of *Psylla* encountered on box bushes in the winter will only be sheltering there. *Psylla buxi* nymphs are often abundant on hedges of their hostplant, causing the shoot-tips to form into little cabbage-like galls of clustered and deformed leaves.

Several other trees and shrubs are hosts to various species of *Psylla* but ash has none, supporting instead the four *Psyllopsis* species, all of which overwinter as eggs and are confined to this tree. One of them, *P. fraxini*, deforms the leaflets, usually causing one side of the lamina to become thickened and rolled inwards towards the midrib. These rolls are yellow-green at first, becoming streaked and blotched with red and purple at maturity. (Similarly coloured thickenings around the midrib, not extending to the margin of the leaflet, are caused by a gall-midge.)

Four British species of Psyllidae are associated with woody Leguminosae (gorse, broom, whin, laburnum). *Livilla ulicis* is a strange, rather beetle-like creature, with the fore wings thick, convex, coriaceous and dark brown. It lives on dyer's greenweed

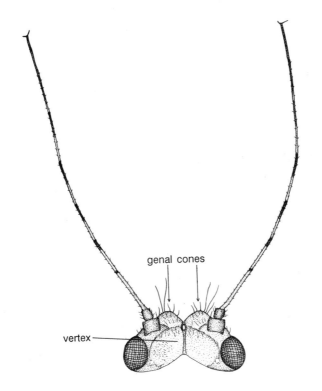

Fig 91. *Psyllidae:* Psylla foersteri, *head.*

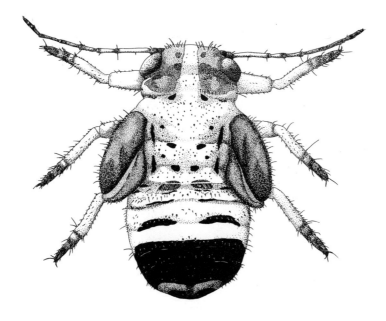

Fig 92. *Psyllidae:* Psylla alni, *nymph.*

(*Genista tinctoria*), broom and perhaps also gorse, from which the name of the species is taken. *Arytainilla spartiophila* and *Arytaina genistae* are common on broom throughout Britain. *Arytaina genistae* overwinters as either egg or adult and has two or three generations each year. It is the only British species of Psyllidae, apart from *Psylla pyricola* and the rare *P. pyri*, with more than one generation a year. *Livilla variegata*, the laburnum psyllid, is a Mediterranean species that was first observed in Britain in 1978. It has since spread rapidly throughout southern England and is continuing to extend its range northwards. *Livilla variegata* is not included in Hodkinson & White's (1979) handbook but Hodkinson & Hollis (1980) gave keys to separate adults of this psyllid, which they called *Floria variegata*, from those of the other three Leguminosae-feeding species. Like *A. spartiophila* and *L. ulicis*, it overwinters in the egg stage.

One other genus, with a single species, has a place in the British checklist: *Spanioneura fonscolombii* is found on box in southern England, where it is much less common than *Psylla buxi*.

Calophyidae

Calophya rhois, a tiny species less than 2 mm long, was once found in the Inner Hebrides. It feeds on the ornamental shrubs *Cotinus* and *Rhus*.

Triozidae

Adult triozids are easily recognized by the distinctive venation of the fore wing, in which Cu, R and M diverge from a single point. All but one of the eighteen British species belong to the genus *Trioza*. The exception is *Trichochermes walkeri*, which has a single generation each year on purging buckthorn, alder buckthorn and various cultivated

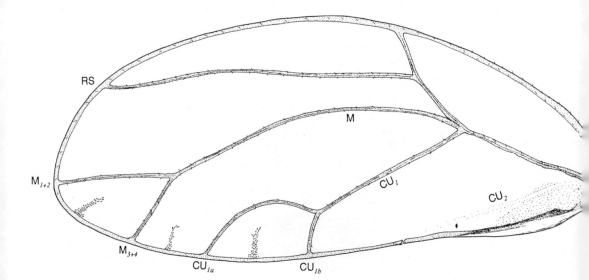

Fig 93. *Triozidae:* Trioza urticae, *fore wing.*

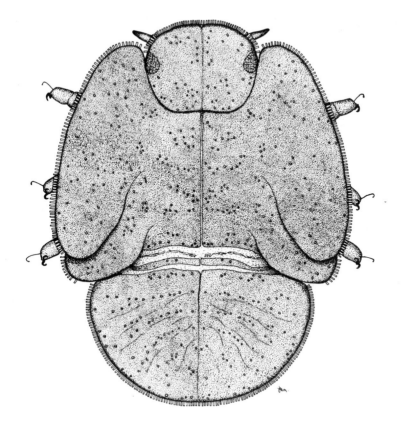

Fig 94. *Triozidae:* Trioza remota, *nymph.*

buckthorns. Nymphs of *T. walkeri* live in distinctive pocket-like galls at the margins of the leaves. Adults have brown fore wings with paler markings. It is not known if eggs or adults of this species overwinter. *Trichochermes* and many other triozids have a characteristic way of twitching their bodies when disturbed.

Trioza species feed on trees, shrubs or herbaceous plants and all of them overwinter as adults except for *T. galii*, which passes the winter as either eggs or nymphs. This species causes the terminal leaves of goosegrass and bedstraws to become pale, incurved and bunched into loose, spherical galls. *Trioza alacris* has one or perhaps two generations yearly on bay laurel, causing the margins of the leaves to become tightly rolled and bright yellow. *Trioza chenopodii*, with two generations a year, crumples the leaf-margins of oraches (*Atriplex*), goosefoot (*Chenopodium*) and sea purslane (*Halimione*). Adults of the summer generation of this species have very noticeably shorter wings than those of the overwintering generation. Such dimorphism is very unusual in Psylloidea. *Trioza urticae* has up to four generations a year, on nettle, deforming the leaves slightly. It is widespread and often abundant. The remaining British species of *Trioza* have only a single generation annually. The uncommon *T. centranthi* deforms the flowers of red valerian and cornsalad, turning them green. *Trioza remota*, on oak, causes small blister-like pit galls to form in the lamina of the leaf. These are convex above and concave below and are about a millimetre in diameter. *Trioza rhamni* has a somewhat similar effect on the leaves of purging buckthorn. *Trioza albiventris* breeds on several species of narrow-leaved willow without causing any obvious deformities. Out of the breeding season it is regularly found sheltering on evergreen plants such as conifers, a habit that it shares with about half the species of the genus.

Homotomidae

Homotoma ficus, an introduced species from southern Europe, is the only representative of this mainly tropical family in the British Isles. It occurs sporadically in the south, where it lives on the undersides of fig leaves. Adults are found from June to October; they are quite large, with a fore wing length of 3.4–4.3 mm, and pigmented in various shades of brown. The broad, flat nymphs resemble those of Triozidae except that the wing-pads are not fused to the thoracic terga. Eggs overwinter and there is a single generation annually.

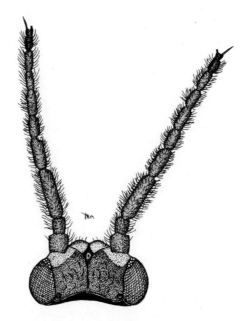

Fig 95. *Homotomidae:* Homotoma ficus, *head.*

Spondyliaspididae

Only one member of this Australian family has become established in the British Isles. *Ctenarytaina eucalypti* is confined to species of *Eucalyptus*, including the commonly grown *E. gunnii* and *E. globulus*, but is not found everywhere the host trees are cultivated. The adult is a tiny insect, about 2 mm long, dark brown with paler legs and antennae. Its fore wings are waxy and whitish between the yellow veins. Adults overwinter but little else is known of the biology of this species in Britain.

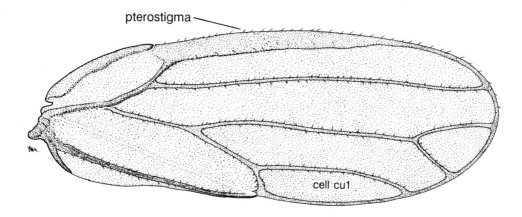

Fig 96. *Spondyliaspididae*: Ctenarytaina eucalypti, *fore wing*.

Aleyrodoidea

Adult whitefly are small, delicate insects with two almost equal-sized pairs of wings coated, like the body, with white wax (Fig. 97). They might be mistaken for small moths or powdery lacewings but for the characteristic hemipteran mouthparts. Adults are difficult to determine to species or even to genus except for the two species of *Aleyrodes* with distinctively spotted wings. Trehan's (1940) paper is of some use in identifying adults. Modern works rely on the much better characters of the puparium, which is the mature last-instar larva (Figs 98–100). Whitefly larvae and puparia are distinguished from immature and adult female Coccoidea by the unique vasiform orifice (Figs 98, 100). This is a median dorsal depression inset from the hind margin of the body and containing the lingula, which bears the anus, and the operculum, a flap that covers the lingula basally and sometimes entirely.

Empty pupal cases, from which the adults have emerged, show the diagnostic details well and need less preparation than unemerged ones for microscopic examination. Puparia can be preserved dry or in alcohol. Mound (1966: 402) and Martin (1987) gave instructions for the preparation of permanent mounts, involving de-waxing and a minimum of maceration in dilute sodium or potassium hydroxide before staining or, with the black pupal cases of *Aleurotuba jelinekii* and *Tetralicia ericae*, bleaching, before mounting in Canada balsam.

Aleyrodidae is the only family of whiteflies. All 13 genera and 19 species that have been recorded living in Britain belong to the subfamily Aleyrodinae. Other species are occasionally imported but do not survive for long. Mound (1966) reviewed the British whitefly known at that date and provided keys based on the puparia to separate the genera and species. *Aleurochiton aceris*, provisionally included by Mound (1966), was confirmed as genuinely native by Martin (1978); *A. acerinus* was added by Dolling & Martin (1985). *Aleurochiton aceris* and *A. acerinus* are confined to Norway maple (*Acer platanoides*) and field maple (*A. campestre*) respectively. *Bemisia afer* was recorded from London (as *B. hancocki*) by Halstead (1981). It was found on a single plant of bay laurel and the colony persisted for several years despite an application of insecticide. Bink-Moenen (1989) added the heather-feeding *Trialeurodes ericae*. Recent changes to the

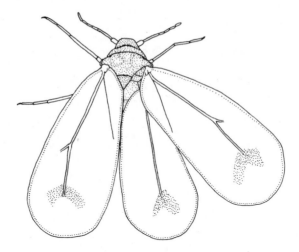

Fig 97. *Aleyrodidae:* Aleyrodes lonicerae, *adult.*

nomenclature of whiteflies can be traced in the World catalogue by Mound & Halsey (1978). *Aleurotrachelus jelinekii* has now been transferred to a new genus, *Aleurotuba*.

The sexes of whiteflies are similar in appearance, except for the genitalia, and there is no seasonal dimorphism. Adult males are often smaller than females and this difference is reflected in the size of the puparium. Parthenogenesis is known to occur in some species. Adults fly short distances when disturbed and also undertake longer dispersal flights. Eggs are laid on the undersides of leaves, attached by short stalks, and are often arranged in arcs or even complete circles as the ovipositing female rotates about her rostrum embedded in the plant tissue. Each female lays a few dozen eggs; the maximum number laid under optimal conditions is 300, over a period of 21 days, in *Bemisia tabaci* (Avidov, 1956) and 100 in *Aleyrodes proletella* (Butler, 1938). The egg batches are conspicuous because they and the surrounding leaf area are dusted with white wax.

The first of the four larval instars has relatively long legs and antennae but probably moves little, since clusters of the older, immobile instars are often encountered. The appendages are greatly reduced in the second and third instars, the legs being represented by short, conical stumps. The legs are somewhat longer, though still non-functional, in the fourth instar. All four instars, like the adults, feed on phloem sap but the fourth-instar larva ceases to feed after a while and becomes physiologically modified (without externally visible changes) into a puparium in which the adult develops. The adult emerges from the puparium through a T-shaped dorsal rent or, sometimes, by pushing off the anterior dorsal part of the puparium as a pair of quadrant-shaped plates.

Life-history studies on some British species of whitefly have been published by Bährmann (1980), Bink-Moenen (1989), Butler (1938), Mound (1962), Trehan (1940) and Wilson (1935). *Aleyrodes lonicerae* and *A. proletella* breed continuously throughout the summer months, but only mated females are found in the depths of the winter. Both of these species are common and widespread in the British Isles. *Aleyrodes proletella*, the 'cabbage whitefly', is often abundant on plants of the cabbage family, flying up in clouds when disturbed. It feeds on a wide variety of herbaceous plants with glossy, hairless leaves in contrast to *A. lonicerae*, which favours leaves with dull, hairy undersides, including those of strawberry, bramble (especially in the winter) and honeysuckle. Adults of *A. proletella* have two dark spots on each wing while those of *lonicerae* have only one.

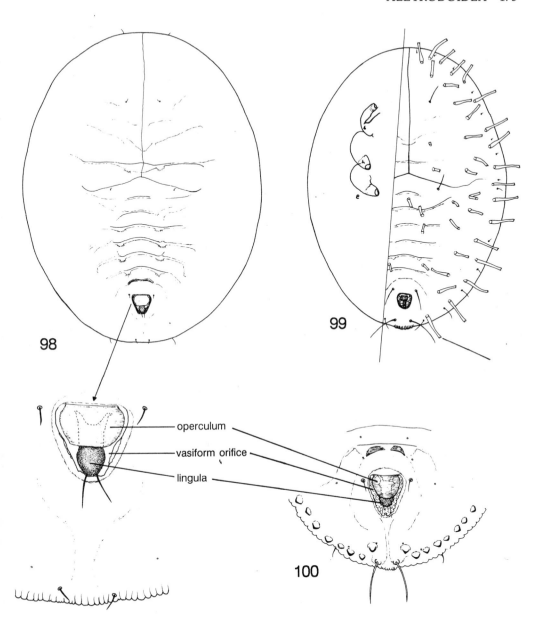

Figs 98–100. *Aleyrodidae. 98,* Aleyrodes proletella *puparium, with detail of vasiform orifice and associated structures; 99. Aleyrodidae:* Siphoninus phillyreae, *puparium; 100,* Trialeurodes vaporariorum *puparium, detail of posterior dorsum.*

Trialeurodes vaporariorum breeds continuously on many different plants in glasshouses and on window-sills. Heavy infestations, if not checked, may kill some susceptible plants such as fuchsias. Another notorious, polyphagous pest is *Bemisia tabaci*. It occurs sporadically under glass and has very occasionally been found out of doors in England. The two *Aleurochiton* species are bivoltine, overwintering as puparia. The other outdoor species are univoltine. Those on deciduous trees overwinter as puparia while those on evergreens overwinter in the younger nymphal stages. On heathers, *Tetralicia ericae* usually passes the winter months in the third instar and *Trialeurodes ericae* in the puparium.

The puparia of the species living on deciduous trees fall with the leaves in the autumn, remaining attached through the winter. In the genus *Aleurochiton* the overwintering puparia are much more thick-walled than the summer ones and are richer in fat. The two types of puparia also differ in some morphological characters (Bährmann, 1973).

Mound & Halsey (1978) gave details of distribution, hostplants and natural enemies of all the world's species of whitefly known at that date.

Adelgoidea

This superfamily is closely related to Aphidoidea and members of both groups are generally similar in appearance. Both of them have apparently originated as phloem-sap feeders on trees in temperate regions and in both groups there is a tendency to develop complex life cycles involving parthenogenesis and host-plant alternation. There are four nymphal instars, as in Aphidoidea, except in apterous Phylloxeridae, which are said to have only three. The chief biological difference between the two superfamilies is the persistence of oviparity in the parthenogenetic females of Adelgoidea and its replacement by viviparity in Aphidoidea. Most of the structural features of Adelgoidea are found also in Aphidoidea. Apterae of Pemphigidae, in particular, have many points in common with adelgoids, in particular the presence of eyes with only three ommatidia and the absence of siphunculi.

Key to the families of Adelgoidea

1. Antennae of apterae and larvae with two rhinaria, those of alatae with three or four (Fig. 103); wings of alatae held in tent-like position at rest, fore wings with Cu1 and Cu2 separate (Fig. 104), hind wings with an oblique vein. (On conifers) .... ADELGIDAE
— Antennae of apterae and larvae with one rhinarium, those of alatae with two (Fig. 107); wings held flat at rest, fore wings with Cu1 and Cu2 with a common stem (Fig. 108), hind wings without oblique vein. (On oak, vine, willow or poplar) .... PHYLLOXERIDAE

Adelgidae

The members of this family are confined to conifers. There are about a dozen British species. Apterae have three-segmented antennae with two rhinaria, eyes with only three facets, a broadly rounded cauda and no siphunculi. Alatae hold their wings in a tent-like position at rest and the fore wings have three oblique veins, all arising separately from the main longitudinal vein. All female morphs are oviparous and a short ovipositor is present. No specialized 'winter egg' is produced; the usual overwintering stage is the first-instar or sometimes the second-instar nymph.

The characteristic life-cycle of the family occupies two years, with one sexual and two asexual generations on spruce (*Picea*) and a series of asexual generations on other conifers. The second asexual generation on spruce live gregariously in characteristic galls on the shoots. In *Adelges cooleyi*, small nymphs overwinter on any of three species of *Picea*. They complete their development to adult fundatrices in the spring and give birth to the gall-forming nymphs which mature into winged migrants. These disperse to

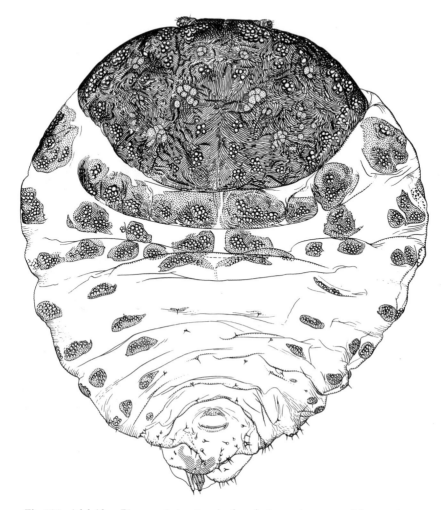

Fig 101. *Adelgidae:* Pineus pini *aptera in dorsal view, microscope slide mount.*

the secondary hosts, which are species of *Pseudotsuga*. One or more apterous, parthenogenetic, generations occur on the secondary host, overwintering as nymphs. They live on the shoots and do not cause galls to form. A winged generation is then produced. These return to the primary hosts, where apterous sexuales are produced. From the union of the sexuales arises the new generation of fundatrices, which overwinter as nymphs and produce the new gall-forming generation in the spring.

This basic life-cycle may be modified by the elimination of host-alternation and sexual reproduction. *Pineus orientalis*, for example, alternates between spruce and pine in the same way that *Adelges cooleyi* alternates between spruce and *Pseudotsuga* but the related *Pineus pini* is restricted to pine. It has forms like the sexuparae of *orientalis* but these do not produce young; sexuales are, therefore, absent and the species is wholly parthenogenetic. *Adelges viridana*, a wholly parthenogentic species living on larch, is related to two holocyclic species, *A. laricis* and *A. viridis*, which both alternate between spruce and larch. Similarly, *A. piceae* is anholocyclic on various species of fir (*Abies*) while its relative *A. nordmanniae* is holocyclic, alternating between *Picea orientalis* and

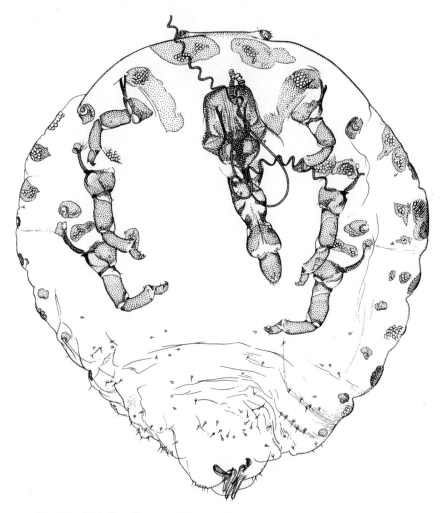

Fig 102. *Adelgidae:* Pineus pini *aptera in ventral view, microscope slide mount.*

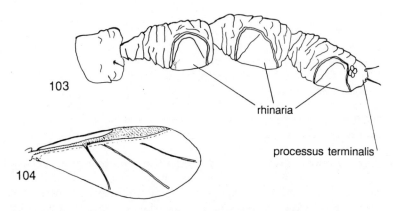

Figs 103–104. *Adelgidae:* Adelges *species alatae. 103, antenna; 104, fore wing.*

Abies species. An unusual anholocyclic life-history is that of *A. abietis*, which galls spruce: the progeny of the gall-producing stage are 'pseudofundatrices' which also live on spruce and produce a new generation of gall-formers directly.

Adelgids are difficult to identify. The host-plant is a very useful guide, as is the form of the galls induced on spruce. Carter (1971, 1976) covered the British species. His nomenclature, particularly above the species-level, differs considerably from that of Heinze's (1962) detailed account of the German Adelgidae.

Phylloxeridae

Six species of this small family have been recorded in Britain. Apterae have only three antennal segments, the last of which bears the only rhinarium; they have three-facetted eyes and a rounded cauda. The fore wings of the alatae are folded flat over the back and have only three oblique veins, the first two of which arise on a common stem from the main longitudinal vein. Siphunculi are absent in both alatae and apterae and all females are oviparous. The sexuales are always wingless and, lacking mouthparts, they cannot feed. British species lack an ovipositor. Sexually produced, overwintering eggs have much thicker shells than those produced asexually in summer.

Phylloxera glabra is the only common species. It lives on the leaves of oaks (*Quercus robur* and occasionally other deciduous oaks). The fundatrices develop in small, marginal leaf-rolls on the undersides of the leaves. Several parthenogenetically reproducing generations of apterae follow, developing on the undersides of the leaves, which develop a yellow-mottled appearance if the infestation is heavy. Eventually, winged sexuparae are produced. Their progeny are apterous sexuales. Female sexuales lay overwintering eggs on the twigs of the host tree.

Two other phylloxerids have been recorded from oak in Britain. *Phylloxera quercus* may have been reported in mistake for *glabra*. In continental Europe it alternates between evergreen oaks (the primary hosts) and deciduous oaks. Its usual primary host, *Quercus coccifera*, is rarely grown in this country and the insect is apparently unable to use *Q. ilex*, the widely grown holm oak. A recently introduced species, *Moritziella corticalis*, lives on the twigs of pedunculate oak and sessile oak and appears to be anholocyclic in Britain, since the winged sexuparae, produced in mid- to late summer, seem not to produce offspring here. The overwintering stage is the first-instar larva.

The serious vine pest, *Viteus vitifolii* (the 'vine phylloxera'), which had died out in Britain, reappeared with the revival of viticulture. The small, crown-like galls on the leaf lamina, produced by the fundatrices, are an obvious sign of the presence of the pest. The cycle in European countries proceeds as continuous asexual reproduction by apterous female 'radicicolae' which live underground near the tips of the roots, whose growth becomes stunted. In the autumn, winged sexuparae emerge from the ground. These sexuparae are of two types: 'androparae', which produce males and 'gynoparae', which produce sexually-reproducing females. The daughters of the gynoparae mate with the sons of the androparae and lay winter eggs on the twigs. Fundatrices hatch from these eggs in the following year but in Europe they usually die without producing any young. Radicicolae are present throughout the winter and continue reproducing asexually; otherwise, the cycle would terminate with the fundatrices. On the American continent, fundatrices produce two types of larvae: 'gallicolae', which make leaf-galls like those of the fundatrix; and radicicolae. Adults of both of these morphs are apterous. The gallicolae produce further mixed broods of gallicolae and radicicolae and the latter eventually produce sexuales as in the European populations. The life-cycles of *V. vitifolii* in Europe and America are dealt with in many standard textbooks. European viticulture was saved

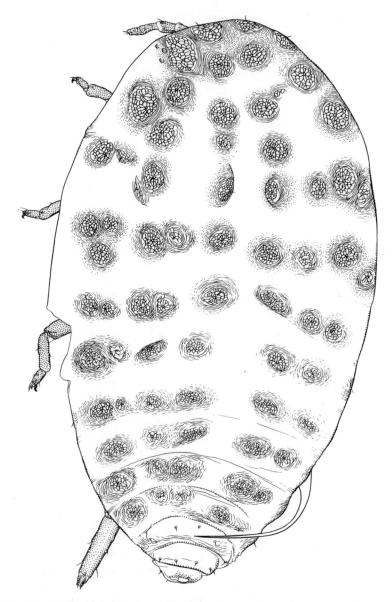

Fig 105. *Phylloxeridae:* Phylloxerina salicis *in dorsal view, microscope slide mount.*

from financial disaster by grafting European vine varieties onto rootstocks of American species resistant to the insect.

The oak- and vine-inhabiting species all belong to the subfamily Phylloxerinae and do not have conspicuous body wax. A second subfamily, Phylloxerininae, is characterized by the presence of wax glands and in life the insects are covered with their white, woolly secretion. *Phylloxerina salicis* occurs sporadically on certain willows (*Salix alba* and sometimes *S. babylonica*), living holocyclically on the trunks and branches. All stages are apterous. Fundatrices emerge from winter eggs in the spring. From them stem several

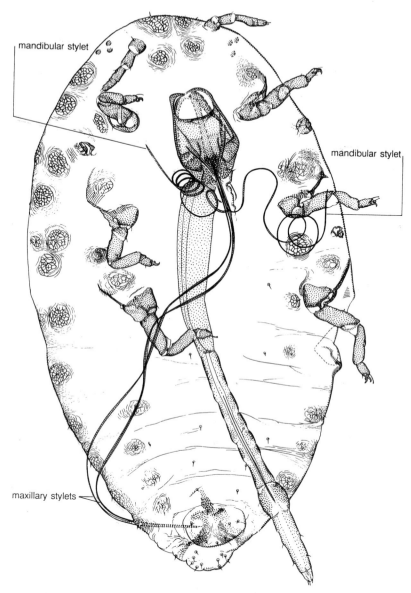

Fig 106. *Phylloxeridae:* Phylloxerina salicis *in ventral view, microscope slide mount.*

generations of asexually-reproducing virginoparae. The females that appear in August and September are 'virgino-sexuparae', whose eggs produce a mixture of males, sexually-reproducing females and further virgino-sexuparae like themelves. A second species, *P. populi*, has been found in Britain on the trunks of Lombardy poplar.

Heinze (1962) gave much taxonomic, morphological and biological information on the European members of the family. Barson & Carter's (1972) account can be used to identify the three British oak-feeding species; the other three are specific to their respective host plants.

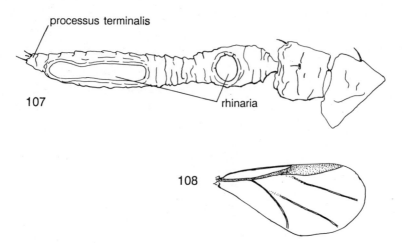

Figs 107–108. *Phylloxeridae: Phylloxera* species alatae. 107, antenna; 108, fore wing.

Aphidoidea

Most aphids are readily recognized as belonging to this superfamily by their general appearance. The presence of a pair of siphunculi (or cornicles) on the fifth or sixth abdominal segment is unique to the superfamily, although siphunculi are absent from a few genera. In all instars of all aphidoids the last antennal segment is divided into a thicker, basal portion and a thinner, apical part termed the 'processus terminalis' (or terminal process). At the junction of the two parts can be found a large, plate-like sensory organ, termed the primary rhinarium. In adults it may be accompanied by a number of smaller, secondary or accessory rhinaria. The processus terminalis is reckoned to begin at the distal margin of the primary rhinarium. Adelgoidea also have a processus terminalis; the two groups may be separated by the characters given in the key to superfamilies of Sternorrhyncha.

Other Sternorrhyncha lack the processus terminalis. In nymphs of Psylloidea the terminal antennal segment may have a short, narrower apical portion but this never has a rhinarium at the base. Adult Aphidoidea can most readily be distinguished from Psylloidea by the structure of the legs, which are slender and usually rather long in aphidoids and never have the posterior pair modified for jumping, unlike the shorter, stouter legs of psylloids. Aphidoid fore wings differ from those of psylloids in that they lack a clavus. The fore wings of aphidoids are conspicuously larger than the hind wings, in contrast to the equal-sized wings of Aleyrodidae. Apterous adult Hormaphididae are strikingly similar to puparia of Aleyrodidae but lack the characteristic vasiform orifice of the latter. Some Aphidoidea superficially resemble Coccoidea and some Coccoidea have a rather aphidoid-like appearance, but the presence or absence of a processus terminalis should resolve any doubts (as should the fact that all Coccoidea have only a single tarsal claw in contrast to the paired claws of other Sternorrhyncha).

Polymorphism is a marked characteristic of Aphidoidea. The major developmental alternatives are between sexually reproducing or parthenogenetic females and between winged ('alate') or wingless ('aptera') forms. All morphs pass through four nymphal instars. In those destined to become alatae, the developing wing-pads are obvious after the second moult. Parthenogenesis occurs universally in the group and the parthenogenetic females are always viviparous, in contrast to the oviparous parthenogentic

Plate 7

Sternorrhyncha. **1**, Nymphs and moult-skins of *Psylla alni* (Psyllidae) surrounded by protective wax threads on alder shoot. **2**, Puparia and emerging adults of *Aleyrodes proletella* (Aleyrodidae) on leaf of *Lathyrus*. **3**, *Phylloxera glabra* (Phylloxeridae) females, nymphs and groups of thin-shelled summer eggs on underside of oak leaf. **4–5**, Aphididae. **4**, *Pterocomma salicis*, with salmon-pink siphunculi and white wax-patches, on sallow stem. **5**, *Macrosiphum cholodkovskyi* parthenogenetic female giving birth to a nymph on meadowsweet. **6**, *Maculolachnus submacula* (Lachnidae) sexually-reproducing females with thick-shelled winter eggs on rose twig.

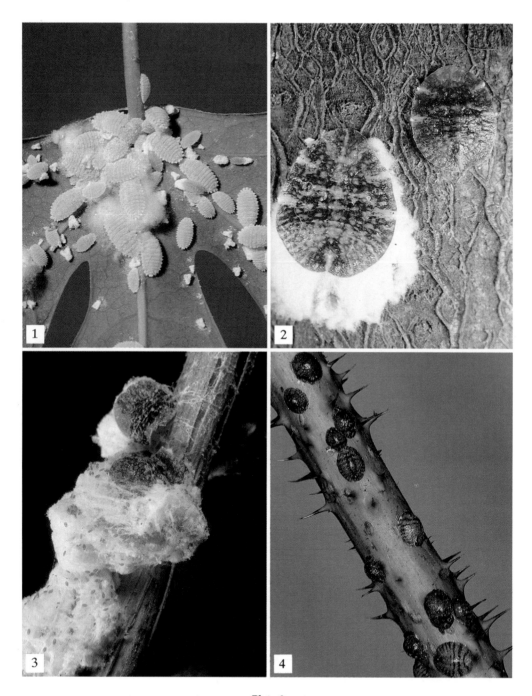

Plate 8

Scale insects. **1**, Colony of *Phenacoccus manihoti* (Pseudococcidae) on cassava leaf in greenhouse. **2**, *Pulvinaria regalis* (Coccidae) mature female with ovisac on stem of bay laurel. **3**, *Pulvinaria vitis* (Coccidae) mature females with ovisacs from which the first-instar 'crawlers' are emerging, on shoot of vine. **4**, *Parthenolecanium corni* (Coccidae) mature females on stem of loganberry.

females of Adelgoidea. Sexual reproduction always involves the production of a few, thick-shelled 'winter eggs' or, in Pemphigidae, only a single such egg. The eggs of Aphidoidea are typically elongate oval and shining black.

The basic annual life cycle of Aphidoidea seems to be determined by the pattern of sap-flow in woody plants. At the beginning of the spring flush, the first generation, the 'fundatrices', hatch from the winter eggs. Clones of parthenogenetically produced aphids build up rapidly, feeding on the nutrient-rich sap of the host-plant. In the autumn, when the sap of the woody host plant is again enriched with mobilized nutrients, there is a second peak of reproductive activity, this time including a sexually reproducing generation, which culminates in the production of winter eggs.

What happens between the two periods of sap-flow differs in different kinds of aphids. For some, the summer is a period of quiescence, when reproduction slows down or stops altogether. The summer resting stage of such species may be either the normal adult parthenogenetic female or a specialized first-instar nymph. Other kinds migrate to herbaceous plants, whose sap is full of nutrients at a time when that of the woody plants is not. A number of species are presumed to have abandoned their ancestral woody hosts altogether, transferring the whole life-cycle, including the sexual stages, to herbaceous hosts on which they live all the year round.

In many species of aphids the development of wings is suppressed in the non-migratory forms, and winged forms arise either in response to overcrowding or in anticipation of the need for migratory flight. The production of sexually reproducing generations or their precursors has been shown in many species to be linked to the number of hours of darkness experienced each day by their mothers and to the temperature at which they developed.

Parthenogenesis enables aphids to reproduce at a very rapid rate. The fecundity of each female is not very great; a few dozen young is the norm. The rapid growth of populations that is often seen is the result of the shortening of the time between generations. With the elimination of the necessity for fertilization, eggs can begin to develop before the female aphid is adult. In fact, each viviparous aphid is born with embryos already developing within her body and with even the rudiments of her grandchildren beginning to form in the ovaries of her embryos.

The typical, complete life cycle, or holocycle, involving both sexual and asexual reproduction in the course of the year, is not universal. Some species have become 'anholocyclic', abandoning sexuality altogether. Only two morphs remain: the apterous and alate viviparous females. Anholocyclic life cycles are particularly common among species that live on herbaceous plants or underground. Such anholocyclic species appear to be derived from holocyclic ones that have abandoned sexual reproduction.

The wingless adult parthenogenetic female (the 'aptera vivipara') is the morph that is most often encountered in the majority of species and most identification guides concentrate on this form. There is no ovipositor but wingless adults may be distinguished from nymphs by their well developed, tail-like cauda, projecting above the anus. The cauda may be triangular, rounded, hemispherical, knobbed, conical or elongate (Figs 113, 119, 130, 124, 132, 133) and it is a useful feature in identification. Females derived from winter eggs ('fundatrices') are rather similar to the other apterous viviparae but usually have shorter appendages. Winged adult viviparous females ('alatae viviparae') have many of the features of apterae, including the cauda, but there are some differences in proportions of the body and appendages and the thorax is modified to accommodate the flight-muscles. In particular, the meso- and metathorax are enlarged, they are more strongly sclerotized than in the aptera, and the thoracic tergites are divided by sutures into a number of separate sclerites.

Egg-laying females ('oviparae'), which are usually encountered only in the autumn, are

184 STERNORRHYNCHA

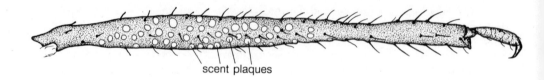

Fig 109. *Aphididae:* Cryptomyzus ribis, *hind tibia and tarsus of ovipara.*

almost always wingless and frequently have numerous scent glands on the surfaces of their often swollen and bowed hind tibiae (Fig. 109). The hind tibiae of viviparous females are straight and slender and lack the scent-glands. Males, which are sometimes apterous, have a median penis near the tip of the body, flanked by a pair of simple claspers (Fig. 110). Sometimes the sexuales (males and oviparous females) are dwarfed and in Pemphigidae they have no mouthparts from birth and cannot feed.

Reliable identification is often possible only with cleared, slide-mounted specimens. Maceration in potassium hydroxide followed by dehydration, clearing and mounting in Canada balsam is a traditional method that has stood the test of time. There are numerous alternatives, of varying degrees of permanence, such as the gum chloral method detailed by Blackman (1974: 122–124).

There are probably more publications on aphids than on any other group of comparable size outside of the Lepidoptera. Blackman's (1974) book forms an excellent introduction to the study of the British fauna. It includes information on such diverse topics as biology, techniques of study and natural enemies as well as an identification

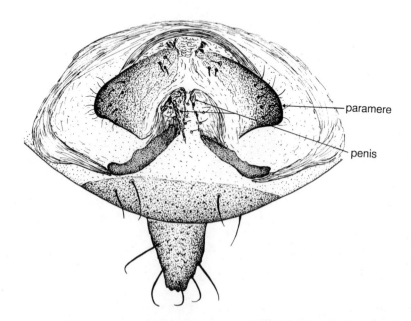

Fig 110. *Aphididae:* Hyalopterus pruni, *abdominal apex of male.*

guide to more than a hundred of the more common species. The most recent comprehensive monograph of the British aphid fauna is that of Theobald (1926, 1927, 1929), which is now greatly out of date and in some places misleading. Hille Ris Lambers's (1933, 1934) notes and Stroyan's (1950, 1955, 1957, 1964, 1972, 1979) addenda go a long way towards remedying Theobald's shortcomings. Some European works that are helpful in dealing with identification of aphids are: Börner (1952, 1953), Hille Ris Lambers (1938–1953) and Shaposhnikov (*in* Bei-Bienko, 1964, English translation 1967). Two local lists from opposite ends of Britain shed some light on the distribution of aphids within the country: Shaw's (1964) Scottish list and Wood-Baker's (1980) account of the Kentish species. The world aphid fauna was catalogued by Eastop & Hille Ris Lambers (1976). Identification of economically important species was dealt with by Stroyan (1952) for Britain and by Blackman & Eastop (1984) for the world. The latter work is arranged according to the crops attacked. A more comprehensive host-plant list for the aphids of the world is that of Patch (1938), indexed by Averill (1945) and updated by Richards (1976). Techniques for the study of living aphids are the subject of Van Emden's (1972) book. Dixon (1973) reviewed some aspects of the biology of aphids and the same author (1985) covered their ecology more fully. Minks & Harrewijn (1987, 1988, 1989) also dealt extensively with many aspects of biology. Polymorphism was reviewed by both Hille Ris Lambers (1966) and Lees (1966).

Key to the families of British Aphidoidea

This key should work for adult viviparae and for most female alatae. It is sometimes difficult to see if the terminal process of the last antennal segment is longer or shorter than half the length of the basal part of this segment and a measuring graticule may be needed to resolve the question. The presence or absence of the 'beak' mentioned in couplet 9 is sometimes difficult to determine. The complexity of the key is due in large measure to the striking differences between the different morphs. It may not be completely reliable for fundatrices or sexuales, or both, in some groups, more especially those of Pemphigidae, Anoeciidae and Hormaphididae.

1. Terminal process of last antennal segment (measured from distal end of primary rhinarium) half as long to several times as long as basal part of the segment (including primary rhinarium) (Figs 116, 129) or, if less than half as long, cauda knobbed (Figs 122–124) and antennae 6-jointed or (apterae) abdominal segments I–VI each with either six pigmented glandular areas (Fig. 119) or with a transverse, pigmented bar (Fig. 120) 2
— Terminal process of last antennal segment less than half as long as basal part (Figs 115, 139, 149–151). If cauda knobbed (Fig. 137), then antennae not more than 5-jointed. If apterae with glandular areas on abdominal segments I-VI, these not pigmented 6
2. Cauda elongate-triangular, strap-shaped or finger-shaped (Figs 127, 131, 133) .... APHIDIDAE (part)
— Cauda short-triangular (Fig. 132), helmet-shaped (Fig. 130), knobbed (Figs 118, 122–124), semicircular or broadly rounded 3
3. Cauda not knobbed 4
— Cauda knobbed (Figs 118, 122–124) 5
4. Siphunculi without reticulate sculpture and not usually stumpy. Antennae with 6 or 4 joints or, if with 5 joints, body with large marginal tubercles on pronotum and elsewhere. Anal plate entire, not bilobed. Secondary rhinaria, when present, round or oval (as in Fig. 128) .... APHIDIDAE (part)

— Siphunculi usually stumpy (Figs 125, 126) or reduced to chitinized rings (Figs 118, 119); if siphunculi several times longer than wide at base then cauda knobbed (Fig. 123). Either siphunculi with reticulate sculpture (Fig. 117), *or* antennae 5-jointed and body without marginal tubercles *or* anal plate deeply bilobed (Fig. 122) *or* secondary rhinaria narrow and transverse (Fig. 121) *or* terminal process of last antennal segment shorter than base of segment 5

5. Siphunculi with reticulate sculpture (Fig. 117) *or* antennae 5-jointed and anal plate entire .... CHAITOPHORIDAE
— Siphunculi without reticulate sculpture (Fig. 126) *and* antennae 6-jointed or if (very rarely) antennae 5-jointed then anal plate bilobed (as in Fig. 124) .... CALLAPHIDIDAE

6. Cauda subtriangular (Fig. 141). Fore wing of alata with last oblique vein (Rs) arising from base of pterostigma (Fig. 140). (On conifers.) .... MINDARIDAE
— Cauda rounded (Figs 136, 138, 142, 146, 148), knobbed (Fig. 137) or very broadly triangular (Fig. 113). Fore wing of alata with last oblique vein (Rs) arising near apex of pterostigma (Figs 111, 134, 143) 7

7. Nymphs and apterous viviparae with head and pronotum not fused and usually clearly separate; if fused (occasional *Anoecia* specimens) then abdomen with large, flattened, marginal tubercles (Fig. 136); if cuticle of head and pronotum membranous then abdomen with wax glands absent or conspicuous and not restricted to a lateral pair on segment VII. Antennae of alatae usually 6-segmented and always with at least one rhinarium on segment III; if 5-segmented then siphunculi absent. Alatae almost always with wings held rooflike in repose; if with wings held flat, then all abdominal tergites with wax glands medially as well as laterally. Cauda never knobbed 8
— Nymphs and apterae with head and pronotum fused (so that eyes are situated at about middle of sides of apparent head) (Figs 137, 138). Without large, flattened, marginal abdominal tubercles. Abdominal wax glands absent, inconspicuous or restricted to a lateral pair on tergite VII only (Fig. 142). Antennae of alatae with 5 or 6 segments; if 6-segmented then segment III without rhinaria; if 5-segmented then siphunculi present. Alatae with wings held flat in repose. Cauda rounded or knobbed 11

8. Hind tarsus very long, more than twice as long as fore or mid tarsus (Fig. 114). On roots of Compositae and Ranunculaceae, attended by ants .... LACHNIDAE (Traminae)
— Hind tarsus not much, if any, longer than the other tarsi 9

9. Apical segment of rostrum without a 'beak' (Figs 144, 145). Siphunculi absent or present only as pores or very small cones, in which case eyes of apterae with only 3 facets (Fig. 147). Antennal segments of apterae variable in number, those of alatae usually 6-segmented, often with annular (Figs 149, 150) or narrowly transverse rhinaria; alatae occasionally with 5-segmented antennae, in which case siphunculi absent .... PEMPHIGIDAE
— Apical segment of rostrum with a short, blunt or longer, narrow 'beak' (Figs 112, 135). Siphunculi present, pore-like, usually on low cones, often surrounded by hairs (Fig. 136). Antennae of normal viviparae 6-segmented, rhinaria of alatae rounded. Eyes of apterae usually with many facets but if with only three then there are large, flat marginal tubercles on most abdominal segments (as in Fig. 136) 10

10. Abdomen with large, flattened, subcircular marginal tubercles (Fig. 136) (not to be confused with the spiracular plates). Pterostigma of alata forming a conspicuous, short, broad, dark spot on anterior margin of fore wing (Fig. 134). On Cornus or roots of grasses or sedges .... ANOECIIDAE
— Abdomen without marginal tubercles. Pterostigma long and narrow (Fig. 111), not forming a conspicuous, dark spot. On conifers and deciduous trees but not on *Cornus* or monocotyledons .... LACHNIDAE

11. Anal plate bilobed. Antennae and legs of adult aptera very short, usually completely concealed beneath body (Fig. 138). Antennae of alata with numerous annular rhinaria (Fig. 139). On birch or in greenhouses on monocotyledonous plants and ferns .... HORMAPHIDIDAE
— Anal plate entire (Fig. 137). Appendages of aptera not concealed beneath body. Antennae of alata without annular rhinaria 12

12. Wax-glands in two groups, on tergite VII only (Fig. 142). Antenna of alata 6-segmented, without rhinaria on segment III. On poplars .... PHLOEOMYZIDAE
— Wax-glands absent or situated laterally on all abdominal segments. Antenna of alata 5-segmented, with a few sensilla on segment III. On oak, birch or alder .... THELAXIDAE

Lachnidae

These are mostly large, hairy aphids. Their pore-like siphunculi are either placed flush with the body surface or elevated on short cones and the cauda is broadly rounded. The terminal process of the antenna is very short and the fourth segment of the rostrum ends in a short, narrow 'beak' which, in the genus *Cinara*, is separated from the fourth segment by an articulation and appears to be a fifth segment. There are 40–45 British species, in three subfamilies. They never live on the aerial parts of herbaceous plants or on the roots of grasses. Host-alternation is unknown in the family.

Conifers are the hosts of the subfamily Cinarinae, which includes about 30 of the British species. Many non-native conifers have one or more species of *Cinara* attacking them. Most cinarines are plump and black or dark brown, but the pine-feeding *Eulachnus* species are slender and green or red-brown, covered with grey wax-dust.

There are six species of Lachninae found in Britain. *Tuberolachnus salignus* is a big, brown, hairy species often occurring in large colonies on willow twigs, where it appears to be anholocyclic. *Maculolachnus submacula* is largely confined to the stems and twigs of roses but in the summer it has occasionally been found at the roots of roses and of some herbaceous plants, perhaps as a result of transport by ants. *Stomaphis quercus* inhabits fissures in the bark of oak and birch and is attractive to ants. Its rostrum is extremely long except in the males, which lack mouthparts altogether. *Lachnus* species are large, brown, long-legged aphids living on the twigs of oak, where they are often tended by wood ants.

About eight species of the subfamily Traminae are found in the British Isles, living on the subterranean parts of various Compositae (Asteraceae) or, in the case of *Protrama ranunculi*, of buttercups (Ranunculaceae). They are attractive to ants and are probably all anholocyclic. Alatae and apterae occur in *Trama* and *Neotrama*. The latter genus also produces brachypterae, with rudimentary wings, and alatiform apterae, i.e. wingless forms with some of the characters of alatae, such as more numerous ommatidia. All three British *Protrama* species occur in the alatiform aptera morph. *Protrama flavescens* also has brachypterae and alatae but no normal apterae; *P. ranunculi* has only alatiform apterae and alatae; and *P. radicis* has both normal and alatiform apterae but true alatae are rare. The flightless morphs of all of these subterranean aphids are white and long-legged. The extremely long second tarsal segment (Fig. 114) of the hind leg is diagnostic of Traminae.

The British Traminae were investigated by Eastop (1953), and the same author published, in 1972, a detailed investigation of the species of *Cinara*. Carter & Maslen (1982) covered all the native and introduced Cinarinae. Heinze (1962) gave a detailed account of the German Lachninae.

Chaitophoridae

These hairy aphids have short or pore-like siphunculi, a rounded or knobbed cauda and a long terminal process on the last antennal segment. Host-alternation is unknown in this family. In Britain, 17 species of Chaitophorinae, in which the siphunculi are conical and reticulately sculptured, live on trees and the nine species of Atheroidinae live on grasses or, sometimes, sedges or rushes. None of them induces the development of galls or other malformations.

188 STERNORRHYNCHA

Chaitophorus species (Chaitophorinae) live on willow, sallow, poplar and aspen. Their males may be winged or apterous. In the other British genus of the subfamily, *Periphyllus*, the males are usually winged. This genus is largely confined to sycamore and field maple

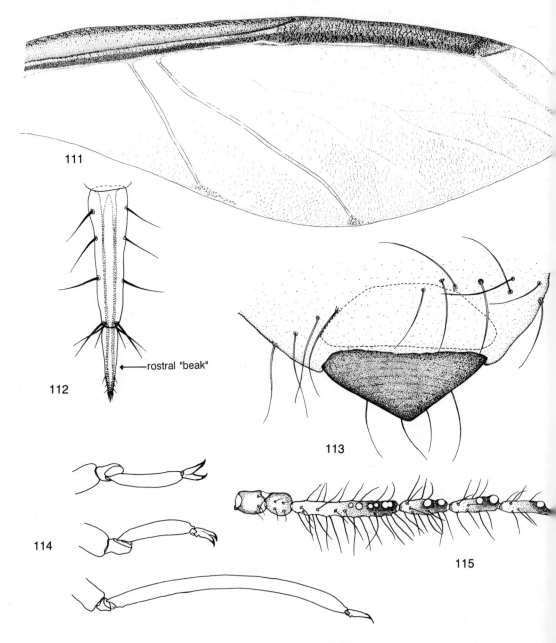

Figs 111–115. *Lachnidae. 111,* Cinara acutirostris, *fore wing; 112,* Cinara schimitscheki, *last segment of rostrum, showing apical 'beak'; 113,* Schizolachnus pineti *alata, apex of abdomen showing triangular cauda; 114,* Protrama ranunculi, *tarsi, showing great elongation of posterior tarsus; 115,* Cinara tujafilina, *antenna.*

but *P. testudinaceus*, the commonest species on field maple, is occasionally found also on horse chestnut. The first-instar nymphs of *Periphyllus* occur in two different forms. In addition to nymphs of normal appearance there are aestivating morphs ('dimorphs'), which spend the summer in a state of suspended development. In *P. acericola* and its closest relatives the aestivating morphs are covered with very long hairs but are otherwise unmodified. The aestivating morphs in *P. testudinaceus* and its allied species have leaf-like hairs around the body margins, on the antennal bases and on the first two pairs of tibiae; the dorsal surface of the head, thorax and abdomen is divided into plates like the carapace of a tortoise. Aestivating morphs of the *acericola* type cluster together, while those of the *testudinaceus* type are solitary. Some species of *Periphyllus* do not produce aestivating morphs, some pass the summer only in this morph and yet others produce a mixture of aestivating and non-aestivating morphs. According to Hille Ris Lambers (1966), the aestivating morphs, when they resume development, mature into apterous or, rarely, alate sexuparae.

Apart from *Caricosipha paniculatae*, which lives on sedges, and occasional populations of *Sipha glyceriae* on sedges and rushes, the Atheroidinae are found only on grasses. Males are always apterous. Viviparae of this subfamily have only five antennal segments whereas those of Chaitophorinae have six, occasionally with segments III and IV incompletely fused.

British species of Chaitophoridae can be identified with Stroyan's (1977) handbook. Heie (1982) united this family with Callaphididae under the name Drepanosiphidae in his monograph of the Scandinavian species.

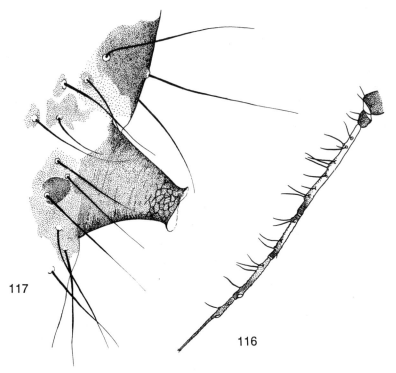

Figs 116, 117. *Chaitophoridae:* Periphyllus californiensis. *116,* antenna; *117,* siphunculus, *showing reticulate sculpture.*

Callaphididae

More than 50 species of this family occur in Britain. They are distributed among four subfamilies. All are monoecious and none causes leaf-rolling or gall-formation. There is considerable diversity of form in the family, making it difficult to define. The siphunculi are usually short and conical but never have the reticulate sculpture characteristic of the Chaitophorinae. The cauda is short and usually, but not always, knobbed. The antennae are six-segmented (except in some apterae of *Callipterinella* and occasional abnormally small specimens of other genera) and the terminal process is very variable in length. In some species all viviparae are alate, including the fundatrices. In others both alate and apterous fundatrices occur but usually, if apterous viviparae occur at all, then all the fundatrices are wingless. Compared with most other Aphidoidea, callaphidids are active insects and some have modifications of the legs that enable them to leap when disturbed.

Callaphidinae, with 22 British species, live mostly on broadleaved trees; exceptions are *Takecallis* species, on bamboos, *Myzocallis myricae*, on bog myrtle and *Ctenocallis setosa*, on broom. Apterae of the last-named species have a series of large, finger-like processes arising both laterally and on each side of the midline of the thoracic and abdominal segments. Male Callaphidinae are winged except for one of the two *Pterocallis* species, in which they are always apterous. Apterous, as well as alate, virginoparae occur regularly only in *Pterocallis* and *Ctenocallis*. In all the other genera only the oviparae are apterous; even the fundatrices are winged. Incompletely alate individuals occur in *M. myricae*, in which brachypters predominate in late summer, and its congener *M. castanicola*, which

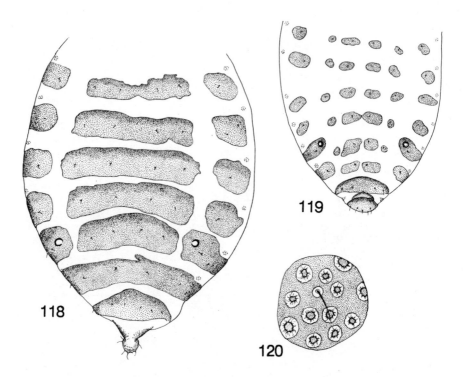

Figs 118–120. *Callaphididae:* Phyllaphis fagi. *118, abdominal dorsum of aptera; 119, abdominal dorsum of alata; 120, detail of wax-plate.*

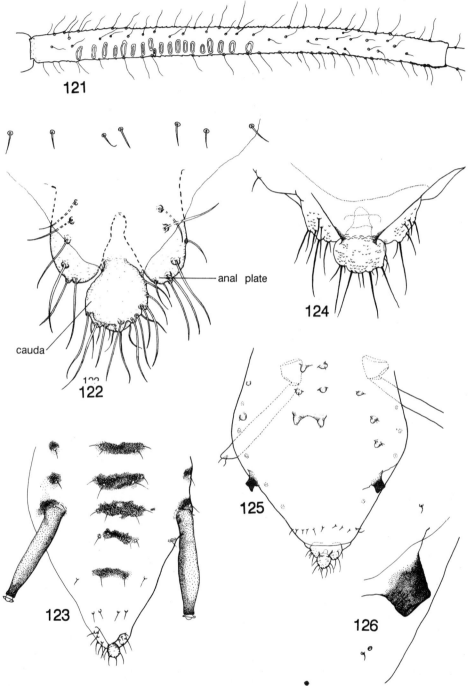

Figs 121–126. Callaphididae. 121, Symydobius oblongus aptera, third antennal segment, showing secondary rhinaria; 122, Eucallipterus tiliae alata, bilobed anal plate and knobbed cauda; 123, Drepanosiphum platanoidis, abdomen; 124–126, Tuberculoides annulatus: 124, enlargement of abdominal apex, showing bilobed anal plate and knobbed cauda; 125, abdomen; 126, enlargement of siphunculus.

may rarely produce a few larviform micropters. *Callaphis* and *Chromaphis*, both living on walnut, have unmodified legs. *Myzocallis* and *Tuberculatus* (including its subgenus *Tuberculoides*) have their anterior femora slightly enlarged. All the other genera have the anterior coxae slightly or, in the case of *Tinocallis*, greatly enlarged for leaping. *Eucallipterus tiliae*, a black and yellow aphid with the wing veins broadly and darkly pigmented, lives on lime (*Tilia*) and large colonies often draw attention to themselves by showering honeydew onto parked cars.

All 16 species of Phyllaphidinae found in Britain live on broadleaved trees, most of them on birch. *Drepanosiphum platanoidis*, a green aphid living on sycamore, belongs to the only British genus of the subfamily with jumping modifications, which involve the enlargement of the anterior femora. It is also the only genus of Callaphididae with long, slender siphunculi. Its honeydew causes the same problem as that of *E. tiliae*. *Phyllaphis fagi* lives on the undersides of beech leaves where it produces, unusually for a member of this family, large amounts of white, woolly wax. All morphs of *Drepanosiphum*, *Euceraphis*, *Clethrobius* and *Monaphis*, with the exception of the oviparae, are alate. Both alatae and apterae occur in the other genera and the males of *Symydobius*, *Betulaphis* and one of the three *Callipterinella* species are also apterous.

The subfamily Therioaphidinae is represented by four species of *Therioaphis*, all of them uncommon. Their anterior coxae are greatly enlarged for jumping. Males, so far as is known, are winged. In one species all virginoparae are alate; in the others both alatae and apterae are known. The host-plants are various herbaceous Leguminosae (clovers, birdsfoot trefoil, melilot, medick, restharrow).

There are ten species of Saltusaphidinae in the British Isles, belonging to three genera. Both alate and apterous virginoparae occur in all of these. Males and fundatrices are apterous. In *Iziphya* the first two pairs of femora and the bases of the corresponding tibiae are enlarged for jumping. The legs of the other two genera are unmodified. *Iziphya leegei* lives on rushes; the others live on sedges of the genus *Carex*.

The species of Callaphididae were keyed by Stroyan (1977). Blackman (1977) added a second species of *Euceraphis* and Polaszek & Cotman (1983) recorded an American species, *Appendiseta robiniae*, living on the introduced tree *Robinia pseudacacia*. Heie (1982) covered the Scandinavian species.

Aphididae

Two-thirds of the British Aphidoidea belong to this family. About 370 species are listed and new ones are added almost every year. The terminal process of the antenna, which is usually six-jointed but may have only five or four joints, is always more than half as long as the basal part of the last joint. All Aphidoidea with an elongate unknobbed cauda belong here but various other forms of cauda are also encountered in the family. Long siphunculi are frequent in Aphididae but also occur in *Drepanosiphum* (Callaphididae). The wings are never folded flat over the back. The subanal plate is always entire, never emarginate or cleft.

The smaller of the two British subfamilies, Pterocommatinae, contains only nine species, in two genera. They are autoecious on willow, sallow, poplar or aspen, often low down on the plants or even below ground level. They are frequently attended by ants. These are large (2.2–4.5 mm) and sometimes strikingly coloured aphids, with a short cauda (less than 1.1 times its basal width) bearing more than 20 hairs. (Aphidinae are never found on poplar or aspen and those members of the subfamily on willow or sallow have a longer cauda with fewer hairs.) Males are winged in some species of Pterocommatinae, wingless in others.

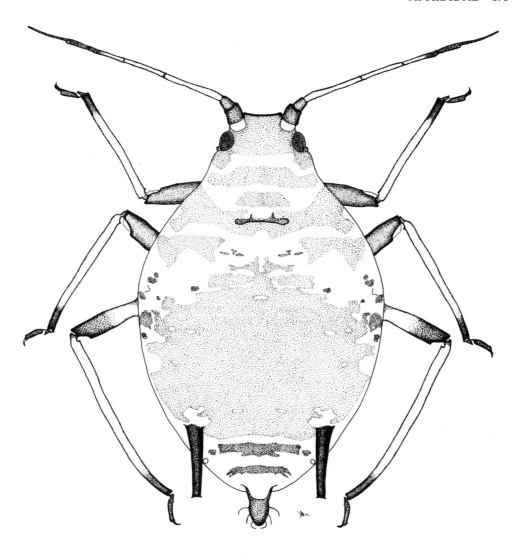

Fig 127. *Aphididae:* Aphis craccivora *aptera.*

Most Aphidinae are autoecious on herbaceous plants and not infrequently anholocyclic. Nevertheless, a heteroecious and holocyclic life cycle seems to be the original pattern for the subfamily. It differs from the heteroecious cycles of Anoeciidae and Pemphigidae in that the return migrants produce only oviparous females. Males are born on the secondary host and make their own return migration.

Aphis fabae, an often devastating pest ('blackfly') of broad bean, has this type of life cycle. Winter eggs, laid on spindle tree, hatch in the spring to produce a generation of apterous, parthenogenetic females, the fundatrices. A few further generations of apterous females follow while the host shrub is in a suitable condition but, as the nutrient content of the sap diminishes at the end of the flush of spring growth, alate females appear in the colonies and fly off to colonize broad bean plants or any other suitable host, including sugar beet, poppy and dock. The offspring of these migrant females are all apterous but

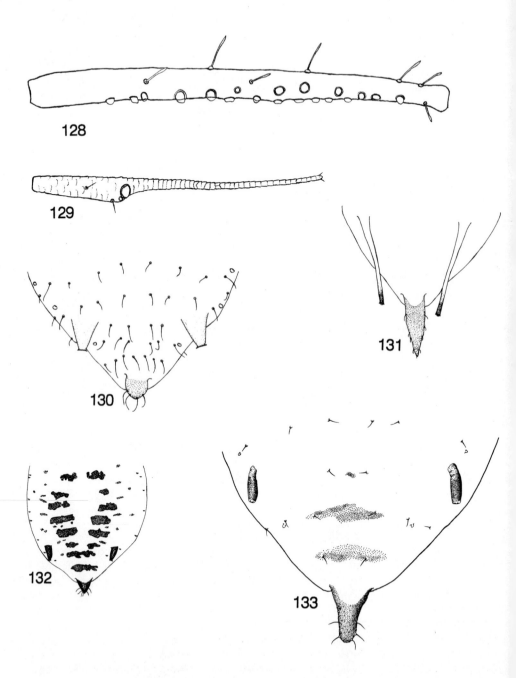

Figs 128–133. Aphididae. *128,* Macrosiphoniella sanborni *aptera, third antennal segment, showing secondary rhinaria; 129,* Aphis gossypii, *last (sixth) antennal segment; 130,* Brachycaudus helichrysi *aptera, abdominal apex; 131,* Acyrthosiphon pisum *aptera, abdominal apex; 132,* Brevicoryne brassicae *aptera, abdominal apex; 133,* Hyalopterus pruni, *abdominal apex.*

overcrowding and deterioration in the nutritional quality of the sap of the secondary host induce the production of an increasing proportion of alatae in the succeeding generations. These alatae fly off to settle on herbaceous plants in a more suitable condition. In the autumn, in response to the lengthening nights, apterae give birth to return migrant females and, later, to males. Both of these new winged forms migrate to spindle, where the autumnal sap flow is in progress. Here, the return migrant females ('gynoparae') give birth to apterous, oviparous females ('oviparae') that mate with the males. Each ovipara lays four to six eggs on the twigs of the spindle bush, where they remain throughout the winter, the new generation of fundatrices hatching in the spring.

Aphis fabae is unusual in that it has three different and distantly related primary (woody) host-plants: spindle, mock orange and guelder rose. It has been divided into four subspecies. *Aphis fabae fabae* typically has the hostplants mentioned in the account above but it can use many other herbaceous hosts in the summer. Another subspecies, *A. f. soldanella*, alternates between spindle and black nightshade or black bindweed (neither of which is accepted by subspecies *fabae*). A third subspecies, *A. f. cirsiiacanthoidis*, can use any of the three woody primary hosts and usually lives on thistles in the summer; it will not colonize broad bean or black nightshade. The fourth subspecies, *mordwilkoi*, can also use any of the three primary hosts but lives in the summer on burdock; it does not feed on thistles. This biological complexity, particularly the unusual circumstance of some of the subspecies being able to use three different primary hosts, has led some aphid workers to suspect that the *fabae* complex originated by hybridization of at least two parent species that had different primary hosts.

There is a group of species closely related to *A. fabae* that are autoecious, holocyclic and host-specific. Their males are apterous and their females have fewer scent glands on the hind tibiae than are normally encountered in their heteroecious relatives. Dock and, sometimes, rhubarb are the are the hosts of the monoecious *A. rumicis*, in which the males and the oviparae are the progeny of apterous sexuparae. *A. viburni*, another member of the *fabae* group, is autoecious on guelder rose, with a slight tendency to alternate to orchids in the summer. It can be persuaded experimentally to accept dock and spindle. Groups of closely related autoecious species centred, like these, on a heteroecious one and restricted to either its primary or secondary hosts, are frequent in Aphididae and the other host-alternating families of Aphidoidea. Another example is the genus *Dysaphis*, in which the cauda is of a characteristic 'helmet-shape'. Most species of this genus crumple the leaves of apple, pear, rowan, hawthorn and related trees. Some of these species are monoecious on the woody hosts but most migrate to roots and leaf-bases of various herbaceous plants in the summer and some live anholocyclically on the underground parts of herbaceous plants.

The most serious aphid pests of cereals (wheat, barley, rye and oat) in Britain belong to the genera *Rhopalosiphum*, *Metopolophium* and *Sitobion*. Each genus has one anholocyclic species living on cereal crops and grasses all the year round and at least one holocyclic species alternating between these and plants of the family Rosaceae (the primary hosts). The species with no sexual stages in their life cycles are *R. maidis*, *M. festucae* and *S. avenae*. *Rhopalosiphum padi* rolls the leaves of bird cherry in the spring before migrating onto cereals and the related *R. incertum* curls the young leaves of apple, hawthorn and rowan before migrating to the roots and tillers of oat and other grasses. *Metopolophium dirhodum* alternates between rose and cereals or grasses, usually annual ones, and *S. fragariae* alternates between bramble and grasses. There is a service, available to farmers, that gives advance warning of outbreaks of cereal aphids, based on information obtained by trapping flying alatae at a network of sampling stations.

In the garden, there are several kinds of aphids that are present in noticeable quantities almost every year. *Macrosiphum rosae* is often the most abundant of the several species of 'greenfly' on rose. It is green or pink, with characteristic black siphunculi, and migrates to scabious or teasel in the summer. The pale yellow *Cryptomyzus ribis* is responsible for the red 'blisters' that appear on red currant leaves in the spring. It spends the summer on woundwort. *Myzus cerasi* is a glossy, black aphid that crumples the leaves of cherry, where it may live all the year round, though it also migrates to speedwell and bedstraw in the summer months. *Myzus persicae* is a pale green or yellowish green aphid whose primary host is peach, on which it forms dense colonies. Almost any vegetable or herbaceous plant can serve as its secondary host, including cabbage, on which it often overwinters parthenogenetically in company with *Brevicoryne brassicae*. The latter is a waxy, greyish-green species confined to cabbage and related plants like swede and mustard. It produces sexual forms and winter eggs on these host plants but its asexual forms frequently overwinter as well.

Several aphids, in addition to *Aphis fabae* and *Myzus persicae*, have very wide host ranges and, consequently, occur on many garden plants. One such is *Brachycaudus helichrysi*. This yellow-green aphid is the commonest one found on plum, its primary host. In summer it lives on chrysanthemum, China aster and many other hosts, especially bedding annuals and rock-garden plants. *Macrosiphum euphorbiae* is a rather elongate, grey-green or pink aphid with long appendages and siphunculi. This originally North American species lives on potato and many other herbaceous plants. It is usually anholocyclic though its primary host is known to be rose. A relative, the large, grey-green *M. albifrons*, is a recent introduction that lives only on lupin, on which it can be a devastating pest. Waterlilies and many other kinds of aquatic plants are the summer hosts of *Rhopalosiphum nymphaeae*. This aphid has the sexual stages of its life-cycle on plum and sloe.

Aulacorthum circumflexum and *A. solani* are yellow-green aphids often found in greenhouses. The former species has a U-shaped black mark on the abdomen; the latter usually has a dark area at the base of each siphunculus. *Aulacorthum circumflexum* is the usual one found on the leaves of lily, crocus and other monocotyledonous plants forced under glass. Both species occur on many broad-leaved herbaceous plants and keep going parthenogenetically through the winter. *Aulacorthum solani* is also often found out of doors on potato and other plants. It is very unusual in being able to produce sexual forms and winter eggs on numerous different hosts. *Aulacorthum circumflexum* never produces sexual forms and is rare out of doors. *Dysaphis tulipae* is another wholly parthenogenetic species that is found on crocus, tulip and other monocotyledonous plants with bulbs or corms. Unlike *A. circumflexum* it does not feed on broad-leaved plants. Its off-white or fawn apterae overwinter on bulbs or corms in the ground or in store and move up onto the leaves and stems in the spring. Its alatae have extensive black markings on the body.

Many aphid species are attended by ants and for some of them this relationship appears to be obligatory.

The subfamily Aphidinae has traditionally been split into two or three tribes or sometimes into three separate subfamilies, in an effort to distribute the large number of genera and species among groups of more manageable size. The British members of the nominate tribe, Aphidini, together with the subfamily Pterocommatinae, have recently been the subject of a handbook by Stroyan (1984). The same groups were covered by Heie's (1986) Scandinavian work. Macrosiphini and the closely related (or synonymous) Myzini are not yet covered by comprehensive identification guides. The economically important species can be identified with the aid of Stroyan's (1952) guide. Heinze (1960, 1961), dealing with some Central European groups, is also helpful. Müller (1975) produced a guide to the identification of alatae trapped in yellow trays in Europe.

Anoeciidae

Nine species of *Anoecia* have been reported from Britain; its subgenera *Paranoecia* and *Neanoecia* are sometimes accorded full generic status. All species occur, for at least part of the year, on the roots of grasses or sedges, where most of them are vigorously attended by ants. Colonies of some species have been found on the roots of dicotyledonous hosts as well, but this is exceptional. *Anoecia corni* and *A. vagans* are dioecious, with dogwood or

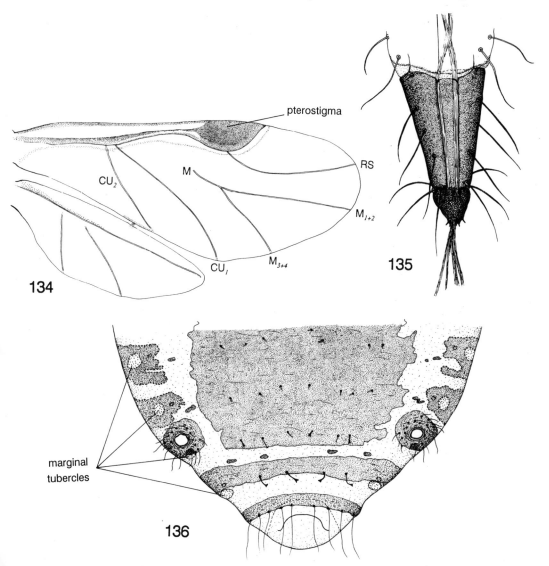

Figs 134–136. Anoeciidae. 134, *Anoecia corni, wings*; 135, *Anoecia vagans aptera, last segment of rostrum, showing 'beak'*; 136, *A. vagans aptera, abdominal apex, showing marginal tubercles, pore-like siphunculi and rounded cauda*.

other species of *Cornus* as their primary hosts. There are several apterous spring generations of *A. corni* on dogwood before alate migrants are produced but the fundatrices of *A. vagans* give birth to alatae immediately. Both species occur in apterous and alate forms on grass roots in the summer and ultimately a generation of return migrants arises. These fly back to dogwood, where they produce both males and sexually reproducing females. These sexuales are wingless dwarfs with the mouthparts normally developed. Each female lays two winter eggs. *Anoecia zirnitsi* and *A. (N.) krizusi* are holocyclic on the roots of grasses and *A. (P.) pskovica* is holocyclic on the roots and runners of sedges. In all of these monoecious species the sexuales, although apterous, are not dwarfs like their dioecious relatives. *Anoecia nemoralis* is usually anholocyclic on grass roots but there is some evidence that it has a partial autumn migration to dogwood. Little has been published on the life cycles of the other species.

Thelaxidae

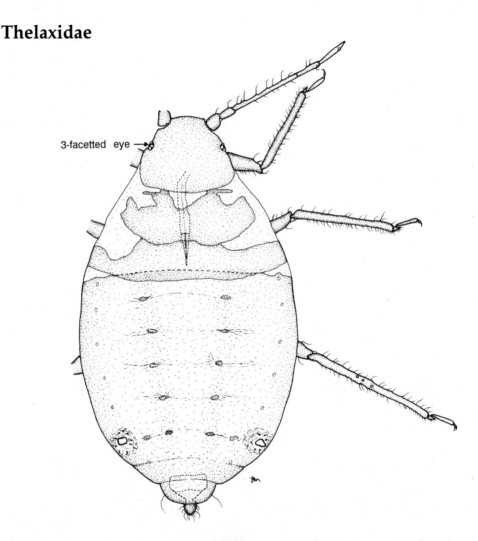

Fig 137. *Thelaxidae:* Thelaxes dryophila *aptera, showing 5-segmented antenna, knobbed cauda, entire anal plate and pore-like siphunculi.*

In this small family the wings of the alatae are folded flat over the back when in repose and the eyes of the apterae have only three ommatidia. The eyes of nymphs and of adult apterae appear to be situated in the middle of the sides of the head due to the fusion of the head with the pronotum. Apterae have well developed legs and antennae. The terminal process of the antenna is short and the siphunculi are pore-like. Nymphs and adult apterae have three to five antennal segments; alatae have five. The anal plate is entire.

There are four British species of Thelaxidae, in two genera. In *Glyphina* the cauda is broadly rounded, whereas it is knobbed in *Thelaxes*. *Glyphina betulae* lives on young shoots of silver birch and sometimes alder. It has a holocyclic life-history. Apart from some alate viviparae, produced in early or mid summer, all its morphs are wingless. Sexuales appear long before the autumn and most colonies have disappeared by September. Each oviparous female lays two eggs. *Glyphina schrankiana*, on various species of alder, has a similar life-history. According to Blackman (1989), who revised the European species of the genus, there are no authentic British records of this second species but a third species, which he named *G. pseudoschrankiana*, lives on downy birch in Britain and elsewhere.

Fundatrices of *Thelaxes dryophila*, on oak, hatch from the eggs in April and give rise to one or two generations of apterous viviparae. Alatae appear towards the end of May or in June. They give birth to small first-instar nymphs, which secrete a fringe of long, thin filaments of wax around the margins of the head and body. These nymphs shelter for some months, without developing further, in the forks of veins and other secluded places on the leaves. In September their development recommences and they mature into sexuales. Polaszek (1986) gave an account of the annual cycle of this species with descriptions and figures of the sexuales. Like the *Glyphina* species, *T. dryophila* is usually attended by ants. Another species of *Thelaxes*, *T. suberi*, has been found on Turkey oak in Britain.

Hormaphididae

This family shares a number of characteristics with Thelaxidae: nymphs and adult apterae have the head and pronotum fused and their eyes have only three ommatidia; there are never more than five antennal segments, even in alatae; the processus terminalis is short; the siphunculi are pore-like or absent; and the alatae hold their wings flat over the back in repose. In Hormaphididae the anal plate is bilobed beneath the knob-shaped cauda in both alatae and apterae. The latter have short legs and antennae, scarcely extending beyond the margins of the body, and might be mistaken for Coccoidea or the puparia of Aleyrodoidea.

Hamamelistes betulinus, living on birch, is the only species of the family living out of doors in Britain, where it has three generations a year and sexuales are unknown. First-instar larvae overwinter on the twigs of the host tree. They develop into apterous adults when the buds open. One or two summer generations then ensue on the undersides of the leaves, causing pale yellow 'blisters' to form in the lamina. The first summer generation is mainly apterous, perhaps wholly so in Britain, and a proportion of the second generation is alate. Alatae or apterae of either generation may give birth to the new overwintering generation. Apterae are largely dark brown, powdered with wax; they lack all trace of siphunculi and have three-segmented antennae. Alatae are brown, with pore-like siphunculi on the green abdomen, and have five-segmented antennae; their legs are much longer than those of the apterae.

Two species of *Cerataphis*, also belonging to this family, have been found in

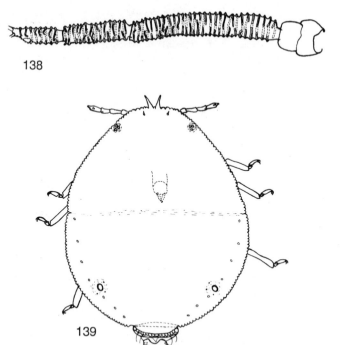

Figs 138–139. *Hormaphididae*: Cerataphis variabilis. *138*, aptera; *139*, alata, antenna, showing short terminal process and annular rhinaria.

greenhouses in Britain. Both are anholocyclic on orchids and other monocotyledonous plants and also on ferns. The better known of the two is *C. orchidearum*. Its apterae are black, almost circular, with very short legs, short, five-segmented antennae and a fringe of white wax-plates all round the crenulate margins of the body. They might be mistaken for puparia of whitefly but for the absence of the vasiform orifice. Alatae also occur; they resemble those of *Hamamelistes* except that the media is forked in *Cerataphis* and simple in *Hamamelistes*.

Mindaridae

The two British species of *Mindarus*, the only genus of this family, were treated by Heie (1980) and were also compared by Carter & Eastop (1973). They are holocyclic on conifers of the genera *Abies* and *Picea* and the life-cycle is similar in both. There are only three generations in the course of a year. The apterous fundatrices produce both winged and wingless viviparae which, in turn, give birth to dwarf, apterous sexuales with functional mouthparts. The wings of the alatae are held roof-like in repose and the fore wings are unique in Aphidoidea in that the last oblique vein (Rs) arises from much nearer the base than the apex of the pterostigma, which is prolonged along the costal margin of the wing almost to meet Rs again where the latter reaches the wing-margin. There are six antennal segments and the terminal process is very short. The cauda is subconical. The siphunculi are indistict and pore-like. The eyes of apterae and nymphs have only three ommatidia.

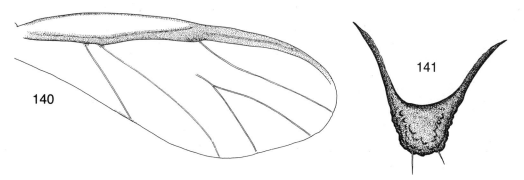

Figs 140–141. *Mindaridae:* Mindarus abietinus. *140, wings; 141, cauda.*

Apart from *Mindarus* and the members of the subfamily Cinarinae (Lachnidae), only two other aphidoids, *Elatobium abietinum* and *Illinoia morrisoni* (Aphididae) occur regularly on the above-ground parts of conifers in Britain.

Phloeomyzidae

This family is probably represented in Britain by only a single species, *Phloeomyzus passerinii*, which lives in fissures in the bark of young *Populus* species, including black, white and grey poplars but not aspen. White wax is produced from two groups of glands on the seventh abdominal tergite and, on the rare occasions when the insect reaches pest proportions, every crevice of the bark is filled with its flocculent masses. There are perhaps five to ten apterous generations in the course of the year. Alate viviparae probably do not occur but males are fully winged and so, uniquely for European Aphidoidea, are oviparous females. Oviparae are also unusual in that their hind tibiae lack scent glands. Each ovipara lays two overwintering eggs. All morphs have functional

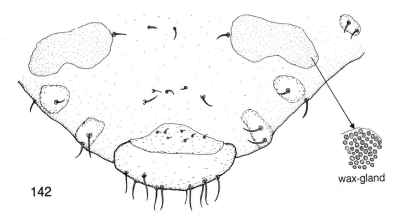

Fig 142. *Phloeomyzidae:* Phloeomyzus passerinii *aptera, abdominal apex, showing pair of glandular areas on seventh segment.*

mouthparts. The wings are folded flat over the back in repose. The forewing has four oblique veins arising posteriorly from the main longitudinal vein and the third of these (M) forks once. The cauda is broadly rounded and the siphunculi are pore-like. The antennae of the sexuales have six segments and those of the apterae viviparae have five or six. The terminal process is short. Secondary rhinaria are absent in all morphs, a condition unique among British Aphidoidea. The suture between the head and the pronotum is obliterated in nymphs and apterae, as in Thelaxidae. The eyes of nymphs have three ommatidia; there are three to five in adult apterae and more in alatae.

Pemphigidae

More than 40 species of Pemphigidae, distributed among three subfamilies, occur in Britain. The widespread production of wax from abdominal glands in this family has earnt some of its members the common name of 'woolly aphids', though many other Aphidoidea produce similar waxy secretions. Host-alternation is a basic feature of the family. The type of life-cycle resembles that of the Anoeciidae rather than the Aphididae, in that return migrant sexuparae give birth to both sexes of sexuales on the primary host. Pemphigidae are unique among Aphidoidea in having both males and oviparae that lack mouthparts throughout their lives (males of *Stomaphis*, in Lachnidae, also lack mouthparts). Because they cannot feed, the sexuales do not grow as they moult. The adults are dwarf and wingless and each female lays only a single egg, almost as long as herself. The cauda is short and the siphunculi are conical, pore-like or lacking. The eyes of apterae normally have only three ommatidia. The terminal process of the antenna is always shorter than the basal part of the last segment.

All but one of the nine species of Eriosomatinae recorded from Britain are holocyclic and heteroecious. The primary hosts are elms, on which they produce a variety of characteristic leaf-rolls or leaf-galls. *Eriosoma (Schizoneura) lanuginosum* deforms the leaves into closed, globular structures the size of walnuts. In the summer, its alatae migrate to the fine roots of apple, pear and quince. Other species spend the summer on the roots of grasses, sedges, labiates, currants or ragworts. Living in enclosed or confined spaces, the aphids produce copious amounts of woolly wax that coats the droplets of honeydew and protects them against being wetted by it. The subterranean generations may be attended by ants. *Eriosoma (E.) lanigerum* is the notorious woolly apple aphid or American blight. It lives anholocyclically on the roots, trunks and branches of apple, pear and *Cotoneaster*. Sexuales and even eggs are sometimes produced on apple but the only form that seems to be able to overwinter is the apterous virginopara. The active first-instar larvae are responsible for most of the dispersal of populations of this aphid both on individual trees and between the trees of an orchard.

More than 20 species of Pemphiginae are known from the British Isles. Most of them are holocyclic and heteroecious. The primary host of most species of *Pemphigus* and *Thecabius* is black poplar (including its more familiar variety, the Lombardy poplar). The fundatrices cause various characteristically shaped galls to develop on the leaves, petioles or, rarely, the shoots of their primary hosts. Apterae of *Pemphigus bursarius*, the lettuce root aphid, may overwinter anholocyclically on the roots of lettuce and other Compositae (Asteraceae) but usually the species produces at least some alate sexuparae that make a return migration to poplar in the autumn. The eggs of their progeny hatch in the spring and the fundatrices provoke the formation of globose galls on the petioles. Alate migrants leave the galls via a slit-like opening in late summer and disperse to the roots of Compositae. This kind of cycle is typical of the pemphigines with poplar as the primary host: their secondary hosts are almost always the subterranean parts of dicotyledonous

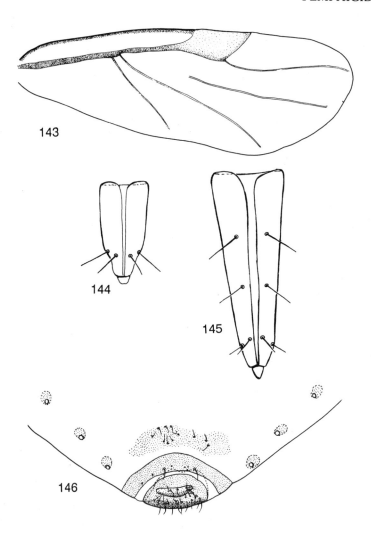

Figs 143–146. *Pemphigidae. 143,* Smynthurodes betae, *fore wing; 144,* Pemphigus bursarius, *last segment of rostrum; 145,* Eriosoma lanigerum *aptera, last segment of rostrum; 146,* Smynthurodes betae *aptera, abdominal apex, showing blunt cauda.*

herbs. *Pemphigus phenax* is sometimes a pest of carrots on which its white, woolly wax is very conspicuous. The spiral petiole galls of *Pemphigus spyrothecae* are often abundant on Lombardy poplar. Some of the daughters of the fundatrices of *P. spyrothecae* are apterous and live all their lives in the galls. They are born as 'soldiers', a special form of first-instar larva with thicker, more muscular hind legs than usual. They remain in this form for several weeks, during which time they will attack any intruder into their gall, piercing its body with their stylets and slashing at it with the claws of their strong hind legs. Eventually they moult into non-soldierlike second-instar larvae and develop normally thereafter. Their daughters and the later daughters of the fundatrices are all alate sexuparae, which migrate to other trees of the primary host species, so there is no host-alternation. The tidal dispersal of first-instar nymphs of *Pemphigus trehernei* is covered in

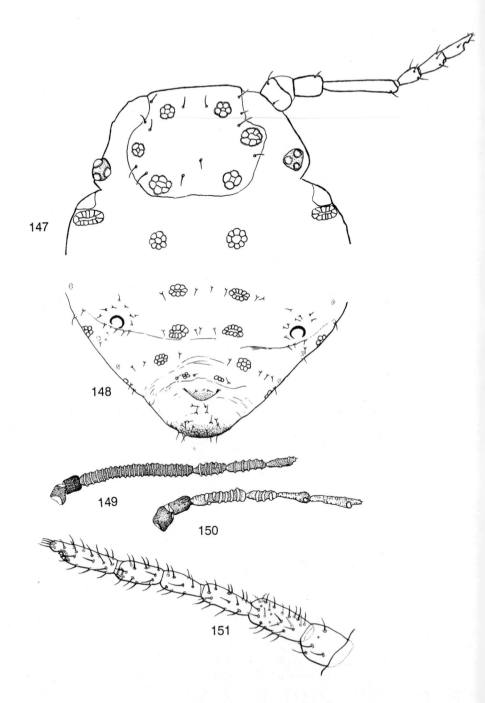

Figs 147–151. Pemphigidae. 147, Eriosoma lanigerum *aptera, head and anterior part of pronotum, showing wax-glands and 3-facetted eyes (triommatidia); 148,* E. lanigerum *aptera, abdominal apex, showing pore-like siphunculi, wax glands and rounded cauds; 149,* E. lanigerum *alata, antenna; 150,* Pemphigus bursarius *alate sexupara, antenna; 151,* Smynthurodes betae *aptera, antenna.*

chapter 8. Other genera of Pemphiginae have as their primary hosts aspen, lime (*Tilia*), ash, hawthorn, *Lonicera* and maple, on which they induce the formation of 'nests' of distorted leaves. Most migrate to the roots or sometimes to the twigs of conifers but *Mimeuria ulmiphila* alternates between the leaves of field maple, which is the primary host, and the roots of elm, where it produces mycorrhizal cysts.

Eleven species of Fordinae are known from Britain. Nine of them belong to a group of genera whose primary host is *Pistacia*, a shrub not found north of the Mediterranean area. Consequently they are anholocyclic in northern Europe. All are subterranean. *Smynthurodes betae* lives on the roots of dicotyledonous herbs but all of the others feed on grass roots. These aphids are all attended by ants, principally the subterranean *Lasius flavus*. *Melaphis rhois* is a tiny species, about 1 mm long, that is anholocyclic on mosses. In North America its primary hosts are various species of sumac (shrubs or small trees related to poison ivy in the genus *Rhus*).

The Scandinavian species of Pemphigidae were monographed by Heie (1980). Furk & Prior (1976) gave keys to the British species of *Pemphigus*, treating each morph separately. Aoki & Kurosu (1986) reported on the soldier morph in *P. spyrothecae*. The literature on aphid soldiers was reviewed by Itô (1989).

Coccoidea

Coccoidea are the most highly modified of all the plant bugs. In the majority of species the newly-hatched first-instar larva and the adult male are the most active stages of the life-cycle. Almost all adult females are immobile throughout the egg-laying part of their lives and those of several families are scarcely recognizable as insects at all (Figs 156, 157, 160–164). Their specializations mainly involve reduction or loss of parts normally present in other Hemiptera, and the development of various sorts of wax-glands. Many of them feed on phloem sap and produce copious amounts of sugary, liquid excreta (honeydew) and their waxy secretions protect them against self-contamination. The honeydew often attracts ants, which then tend and protect them as they do aphids. Some species depend wholly on ants to remove their honeydew and drown in it if ants are not in attendance. Others, though not wholly dependent upon ants, thrive better if they are present to clear ejected honeydew away from the colonies. Certain subterranean kinds are known only from ants' nests. While some species live in confined situations underground, in leaf litter, beneath the leaf-sheaths of grasses and sedges or in fissures of bark, others live exposed on the undersides of leaves or on smooth twigs. These latter kinds are often able to squirt honeydew away from themselves, though they may release it slowly in response to the attentions of ants.

In northern Europe, only the immature stages of Aleyrodoidea and wingless females of Hormaphididae are likely to be mistaken for scale insects, but the characteristic vasiform orifice of the former and the paired tarsal claws of the latter show that they are not Coccoidea. The tarsi of all coccoids, if they are present, invariably each bear a single claw in all stages of both sexes. All other Hemiptera have paired claws on at least one pair of legs or no claws at all.

Male coccoids are sometimes wingless but more usually have the fore wings sufficiently well developed to enable them to fly (Fig. 158). The hind wings are reduced to a pair of short straps, known as hamulohalteres, sometimes with 1–3 apical hooks coupling them with the basal angles of the fore wings in flight. They often resemble small Diptera. Males are incapable of sustained flight and do not travel far. Lacking mouthparts and gut, they cannot feed and do not live long. Some Coccoidea reproduce parthenogenetically and in these species or races males are rare or lacking.

Females are always wingless. They are slow-moving or immobile as adults, often lacking legs, and sometimes their antennae are reduced to minute tubercles. Protection is almost always provided for the eggs: in some groups it takes the form of an ovisac of felted threads or fluted plates of wax and may be carried about attached to the abdomen of the female if she is still mobile; in others the eggs are protected in a chamber within or beneath the toughened body of the female or inside a horny scale that she secretes about herself. Some margarodid females develop a pouch, called a marsupium, by invagination of the ventral wall of the abdomen.

The eggs may hatch within a few hours of being laid (ovovivipary) or after several days or weeks; in some species it has been suggested that they overwinter and hatch in the spring. Young first-instar larvae usually remain protected by the ovisac or scale for a maturation period of several hours or days before venturing into the outside world. They always have well-developed legs and antennae and they are the most active stage in the life cycle apart from the adult male; in some species the later larvae and adult females are completely immobile. For a few hours the little larvae, termed 'crawlers', wander about in search of a suitable place to settle and begin feeding. During their wanderings they may be blown about by the wind and so dispersed to new localities. Some are phoretic on flies, locusts and other insects.

In female coccoids there are two or three larval instars, the last of which moults to become the adult. In males there are two larval instars followed by two non-feeding stages termed prepupa and pupa. The second-instar male larva secretes a coccon of wax threads or a puparium of glassy plates or a scale in which the two subsequent stages are passed. The adult male frequently remains in this protective covering for some time, awaiting an opportune time to emerge. The prepupae of Margarodidae are mobile but those of other families, and all pupae, are not.

Apart from a few species that live inside the leaf-sheaths of grasses and some others with a strongly sclerotized dorsum, all larval and adult female coccoids are covered with a protective coating of wax. This wax may be powdery, as in Margarodidae, Pseudococcidae and some Eriococcidae; it may be in the form of a woolly cocoon, as in most Eriococcidae; or a number of large, separate plates, as in Ortheziidae; it may coalesce into a uniform, translucent layer (called a test) separate from the body, as in Asterolecaniidae; it may be uniformly distributed over the body as in most Coccidae; or it may be compounded with a secretion of the gut and often also with cast-off exuviae, to form a horny shell enclosing the body but separate from it, as in Diaspididae. Among the Coccidae there are some species with powdery wax and some heavily sclerotized ones with an inconspicuous wax covering. In addition to the general coating there are often glassy or cottony threads of wax projecting from the body and, in adult females, often a felted, woolly ovisac as well. All of this wax is secreted by certain specialized setae or from a variety of cuticular pores whose type, size and distribution provide important taxonomic characters at all levels.

Coccoidea or their products have been used in the manufacture of dyes, lacquers, medicines, cosmetics, ornaments, food ('manna') and candlewax. The secretions of the Asian *Laccifer lacca*, a member of the family Kerriidae, are the raw material from which shellac is made, and the bodies of *Dactylopius* species (Dactylopiidae) yield the red dye cochineal. Both margarodids and kermesiids have also been used to produce dyestuffs.

Identification of Coccoidea relies heavily on the characters of the young adult female, before her body has become distorted either by being modified into a protective covering for the eggs or by shrivelling after they have been laid. It is necessary to remove the wax and to macerate, clear and stain the insect to reveal the details of its structure well enough for accurate identification to be made. In the small British fauna, the external appearance of the mature female, or of the scale enclosing her body (in Diaspididae), is often sufficiently distinctive to permit identification at least to genus. Green (1927, 1928) provided a brief introduction to the British fauna. Newstead's (1901, 1903) monograph is still useful, particularly for its illustrations and biological data. Williams (1962, 1985) revised the British Pseudococcidae and Eriococcidae respectively. The only concise and comprehensive accounts (in English) of the European species are Kosztarab & Kozár's (1988) account of the Central European fauna and the translation (1967) of Danzig's (1964) annotated keys to the Coccoidea of the European USSR. Schmutterer's (1952a,b,c) careful biological studies in northern Germany include many species found in Britain and correct some errors of earlier workers. Despite modern plant quarantine procedures, coccoids from many parts of the world are introduced to the British Isles from time to time and some, including a number of pests, have become established under glass. Coccoidea found living under artificial conditions may belong to families not included in the key or to genera and species not covered by British monographs. MacGillivray's (1921) identification guide may be of use in identifying such species, but it should be used in conjunction with more modern literature, which is gathered into the briefly annotated bibliography of Morrison & Renk (1957) and its three supplements by Morrison & Morrison (1965), Russell, Kosztarab & Kosztarab (1974) and Kosztarab & Kosztarab (1988). Miller & Kosztarab (1979) reviewed progress in research up to that date, under the headings of Ecology, Control, Genetics, Endosymbionts, Sperm (the spermatozoa of Coccoidea are of a very unusual type) and Systematics.

Key to families of indigenous British Coccoidea – adult females

1. Body enclosed by a horny scale and easily separable from it. (Figs 162–164) .... DIASPIDIDAE
— Body with or without scale-like development of the dorsum but never separable from it 2

2. Posterior end of abdomen divided by a cleft, with anus situated at anterior end of cleft and covered by a pair of triangular anal plates (Fig. 157) or, if anal plates lacking (*Physokermes*), animal globular or kidney-shaped and living on conifers .... COCCIDAE
— Anal cleft and plates lacking; if body globular then not on conifers 3

3. Body lacking apparent segmentation, flat and disc-shaped or globular; without powdery, woolly or plate-like wax 4
— Body clearly segmented, usually at least one and a half times as long as wide and covered with one of the following: powdery wax, a woolly cocoon or a tile-like covering of wax plates 5

4. Body disc-shaped, covered with a translucent secretion and with radiating, glassy threads of wax (Figs 160, 161). (Usually in shallow pits on twigs of Oak; sometimes on low plants.) .... ASTEROLECANIIDAE
— Body globular, without wax threads (Fig. 156). (On Oak.) .... KERMESIDAE

5. Dorsum completely or largely covered with large plates of wax. (Fig. 152) .... ORTHEZIIDAE
— Waxy covering of body powdery or filamentous but never forming dorsal plates 6

6. Anus not surrounded by a bristle-bearing ring (Fig. 153). (British species on conifers or birch.) .... MARGARODIDAE
— Anus surrounded by a ring bearing six large bristles. (British species never on conifers and rarely on birch.) 7

7. Abdomen without ventral adhesive organs (Fig. 155) .... ERIOCOCCIDAE
— One or more abdominal segments often with a ventral, median, adhesive organ (circulus) (Fig. 156) .... PSEUDOCOCCIDAE

NOTE. The last three families are difficult to separate without making microscopic preparations and not all Pseudococcidae have circuli. The anal ring is more readily seen on a wax-free slide mount, where the abdominal spiracles of Margarodidae (lacking in the other two families) should also be visible. Eriococcidae and Pseudococcidae can be separated microscopically as follows:

1. Cuticle of dorsum with wide-bore tubular ducts with broad, cup-like inner ends; trilocular pores absent .... ERIOCOCCIDAE
— Cuticle of dorsum with narrow-bore tubular ducts lacking cup-like inner ends though sometimes with rings surrounding their outer apertures; trilocular pores usually present at least near coxae .... PSEUDOCOCCIDAE

Ortheziidae

The members of this family are the least modified of all Coccoidea. The female possesses abdominal spiracles and an anal ring bearing setae and pores. In this sex there are three larval instars before the adult and all stages are mobile, with legs and antennae that lengthen at each moult to maintain their size in proportion to that of the body. In both larvae and adult females the body is covered wholly or partly by several rows of large, white or grey plates of wax that give the insects a very characteristic appearance. In the adult female a group of these plates, secreted by the posterior abdominal segments, extend posteriorly beyond the body and coalesce to form a pouch (ovisac) for the reception of the eggs, of which there may be a few dozen to a few hundred, depending on the species. About a week after oviposition, the eggs hatch and the larvae force their way out of the anterior end of the ovisac. At each moult of the female larvae and at the first moult of the male, the skin to be shed splits all round the margin, separating into two halves. The dorsal half is usually carried around for a while until new wax-plates have developed on the initially naked new cuticle beneath.

Males are indistinguishable from females until the end of the second larval instar but then the paths of development of the two sexes diverge. The second-instar male larva constructs a loose cocoon of wax threads in which it undergoes the remaining stages of its development to adulthood. Moulting of the second and succeeding instars in this sex involves a splitting of the old cuticle anteriorly and an active sloughing of the exuviae from the anterior to the posterior end. Males live for up to five days if unmated, less if mated. Copulation may last for a few minutes or up to several hours.

Sikes (1928), in a detailed study of *Orthezia urticae*, found that larvae and adult females feed on the stems and undersides of leaves of various herbaceous plants but not of grasses. Overwintering larvae shelter among leaf-litter below the host-plants. Adults are present in the field from March to June. Mated females descend into the leaf litter but climb up the host-plants again when the crawlers are ready to emerge from the ovisac.

Males of *O. urticae* have colourless, glassy wings, compound eyes and a tuft of wax filaments arising from the end of the abdomen. The short hind wings are attached by three apical hooks to the fore wings. The body length of the male, 2.5 mm, is not much less than that of the adult female (2.8–3.0 mm long, 2.3–2.5 mm wide). In the population studied by Sikes, males were rare in comparison with females and she suspected that facultative parthenogenesis occurred.

Ortheziidae are unique among the outdoor coccoids studied by Schmutterer (1952*a,c*) in that they have a variable number of generations in a year, typically two to three, overlapping, and overwinter in almost any instar. There are half a dozen species in Britain. *Orthezia urticae* lives on a wide range of herbaceous plants and Harrison (1916*a,b*)

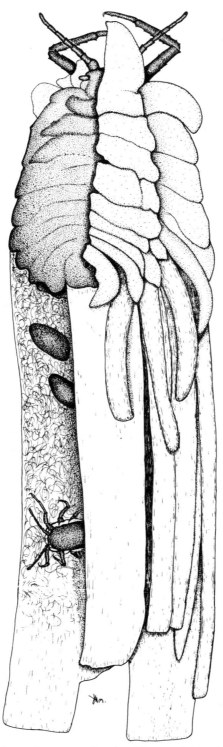

Fig 152. *Ortheziidae:* Orthezia *species, mature female with wax-plates removed from left side to reveal segmentation and to show eggs and first-instar larva (crawler) in brood-chamber.*

even found it abundantly on thrift and sea lavender in a salt marsh, where it must have been subjected to submersion by the tide. Other species are found in leaf litter and under moss, stones and logs. *Arctorthezia cataphracta* is sometimes abundant under stones in the high glens of Scotland. It has been found with its mouthparts inserted into the stipe of a basidiomycete fungus (Thorpe, 1968). *Newsteadia floccosa* has been found in association with various plants, including fungi. The mycorrhizal roots of bilberry are one of its favourite foods.

Margarodidae

Abdominal spiracles are present in all larval instars of Margarodidae and in adults of both sexes, a characteristic that they share only with Ortheziidae. They lack the distinctive wax plates of the latter, having only a powdery covering of wax like that of mealybugs. The spiracles are small and may be difficult to find even in slide-mounted specimens. The insects might be mistaken for mealybugs but for the absence of a setose anal ring. There are only two native British species, *Matsucoccus pini* and *Steingelia gorodetskia*.

Boratynski (1952) gave a detailed account of the life-cycle of *M. pini* in Britain. It reproduces parthenogenetically, males being completely unknown, and there are two generations a year. For most of the year the insect lives concealed beneath flakes of bark on the trunks and limbs of Scots pine. On hatching, the first-instar larva is narrowly oval and about 0.4 mm long. The antennae are six-segmented with segments 3 and 5 very short and 6 the longest. Legs are well developed. There are two thoracic and seven abdominal pairs of spiracles, all of which persist in the later instars and in adults. When fully fed the first-instar larva is plumply pear-shaped and about 0.7 mm long. This is the stage that overwinters. At the first moult the skin splits obliquely at the anterior end and is slowly forced off as the second-instar larva grows. This instar is very different in appearance from either the first instar or the adult. Legs are totally absent, the antennae are reduced to flat plates and the body is almost circular in outline, with only a trace of narrowing posteriorly. The stylets arise almost centrally in the body. The length increases in this instar from 0.8 to 1.6 mm. When the larva is fully fed its skin splits transversely along well defined lines at the posterior end of the body and the adult female emerges. She is 2–4 mm long and only about one-quarter to one-third as wide, with well developed legs and antennae but without functional mouthparts. A white, fluffy ovisac develops beneath her body and 30–100 eggs are laid into it. Adult life lasts for only four days, the first of which is spent wandering in search of a suitable oviposition site. Adult females of the first generation appear in May and those of the second in August.

Steingelia gorodetskia is less well known than *M. pini*. It is bisexual and univoltine on birch in damp places. The three stages of development of the female are similar to those of *M. pini* and adults of both sexes appear in May or June. Male larvae in the second instar have well developed legs and antennae and gather in groups, usually among leaf litter beneath the trees, to produce flocculent masses of cocoons. The adult male has seven ommatidia in a curved row down each side of the head. Its antennae are long and 10-segmented. It is much smaller than the female, being only about 1 mm long with a wingspan of 2.75 mm. There is a pair of caudal wax filaments about 2 mm long. In general appearance it is more like a pseudococcid male than a typical margarodid; most margarodid males have compound eyes like those of ortheziids and are not so small. The female is 4.5–6.0 mm long and about one-quarter to one-third as broad, with well developed limbs. She becomes enveloped in her fluffy ovisac, which is usually formed on the ground under the trees.

A few species have been introduced into Britain from warmer climates. The most

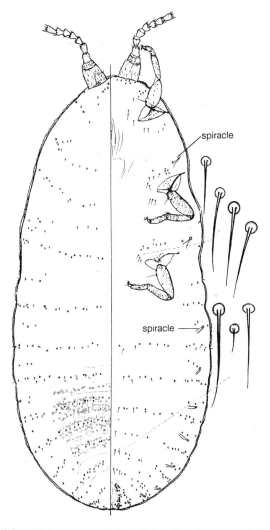

Fig 153. *Margarodidae:* Matsucoccus pini *adult female, microscope slide mount.*

important of these is the cottony cushion scale, *Icerya purchasi*, a notorious pest in the open in warmer countries and in heated greenhouses in cooler ones. If detected in Britain, its presence should be reported to the Ministry of Agriculture and measures to exterminate it should be undertaken. The adult female has functional mouthparts and produces a long, longitudinally fluted ovisac posteriorly. Her body is between 3.3 and 8.2 mm long and much broader in proportion than the native species. Abdominal spiracles are present in segments 6 to 8. The male has compound eyes and two short caudal wax filaments. His body is about 2.7 mm long and his wingspan is about twice as great as this. There are three larval instars in both sexes and all are active, with well developed legs and antennae. The short-lived male third-instar larva lacks mouthparts and constructs a loose, flocculent cocoon. Males are infrequent and the apparent females are in fact self-fertilizing hermaphrodites. This species attacks a wide variety of plants including *Citrus*, mango, guava and *Acacia*. At one time it was so abundant in orchards in

California, to which it had been accidentally introduced from overseas, that dense masses of the scales completely covered the orange trees and killed them by removing enormous quantities of sap. It looked as though the California *Citrus* industry faced extinction. The introduction of a predaceous ladybird, which rapidly brought the population of *Icerya* down to a tolerable level, has passed into history as one of the classic examples of biological control.

The second-instar larvae of some subterranean Margarodidae are globular and 2–4 mm in diameter. They enter a long resting phase (and have been termed cysts for this reason). These resting larvae are sometimes dug up and strung together to make necklaces of 'ground pearls'.

Pseudococcidae

Adult female mealybugs are distinctly segmented, covered in mealy wax and usually 2–5 mm long and about half as broad. Legs are present in all native and most introduced species and the insects are usually mobile at least until the ovisac is formed. The anus is surrounded by an anal ring, which bears several pairs of bristles, and there is no anal cleft. The cuticle bears numerous pores and tubular ducts. The ducts lack the cup-like internal ends characteristic of the Eriococcidae, though they may have ring-like expansions at their external apertures. Trilocular pores are usually present, as are specialized structures called 'cerarii' (pairs or clusters of lateral spines with associated groups of trilocular pores) that secrete short, waxy outgrowths or filaments that radiate from the body margins. Other characteristic but not universal features are one or two pairs of large, dorsal pores, called ostioles, whose possible functions are discussed by Williams (1978: 9–12), and up to four 'circuli' (ventromedian structures that enable the animal to adhere to the substrate). Superficially similar animals are Margarodidae and Dactylopiidae (the latter only on cacti), both of which families lack the setiferous anal ring, and Eriococcidae, which never have trilocular pores and whose tubular ducts have cup-like internal ends.

There are three larval instars in female pseudococcids and they are almost always mobile, albeit sluggishly. They have well developed legs and antennae. The female's ovisac is composed of waxy threads and may either envelop the animal or be carried posteriorly. In a few species, in which the eggs hatch almost as soon as they are laid, the ovisac is reduced to a few, loosely tangled filaments. Most mealybugs lay between 30–200 eggs but the biggest females of *Phenacoccus aceris* may produce several thousands.

The male second-instar larva constructs a cocoon of wax threads in which it undergoes the transition to prepupa, pupa and adult. In all moults of both sexes the old skin splits anteriorly and is sloughed off posteriorly, as in most Hemiptera. The male exuviae of the second instar larva and prepupa are thrust out of the cocoon. Male pseudococcids are usually about 1 mm long with one or two pairs of caudal wax filaments of comparable length to the body and some much shorter ones. The head is distinctly separate from the thorax and bears a pair of small ocelli at about the same level as the ten-segmented antennae. There are four larger, isolated ommatidia, one above and one below each ocellus. Adult males live for only a few days and may mate with several females in that time. In many species males are lacking and reproduction is parthenogenetic.

The outdoor species, in the main, have a single generation annually and overwinter as second- or third-instar larvae or, more rarely, as eggs in the ovisac. *Atrococcus cracens* is unusual in having two generations a year. Species living under glass generally have three or more generations in the course of a year, depending on the temperature.

In his revision of British mealybugs, Williams (1962) recognized 27 species found under

PSEUDOCOCCIDAE 213

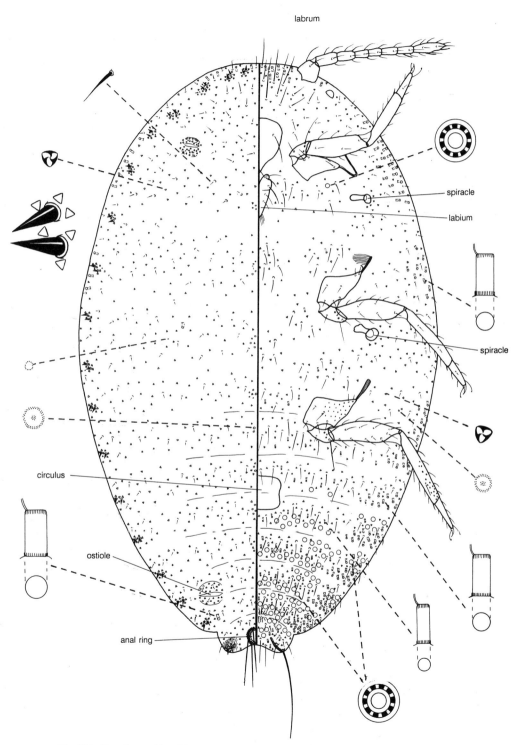

Fig 154. *Pseudococcidae:* Planococcus citri *adult female, microscope slide mount.*

natural conditions and a further 13 living only under glass. Two-thirds of the native species are associated with grasses, some living on the roots in association with ants, some on the stems and leaves and some, mainly species of *Trionymus*, inside the leafsheaths. *Trionymus newsteadi* lives on beech and *Phenacoccus aceris* is found on the trunks, branches and twigs of many trees and shrubs; it is common on gorse throughout the British Isles. No other native species is associated with woody plants. *Atrococcus* species live on various herbaceous plants but rarely on grasses. None of the native species is of much economic importance but several of the introduced ones can be serious pests in greenhouses. *Pseudococcus affinis* is the commonest of these and attacks a wide range of plants. Some species live on the roots or stems of cacti and *Vryburgia lounsburyi* can be troublesome on the roots of bulbous plants like *Crinum* and *Nerine*.

Eriococcidae

Members of this family closely resemble the pseudococcids and sometimes they, too, are referred to as mealybugs. Usually, though, they are called felted scales or felted coccids, a reference to the structure of the ovisac. Adult females are segmented, usually covered in mealy wax and generally shorter and proportionately fatter than pseudococcids. They have more or less numerous, wide-bore tubular ducts pores with cup-like internal ends and never have trilocular pores. There are never any dorsal ostioles or circuli.

Legs and antennae are well developed except in *Cryptococcus fagisuga*, which is less than 1 mm long and has short antennae, mere vestiges of the metathoracic legs and no trace at all of the other two pairs. Some authorities prefer to place this genus in a family of its own, Cryptococcidae. *Eriococcus* species are completely enclosed by the ovisac of densely felted wax threads at maturity with the exception of *E. spurius*, which is ovoviviparous and secretes an ovisac all round the sides of the body but is not enclosed by it dorsally. All species of this genus are mobile until the ovisac is formed. Colonies of *Cryptococcus fagisuga* cover themselves with copious fluffy wax threads and no separate ovisac is made. In *Pseudochermes fraxini* a covering of wax threads cemented with excreta is developed as early as the first instar and serves as the ovisac.

There are only two larval instars in female eriococcids, not three as in pseudococcids. Moulting in both sexes and development of the male proceed exactly as in Pseudococcidae. Adult males resemble those of pseudococcids except that the true ocelli are nearer to the dorsal ommatidia than to the ventral ones, instead of being equidistant from both, and the caudal setae are much shorter than the body. Male eriococcids are usually macropterous but in *Eriococcus spurius* both macropters and brachypters occur. In *Pseudochermes fraxini* they are minute, degenerate and wingless, while *Cryptococcus fagisuga* has no males at all and reproduces parthenogenetically. Females of most species lay a few dozen eggs, *C. fagisuga* averaging 19, but *E. spurius* lays several hundreds. Some *Eriococcus* species are thought to overwinter as eggs in the ovisac but it is known that *E. spurius* overwinters in the second instar, as do females of *P. fraxini*. Males of the latter species pass the winter as prepupae or pupae in the cocoon. *Cryptococcus fagisuga* overwinters in the first instar. Apart from *Eriococcus munroi*, which is bivoltine, those species whose biology is known have only one generation a year.

Most British *Eriococcus* species feed on grasses, usually on the leaves or stems, but some live underground on the rhizomes. *Eriococcus munroi* is generally found on dicotyledonous herbs, but produces its ovisacs in hollow grass stems or similar sheltered places; *E. devoniensis* lives on the shoots of *Erica tetralix* (cross-leaved heath), causing them to bend round in a circle with the insect on the inside curve; and *E. spuria* lives on the bark of elm branches, but is rarely seen except in the years when it stages a sudden outbreak. This last species is tended by ants for its honeydew, unlike the other British members of the family. *Pseudochermes fraxini* is often abundant on branches of ash; it has also been found on lilac and aspen. *Cryptococcus fagisuga* occurs in colonies on the trunks of beech. Trees vary in their susceptibility to attack and some cultivated clones of beech are immune (Wainhouse & Howell, 1983). It can be a serious pest in plantations as its presence facilitates infection by the fungal agent of beech bark disease, *Nectria coccinea*. A few eriococcids have been found on introduced plants but only one has been known to persist for any considerable length of time. This is *Ovaticoccus agavium*, established at Kew for many years on species of *Agave* and *Aloe*. Williams (1985) provided a key to the British species.

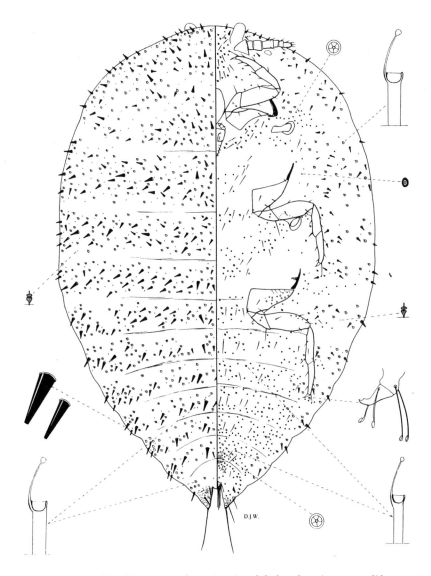

Fig 155. *Eriococcidae:* Eriococcus devoniensis *adult female, microscope slide mount.*

Kermesidae

The mature female of *Kermes*, the only British genus of this family, swells into a globular or heart-shaped form and the body wall hardens to form a two-celled chamber 3–5 mm in diameter. As the eggs are laid, the soft parts of the body shrivel up against the dorsal wall of this egg-chamber, which ultimately contains several hundred to a few thousand eggs. The first-instar larvae emerge via a single, ventral aperture and seek suitable sites in which to settle.

Once settled, each larva becomes enveloped in a matted mass of wax threads cemented with faeces. Second-instar larvae differ between the sexes: male larvae retain functional legs and well developed antennae but these appendages are greatly reduced in the female larvae which, once settled, never move again in their lives. Second-instar larvae remain enveloped throughout the winter. In spring, the fully fed male larva frees itself from its covering and, after wandering for a few hours, constructs a cocoon of loose threads in which it undergoes the remaining two moults, as in mealybugs. Female second-instar larvae moult in late April or early May to become adult. For about a fortnight the young females grow rapidly, excreting copious amounts of honeydew; they are attended by ants during this period. In all moults except the final one of the female the old cuticle splits anteriorly and is actively sloughed off rearwards. At the final female moult the old cuticle splits down the ventral midline and its two halves, hinged together, remain enclosing the adult dorsally until her growth forces them off. Males are macropterous and of the same general type as those of Pseudococcidae and Eriococcidae. The young adult female has an anal ring, as in these two families, but the ring lacks the long, paired bristles. She has stumpy but distinctly segmented antennae and short, one-segmented legs.

Three species of *Kermes* are known from Britain and all occur on oak. *Kermes quercus* lives in crevices of the bark and among crowded adventitious buds on the stems of young trees while *K. roboris* prefers small twigs, on which the mature females resemble little galls. Both species were described and illustrated by Balachowsky (1950). Sternlicht (1972) described a third species, *K. williamsi*, from Britain, allied to *K. quercus* and found in the same situations.

Fig 156. *Kermesidae:* Kermes quercus *mature female on Oak twig.*

Coccidae

One of the largest and most diverse families of the Coccoidea, this group has often been called Lecaniidae in the literature. Where a 'scale' is produced, it is the result of a thickening of the dorsal body-wall, not a separately secreted structure. A powdery or sometimes uniform and translucent waxy coating is usually present but some Coccidae produce a felted ovisac like that of Eriococcidae. In young adult females all abdominal segments are fused with the thorax and no segmentation is apparent. The anus is situated at the anterior end of an anal cleft separating two posterior lobes of the body; it is surrounded by a setiferous ring and overlain by a pair of large, usually triangular anal plates.

Males resemble those of pseudococcids and are usually fully winged. The mode of pupation is unique. The integument of the second-instar larva becomes fragmented into a few, large, glassy plates forming a puparium (Fig. 159) within which the transition to adulthood, via prepupa and pupa, occurs. The posterior plate of this puparium is hinged upwards by the emerging male, who comes out backwards.

There are usually only two larval instars in the female, the same number as in the male. In other families of Coccoidea it is usual for females have one more larval instar than males and some female Coccidae do, in fact, have three.

In the majority of species there are well developed legs and antennae and locomotion is possible in larvae and in adult females until they become grossly swollen with eggs. Some species, like *Pulvinaria ribesiae*, live as larvae on the leaves of deciduous plants and the mature larvae migrate onto the stems before leaf-fall. Pupae and adults of such species are never found on the leaves of their host plants. In *Physokermes*, the appendages of the

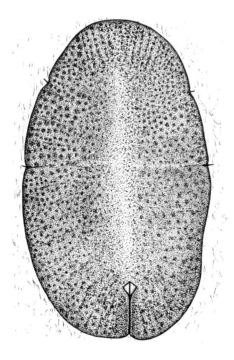

Fig 157. *Coccidae:* Coccus hesperidum *adult female.*

female larva become strongly reduced at the first moult and no locomotion occurs after this.

The type of protection afforded to the eggs varies in the different genera. In some, like *Eulecanium* and *Parthenolecanium*, the soft parts of the body shrivel up inside the thickened integument to make a chamber for the eggs. An extreme development of this procedure is seen in *Physokermes*, where up to 3000 eggs are accommodated in an inflated, two-chambered egg-chamber like that of *Kermes*. A second group of genera, including *Pulvinaria* and *Eriopeltis*, construct an ovisac of felted, waxy threads beneath or behind the body. In the ovoviviparous genera *Coccus* and *Eucalymnatus* a small brood-pouch is formed in the vicinity of the genital opening and the young larvae remain in this for a short period before emergence. Whatever the facilities for protecting them, total number of eggs produced by one female is usually several hundred to a few thousand, though a group of genera with the second type of provision for protecting the eggs (*Luzulaspis*, *Lecanopsis* and *Parafairmairia*) lay rather fewer, *Parafairmairia gracilis* not exceeding a hundred.

Parthenolecanium corni is known to growers of fruit and cobnuts as the brown scale. It lives on stems, shoots and leaves of a very wide range of host-plants including hazel, various currants, blackberry, plum, peach, ash, lime (*Tilia*), broom, grape vine and rose. The second-instar larvae overwinter and there are three larval instars in the female. Adults appear in April. There are both bisexual and parthenogenetic strains. Males, when present, are fully winged. Females vary in size depending on the species and condition of the host-plant; the largest ones may be more than 5 mm in diameter and lay 4000 eggs.

Pulvinaria vitis, the woolly vine scale, is a pest of grape vine, peach, currants, gooseberry and *Pyracantha*. It is common on birch and hawthorn and not infrequent on rowan, willows, alder, quince, spindle, poplars, apricot, sour cherry, blackthorn,

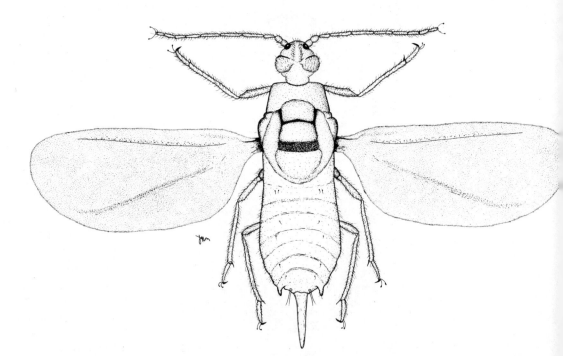

Fig 158. *Coccidae:* Pulvinaria *species male.*

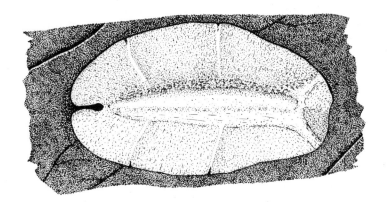

Fig 159. *Coccidae*: Lichtensia viburni, *male puparium*.

whitebeam and many other woody plants. Both sexually reproducing and parthenogenetic strains of the species exist and the young female, already fertilized if she belongs to a sexually-reproducing strain, is the usual overwintering stage. There is a single generation yearly, as in most British Coccidae, but a second generation may occur at higher temperatures. A conspicuous, white, flocculent ovisac develops in May or June and its development forces the mother scale to hinge forwards on her stylets. The eggs hatch in June or July and new adults are present by September or October. Males are about for only a short period and die soon after mating. Several kinds of woolly scales have been named according to the plants on which they live, for example, *P. ribesiae* on currants and *P. betulae* on birch, but in Britain there seems to be only a single, polymorphic species with many different hosts.

In recent years *Pulvinaria regalis* has appeared in Britain. It can form dense colonies, conspicuous when the white ovisacs are developed, on the trunks of horse chestnut, lime, plane, sycamore, various maples and other trees, usually in gardens or streets. Sweet bay (bay laurel) is a favoured host.

The soft scale, *Coccus hesperidum*, is oval, flat and 2–5 mm long when full grown. It has no ovisac and its waxy covering is translucent. It can be a serious pest of ornamental plants in houses and greenhouses and on evergreen shrubs, especially bay and myrtle, in the open in southern Britain. Its pale, yellow-brown, translucent body is very inconspicuous on the stems and the undersides of the leaves of its hostplants. Even heavy infestations may go unnoticed until the plant begins to look sickly or becomes sticky with honeydew and blackened by the associated fungi. Like most Coccidae, it may be visited by ants for the honeydew. There are several generations a year and males are rare.

In heated greenhouses, *Saissetia* species may be a problem on all sorts of plants. Males are rare or absent in this genus. Mature females are dark brown, almost hemispherical and about 3 mm across. Several species of *Ceroplastes*, too, are found under glass. Mature females of this genus are thickly covered with a homogeneous coating of wax, making them about as high as they are long.

There is a well-defined group of genera, including *Filippia* and *Eriopeltis*, whose British members, with one exception, live on grasses, sedges or broad-leaved rushes, usually low down on the plants. The exception is *Lichtensia viburni*, which lives on the leaves or young stems of ivy or, sometimes, of *Viburnum tinus* ('laurustinus'). *Lecanopsis formicarum*, living on the roots of grasses, is frequently attended by ants. Several species of this group are said to have three larval instars in the female sex. The third-instar larva of *L. formicarum*, which has much reduced legs and antennae, overwinters in a casing made from secretions from the gut. Most members of this group of genera are said to have only two larval instars in both sexes and to overwinter as eggs. The cottony ovisacs of *Eriopeltis* species are sometimes seen in conspicuous groups on grasses. Manawadu (1986) reported three species of this genus from Britain and provided keys to both sexes. Borchsenius (1957) gave a comprehensive account of the whole family in the USSR.

Asterolecaniidae

Asterolecaniids, or pit scales, have a number of features in common with diaspidids but the two families do not appear to be closely related. A glassy scale incorporating faecal material is secreted by glands similar to those in the integument of diaspidids but it is discarded at each moult, splitting anteriorly and being jettisoned backwards. It is not incorporated into the scale of the next instar. The soft parts of the adult female shrivel into the anterior part of the cavity beneath the scale, vacating a space that serves to protect the eggs, as in diaspids. The posterior segments are not united into a pygidium. There are no anal plates and no anal cleft. A setiferous anal ring is sometimes present.

Males, when present, resemble those of diaspidids, with little indication of a suture between head and prothorax and no caudal wax filaments.

There are two larval instars in both sexes, the legs and antennae becoming atrophied at the first moult. The last two moults of the male take place beneath the scale of the second instar. The most characteristic feature of larvae and adult females is the presence of unique cuticular glands which encircle the lateral margins of the body and also occur scattered on the dorsum. These glands are bilocular and are usually described as being 8-shaped. They secrete a fringe of radiating, glassy filaments all round the body.

Asterodiaspis (formerly *Asterolecanium*) is the nearest approach to a gall-forming coccoid to be found in Britain. All three species recognized by Boratynski (1961) occur, often together, on the twigs of oak. Each insect causes a shallow pit, surrounded by a raised rim, to develop around it. Adult females are virtually circular in outline and usually between 0.8 and 1.6 mm in diameter. Adult females overwinter, producing about 100 eggs in May or June. Adults of the new generation appear in August. All species are univoltine. Considerable damage can be caused by heavy infestations distorting the twigs and stunting their growth, but blue tits and other Paridae usually keep the insects in check by winter predation. *Planchonia arabidis*, the only other indigenous asterolecaniid, lives on dicotyledonous herbs. Its life-cycle is unknown. *Bambusaspis bambusae* has been recorded under artificial conditions on bamboos.

Diaspididae

This large family abounds in tropical species but has only a dozen native British ones, several of which may be troublesome in gardens and orchards. More than 40 additional species have occurred here under glass and some of these have become established as glasshouse pests. Unlike most Coccoidea, Diaspididae feed on parenchyma, not vascular tissues, and consequently do not produce honeydew. They are, therefore, never attended by ants.

All native species have a single generation yearly. There are two larval instars in both sexes. In second instar larvae and adult females the antennae are reduced to rudiments and legs are absent. Females and larvae have the posterior five abdominal segments fused into an unsegmented pygidium. A thin, ventral scale and a much thicker dorsal one are

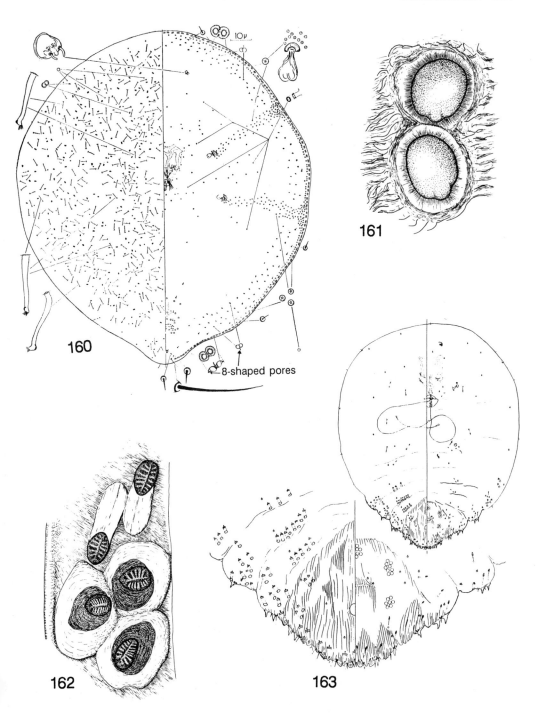

Figs 160–163. *Coccoidea.* 160–161, Asterolecaniidae: *Asterodiaspis quercicola*: 160, mature females in pit-galls on Oak twig; 161, mature female, microscope slide mount; 162–163, Diaspididae: *Carulaspis juniperi*: 162, scales of three mature females and two male puparia; 163, female, removed from scale, microscope slide mount with detail of pygidium.

constructed from secretions of cuticular glands and of the malpighian tubules. These scales are quite separate from the body of the insect, surrounding it like a loose shell. Ecdysis is usually accompanied by the separation of the dorsal and ventral scales laterally and posteriorly with the addition of new material to the free edges of the existing scale. The scale of the second-instar larva thus incorporates that of the first-instar larva and both are incorporated into the scale of the adult female. In Aspidiotini the dorsal and ventral scales separate completely and new material is added all round the edges of the old scale so that, in adult females of these roughly circular forms, the two or three elements of the scale are disposed concentrically.

No ovisac is secreted; the space beneath the scale of the mother provides protection for the eggs. In certain genera found in Britain only under artificial conditions (*Gymnaspis, Fiorinia, Lopholeucaspis*) the female remains within the second exuviae; although the cuticle of the second larval instar separates from that of the adult it is not shed and no further material is added to the scale. This kind of female is known as the 'pupillarial' type. The space between the female's body and the wall of the puparium serves as the brood pouch. The pupillarial type of female usually lays fewer than 50 eggs and the others rarely more than 100. Female scales have a horny texture and may be circular or elongate but male second-instar scales are usually whitish, waxy and oblong, with a dorsal keel.

The male prepupa and pupa develop within the second-instar scale. Their exuviae split anteriorly and are shed backwards. They are usually thrust out from the scale but in *Lepidosaphes* they accumulate around the posterior end of the body. Male diaspidids are, on average, the smallest of all coccoid males. They are usually winged but, even so, may not be capable of flight. The head is fused with the pronotum. The abdomen usually terminates in a long, median spike and wax filaments are absent.

The oystershell scales (*Quadraspidiotus* species) live on the branches of various trees. *Quadraspidiotus ostreaeformis* is sometimes found on fruit-trees but its original host is thought to be birch. The adult female is roughly circular, with the concentric scales typical of Aspidiotini, and about 2 mm in diameter. Adults appear in May or June and crawlers in July. Second-instar larvae overwinter.

Chionaspis salicis, the willow scale, is found on willows, Ash and various other trees, including ornamentals. Eggs overwinter beneath the dead scale of the mother and crawlers emerge in late spring or early summer. Males, unusually for Diaspididae, are not always macropterous; apters and brachypters appear a week or two before macropters. The oval scale of the female resembles that of *Quadraspidiotus* but the exuviae of the earlier instars are attached peripherally, not centrally, to the scales of the later stages.

Aulacaspis rosae lives on the stems of both wild and cultivated roses and may be a pest in nurseries, particularly under glass. The circular, white scale of the female, 2.0–2.5 mm in diameter, resembles that of *C. salicis*, with the brown exuviae of the larvae attached marginally or submarginally. The male puparium is white, oblong and tricarinate, with the exuviae of the first instar larva attached at the anterior end. The fully winged, red or orange males appear in May. They are fully winged and the abdomen is terminated by a median spike longer than itself. Such elongate genitalia are characteristic of all male diaspidids. Eggs are laid in August and crawlers appear in September. Females overwinter as first-instar larvae but males undergo the first moult before winter sets in. Development of the male scale to form a puparium is not completed until the spring.

Lepidosaphes ulmi, the mussel scale, is a pest of apple trees. Like the oystershell and willow scales, its dense colonies may completely cover the woody stems of the hosts. Adult females are covered by a row of three scales of increasing size, the composite structure being about 3.5 mm long and half as wide. Eggs overwinter beneath the dead mother scale and crawlers appear in May or June. Woody plants of many families are attacked and on most of them, including fruit trees, the species reproduces

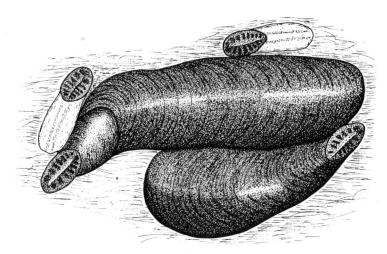

Fig 164. *Diaspididae:* Chionaspis salicis, *scales of two mature females and two male puparia.*

parthenogenetically. *Lepidosaphes ulmi* is often found on bilberry, where winged males are freely produced. The scale of the second-instar male resembles that of the female but is only about 1 mm long. This species is rarely found under glass.

Other species that are likely to attract attention are *Carulaspis* species on juniper, cypress and other Cupressaceae (Boratynski, 1957) and *Diaspidiotus bavaricus* on heather.

The German species of Diaspididae were covered by Schmutterer (1959). Balachowsky (1954) dealt with the entire Palaearctic fauna of Diaspidini. Rosen (1990) provided an introduction to the morphology, systematics and biology of the whole family.

Families occurring in Britain only under glass

Dactylopiidae (cochineal insects) superficially resemble pseudococcids in both sexes. *Dactylopius tomentosus* has been found in Britain on cacti of the genus *Opuntia*. The mature female is easily distinguished from the mealybugs by the lack of a setiferous anal ring. All stages, when damaged, exude body fluids stained dark red with a pigment which is the raw material of the the dye cochineal.

Aclerdidae differ from Coccidae in the possession of a single anal plate intead of a pair. *Aclerda tokionis* has been found in Britain on at least one occasion beneath the leafsheaths of bamboos under glass. The larvae are elongate and long-legged. Adult females are legless and their antennae are reduced to tubercles. They are more than twice as long as broad and lack obvious segmentation.

Conchaspididae owe their presence on the British list to the discovery of *Conchaspis anagraeci* in an orchid house in the nineteenth century. There are three female and two male larval instars in this family. The female has two pairs of thoracic spiracles and no abdominal ones. There is no tibiotarsal articulation in the prothoracic legs. The abdomen is clearly segmented and ends in a segmented pseudopygidium. The body is covered by a scale that does not incorporate the larval exuviae. Males of this family undergo the pupal metamorphosis in a felted puparium. The scale of the mature female of *C. anagraeci* is roughly circular, 2 mm in diameter and 0.5–1.0 mm high, white, and ornamented with a star-like pattern of ridges radiating from a central boss.

Halimococcidae are represented on the British checklist by *Colobopyga kewensis*, described from Kew, where it was found on a palm of the genus *Howeia*. The female remains within the scale of the last-instar larva, like the pupillarial forms of Diaspididae. This scale is 0.7–1.0 mm long and coated with thick, white, felted wax. The crawlers emerge from a posterior, ventral orifice in the puparium. Males, which are apterous, are freely produced.

17

TECHNIQUES FOR COLLECTING AND PRESERVING HEMIPTERA

The value of a specimen in a collection depends on its state of preservation and on the amount of information that is preserved with it. Conversely, the value of field or laboratory observations is enhanced if voucher specimens are preserved so that their identity can be checked should any subsequent doubt arise. The investigator should ask himself: how can I ensure that my results can be checked by others? The collector should ask himself not only: why am I collecting this material? but also: how can I make it of most value later on?

This latter question is best answered by using the collecting methods that cause the least damage to the material and the least disturbance to the habitat and by recording as much relevant information as possible about the circumstances of capture. There is still much to be found out about the habitat, host-plant and prey requirements of the British Hemiptera, especially Heteroptera and Auchenorrhyncha. Thrashing about with a net in mixed vegetation, although often enormously productive of specimens, stirs up clouds of insects that settle haphazardly on any available plant, whether or not they would normally be found on it. Such a procedure may be useful in alerting the biologist to the presence of an interesting species, but yields no further information than locality, date, time and the general type of habitat. Claridge & Wilson (1976, 1981), working on tree-dwelling typhlocybine leafhoppers, found that the only satisfactory method of collecting them with dependable host-plant data was not to catch the readily disturbed and highly mobile adults but to pick leaves on which the less mobile nymphs were feeding and to rear them through. The relative (and at times total) immobility of Sternorrhyncha is undoubtedly the reason why their host plant preferences are much better known than those of the larger and more visually attractive Auchenorrhyncha and Heteroptera.

Collecting techniques

The most generally useful piece of collecting apparatus is a short-handled net of about 40 cm diameter with a bag of strong, white cloth. The bag should be somewhat deeper than it is wide, to impede the escape of the more mobile quarry. For a quick preliminary survey, it can be swept back and forth among herbaceous vegetation in such a way that the rim dislodges insects from the plants so that they fall into the bag. This kind of use rapidly wears out the bag where it is attached to the rim, so it is advisable to reinforce it in this region by sewing on a strip of robust tape. Some users prefer a triangular or D-shaped rim to a circular one for use in very short vegetation such as grazed turf, using it with the straight side close to the ground. The net can be used selectively on suitable vegetation by slipping it over a whole plant or a branch of a tree or shrub and vigorously shaking the latter to dislodge any insects on it. When using the net in this way on a woody plant, the insects can be shaken off by dealing the branch a sharp blow from a stout stick close to the point where it enters the net. Flat beating trays are usually more trouble than they are worth because they are difficult to use in even slightly windy weather and do not hinder the escape of active insects like mirids and cicadellids. They do, however, make it possible to work large amounts of overhanging branches on reasonably still and cool days when the insects are least likely either to be blown off the tray or to take to the wing.

A major advance in sampling the arthropod fauna of trees whose foliage cannot be worked from the ground is the canopy-fogging technique. This procedure involves blowing a fog of insecticide into the crown of the tree and collecting the insects that fall down on sheets or trays spread out beneath. Fogging yields huge amounts of material and it is usual to cover only a small part of the area beneath the tree with collecting trays in order to avoid having to deal with large numbers of insects. The insecticide used must be one of very low toxicity to vertebrates because of the danger to insectivores that feed on the unwanted part of the catch. It should also be of such a kind and concentration as to induce uncoordinated activity rather than immediate death, so that the insects do not fall motionless onto leaves and branches instead of struggling and dropping onto the collecting trays. Pyrethroids are the preferred insecticides on both counts. Stork & Hammond (in prep.) give an account of a programme of fogging and the techniques used in an oak plantation in southern England.

A very effective mechanised method of collecting from the herb layer and the upper litter layers in grassland is vacuum sampling. The Dietrick vacuum sampler is one of the better models. It extracts a very high percentage of the fauna of a circumscribed area and collects it into a bag without passing it through a hose. The catch, even of very small and delicate organisms, is recovered alive and in good condition. Various types of vacuum sampler, and many other collecting techniqes, are outlined by Southwood (1966).

Of the various kinds of stationary traps, with or without baits and attractants, only coloured pan traps are much used by hemipterists. They consist of plastic trays of water containing a few drops of detergent. Certain colours, principally yellow (not green), attract flying aphids and some other insects. The insects attempt to settle on the yellow surface and drown in the water. Obviously, such traps only work in the daytime and must be regularly inspected. They can only be used in places where disturbance by humans and other animals is minimal. Light traps produce disappointingly small quantities even of those kinds of Hemiptera that are known to fly at night. To judge from the large numbers of Heteroptera that are caught at light in the tropics, British conditions are rarely such as to encourage the native species to fly. Pitfall traps, much used by coleopterists, and Malaise (flight interception) traps beloved of hymenopterists produce a few bugs but not in sufficient numbers to make it worth while using them specifically for this group. Powerful suction traps that continuously sample large volumes of air and direct airborne insects into a collecting jar of alcohol are too cumbersome and expensive for the casual user but are used professionally to sample flying aphids.

Berlese or Tullgren funnels, used for extracting soil microfauna, operate on samples too small to be useful for most Hemiptera but a scaled-up version can be used for collecting such families as Ortheziidae and Lygaeidae from large samples of leaf litter. A garden sieve and a plastic dustbin of similar diameter are the basic components. A cone of plastic sheeting forms the funnel. Its wider end is wedged between the rim of the bin and the sieve and its narrower end leads down towards a jamjar with a centimetre or so of preserving fluid in it, standing on the bottom of the bin. About a kilogram of litter can be spread on the sieve and a 100 or 150 watt light bulb or even a small electric fire is suspended above the surface of the litter. The larger, more active organisms are driven downwards and accumulate in the jar. The smaller and more sluggish ones, for the most part, dry up and die in the litter. Litter can, of course, be sorted by hand on a white sheet or tray or sieved onto such a surface. Sieving tends to damage the appendages of Heteroptera but, like hand-sorting, it does have the advantage of giving the collector the option of preserving his material dry or even of keeping it alive for further study or rearing immatures. Attempts to collect live material from funnels into dry containers are often thwarted by web-spinning spiders and other predators.

The most effective way of collecting specimens with useful data is simply to look for them in their natural habitats without using any mass-collecting or extraction methods at all. It is the only method normally used by collectors of most kinds of Sternorrhyncha and has much to recommend it to people interested in all terrestrial Hemiptera but most especially those associated with plants. The collector will soon become acquainted with the subtle or not-so-subtle effects of feeding on plants. A slight tattering or peppering with black dots or holes of the young terminal leaves of a shoot, for example, indicates that a mirid has been feeding there and may still be present among the leaves or close by. White mottling of the upper surface of leaves betrays the presence below of mesophyll-feeding Hemiptera, usually typhlocybine leafhoppers. Bunched or distorted leaves or inflorescences on many different kinds of plants may indicate the presence of aphids or psyllids and some of these distortions take a more precisely defined form characteristic of a particular species.

Many bugs, however, produce no obvious symptoms in their hosts and must be sought on apparently healthy plants. A whole miniature world awaits the naturalist who is prepared to spend even a few minutes gently examining the undersides of leaves on the lower branches of trees and many of the inhabitants of this world are Hemiptera of one kind or another. Aphids and their natural enemies predominate but psyllids, whiteflies, coccids, leafhoppers and certain Heteroptera, as well as the ubiquitous Psocoptera, will be encountered sooner or later. The seeker after coccids, in particular, would do well also to examine twigs and young stems for their specialised fauna. Tree trunks, especially when covered with moss and lichens, harbour a fauna of their own in which certain Microphysidae and Miridae play a role as predators. The sharp and practised eye is the most appropriate means of detecting their presence as they rest there or slowly search for prey. Buds, flowers and fruits of various kinds are exploited by a wide range of Hemiptera as they are sites towards which nutrients are channeled by the plant. Phloem-sucking insects are especially attracted to them and to the flower- and fruit-stalks along which the nutrients are carried.

Qualitative samples of insects living on or close to the ground surface are best obtained by the unsophisticated technique of 'grubbing'. The collector kneels or lies on the ground and parts the vegetation by hand, searching for insects that live beneath the plants. A very large and varied invertebrate fauna is revealed by this method. Vole runs in grassland and the ground beneath mats of vegetation beside paths are particularly productive of interesting species and so are patches of bare ground, such as scrapes made by rabbits, on sandy and chalky soils. A number of species that are close to the limits of their ranges in Britain are able to survive only in dry, sparsely vegetated areas where the sun warms the soil, like sand dunes and cinder tips, and grubbing is the only practicable method of collecting in such habitats.

Pond nets for aquatic insects are too well known to need any description or comment apart from the observation that a simple bag on a stick is quite sufficient without any elaborate plankton jar or other attachment. Various kinds of trawls and dredges are available from biological suppliers and may be of use in specialist surveys. A lightweight plastic tray is useful for sorting the contents of the net if a lot of weed is brought up.

Handling the catch

Material is best kept alive and dry until the time comes for it to be preserved. If a large enough supply of glass specimen tubes with plastic tops can be carried, each specimen can be kept separately, avoiding the risk of mutual interference and damage. The tubes, particularly if they contain pieces of plant material, are apt to 'sweat' in hot weather and the imprisoned insects may be wetted by the condensation and damaged by it. This problem can be avoided by keeping the tubes in the shade and by lining each one with a piece of tissue paper. Alternatively, they can be stoppered with cotton wool, which allows sufficient ventilation to prevent the development of condensation. It is particularly important not to let Miridae and typhlocybine Cicadellidae get wet as they are delicate and easily damaged. Aquatic bugs should never be transported in tubes of water. Adults can survive perfectly well in dry tubes for hours or even days. Nymphs are more difficult to transport. They may survive if accompanied by a piece of damp (not wet) tissue paper in a stoppered tube, but not for long.

Sternorrhyncha are best collected into tubes of preservative fluid or on pieces of their host plants in polythene bags. Samples of host plant material required for identification should also be collected into polythene bags and pressed between newspapers later.

Some mass collecting methods automatically involve drowning the catch in alcohol or other preservative fluids. When samples in fluid have to be transported, they should be decanted into small tubes which are then completely filled with liquid before being stoppered. If the preservative is in short supply, the material can be wedged in the bottom of the tubes with wads of cotton wool and the surplus fluid drained out before stoppering. Small amounts of fluid slopping about in large containers should be avoided as they can do considerable damage to the specimens.

When collecting large numbers of small specimens it may be convenient to use a 'pooter' or aspirator. The type of pooter in which the catch is sucked into a specimen tube (usually 3" × 1") is ideal, as the collecting chamber can easily be changed for another when it becomes too crowded for any more material to be admitted safely. With large numbers of specimens in a tube, condensation is an even greater problem than with single specimens and a tissue-paper lining is advisable both for the purpose of soaking it up and to give the insects a foothold so that they do not get damaged by being jumbled together.

Dry material, if it is not to be kept alive, should be killed once it has been brought in from the field by introducing into the collecting tube a slip of paper on which has been pipetted a single drop of ethyl acetate (amyl acetate and diethyl ether are possible alternatives). If time permits, the material should be mounted straight away. If not, it can be layered. Layering requires a cardboard box with a slip-on lid, some layers of cellulose wadding and soft paper and a preventative against mould. The anti-mould agent (crystals of thymol are preferred) is sprinkled in the bottom of the box and covered with a layer of wadding. The insects are placed between sheets of paper, on which the relevant collecting data are recorded, alternating with layers of wadding. If the specimens are to be kept in a liquid preservative it is advisable to kill them with ethyl acetate first. This is particularly important with Miridae, which shed their legs if dropped into alcohol while still alive.

Dry preservation

It is best to mount specimens as soon as they are dead and before rigor mortis sets in. If this cannot be done then either the material should be kept in a relaxing box overnight for attention the following day or it should be stored dry and relaxed for a day or so when time is available to handle it. Relaxing boxes or jars are simply high-humidity chambers with an antimicrobial agent incorporated so that the specimens take up water from the damp atmosphere without rotting. Traditionally, a laurel jar is used for this purpose. An inch or so of finely chopped young leaves of cherry laurel (*Prunus laurocerasus*) is placed at the bottom of the jar and covered with a layer or two of paper, on which the insects to be relaxed are placed. A tight-fitting stopper for the jar is needed to keep up the humidity. The cyanide vapour given off by the leaves, which is the anti-rotting agent, is strong enough for a new laurel jar to be used as a killing bottle, but condensation problems can be a nuisance. A good laurel jar will last a year or more but will need a few drops of water to be added from time to time. A more mundane relaxing chamber can be made from a plastic lunch box with a few crystals of thymol (anti-mould and also slightly toxic to entomologists) sprinkled in the bottom and covered with a layer of wadding that can be wetted with water as required.

All dry preservation methods use pins and these should be of stainless steel if at all possible. Nickel-plated or lacquered brass pins corrode in contact with the body contents of insects and also react with the glue or paste used to stick paper linings in drawers and boxes as well as with some fumigants. Even where the pin does not pass through the insect, in double mounts, body fats 'sweat' out of the body and soak into card or polyporus, eventually reaching the main pin, unless very thorough measures are taken to de-grease the specimen.

Large specimens can be pinned directly, on long pins. Because many Heteroptera are apt to break in two at the weak junction between the pro- and mesothorax, care should be taken to see that the pin enters through the posterior lobe of the pronotum and leaves via the mesosternum, holding the two parts together. The pin should be inserted slightly to one side of the midline to avoid damaging median structures. The specimen should be placed high enough on the pin to allow space beneath for labels and for the pin to be gripped by pinning forceps.

Small specimens may be mounted in any of three ways: double-pinning (staging), card pointing and carding. Double-pinning requires the specimen to be pinned directly with a fine, stainless steel micro-pin which is then stuck into a stage – a strip of polyporus or polyethylene foam – which is itself mounted on a full-sized pin that bears the labels.

Card points are narrow triangles of card, or rectangles cut to a point at one end, that are mounted on full-sized pins. The extreme tip of a point is dipped in a very small amount of strong, water-souble bone- or fish-glue and applied to the side of the thorax of the specimen. Tiny amounts of glue dry very quickly and this method is no more time-consuming than micro-pinning.

Carding involves sticking the specimen on a flat piece of card with the legs and, where appropriate, the antennae spread out and stuck down with the adhesive. Water-soluble glue or gum should be used so that the insect can be floated off again in water if it is necessary to examine its underside or to dissect it. Gum tragacanth or gum arabic are preferable to glues as they dry almost matt, do not cause the card to curl upwards so badly as they dry and do not discolour or crack with age. As with other kinds of dry mounts, the pins of carded specimens should be long enough to allow space for labels and forceps beneath the card.

The body fat in dry-mounted specimens tends to 'sweat' onto the surface, making the specimens greasy and causing dust to adhere. It is advisable to de-grease dry-mounted material once it is air-dry by soaking it for a few days in a fat solvent such as xylene, benzene, toluene, petroleum ether, petrol or carbon tetrachloride. All of these substances are dangerously flammable or poisonous or both and should be handled with great care and common sense. They should never be used in open dishes or near naked flames or possible sources of sparks. Some perfectionists like to card all their specimens, even large ones, mounting them first on rough pieces of card and then, after drying and de-greasing, transferring them to clean, white Bristol board with a minimum of adhesive. De-greasing may slightly alter the appearance of some waxy insects, notably Tingidae, as many surface waxes are soluble in the de-greasing chemicals. Generally, though, a de-greased specimen retains a nearer resemblance to its living appearance than a greasy one.

Freeze-drying in a vacuum chamber preserves colours, particularly greens, much better than the slower process of air-drying. Some institutions have begun to use the newer technique of critical-point drying (Gordh & Hall, 1979). Both processes require expensive equipment and leave the specimens very brittle.

Genitalia preparations

Male genitalia provide important diagnostic characters in many families of bugs. It may save time later if the genital capsules of males of critical groups are removed at the time of mounting and glued to the cards on which the insects are mounted or stored in gelatine capsules or microvials attached to the pins. It is very important, of course, to keep the genitalia associated with the rest of the specimen. If the genitalia were not removed at the time of mounting, the specimen must usually be relaxed before this operation can be performed. It is sometimes possible to knock the whole abdomen off a dried specimen without relaxing it but often a leg or two, or even more, will come away at the same time. The specimens to be dissected should be left overnight in a relaxing box or, if time is short, the material may be 'steamed' over boiling water or even immersed in hot water for a few minutes. The genital capsule can be removed from the relaxed specimen with fine ('watchmakers') forceps and needles (micropins pushed through matchsticks or inserted into the melted ends of glass rods or tubes). The same instruments can be used to tease the constituent parts of the genitalia apart for examination.

In many cases, satisfactory dissections can be done under water or 20 per cent alcohol in a watch glass or other small dish, with a wetting agent added if necessary, without further treatment. For the best results, the genitalia or the whole abdomen should be macerated by immersion in a ten per cent solution of potassium hydroxide (sodium hydroxide will do) to remove the soft tissues. This reagent is a caustic alkali and great care should be taken to avoid spillage or splashes onto face, hands and clothing. Any such splashes should immediately be mopped up or washed off with copious amounts of water. The alkali may be used cold, in which case the process of maceration takes several hours, or hot, when it takes only a few minutes. Hot alkali, of course, is even more dangerous than cold. The alkali should never be heated directly as it may boil, with disastrous results. A water-bath can be used, the material to be treated being placed in a small amount of the solution in a tube which is stood in a beaker of boiling water. When maceration is complete, the specimen assumes a characteristic, transparent, amber colour. It is transferred to a dish of clean water and the macerating solution is safely disposed of (down the sink, with plenty of water) before the specimen is attended to. Some workers like to neutralise the alkali at this stage, adding a drop or two of acetic acid to the dish but, if enough water is used, dilution is usually sufficient to prevent further softening. The macerated tissues can be expelled from the specimen by repeated gentle pressure and any dismemberment necessary to expose the diagnostic features can then be done. The dissected material can then be cleared by transferring it for a few minutes to a dish of glycerine before returning it to clean water for examination. It is rarely necessary to use any stains unless maceration has been overdone. If staining is necessary, lignin pink and chlorazol black are useful general stains for chitinous structures.

The aedeagus in some Heteroptera is an eversible sac and only careful manipulation, practice, experience and patience or, occasionally, luck and osmosis, can be used to evert it. Osmotic eversion, if it is going to happen at all, should begin as the specimen is transferred from the macerating solution to water. It is as well to watch the aedeagus for half a minute or so, refraining from any manipulative activity, to give this process a chance to start.

Dissections may be stored dry, particularly if maceration has not been necessary. The various pieces can be glued to the same card as the specimen or a separate card on the same pin. Otherwise, they can be placed on a card or a strip of plastic in a drop of warm glycerine jelly, which hardens as it cools and dries and can easily be dissolved in warm water at any time. Alternatively, they can be stored in a drop of liquid glycerine in a microvial attached to the specimen by being pinned through the stopper. Glass vials are expensive and may be difficult to obtain. Van Doesburg (1980) describes a method of making plastic ones. Now that stereoscopic microscopes are commonplace it is advantageous not to mount complex, three-dimensional structures like genitalia on microscope slides, where they can be viewed in only one plane, not necessarily the most satisfactory one.

Preservation in fluids

Preservation in liquids is often the only practical method of storing soft-bodied Hemiptera, particularly the immature stages. The most widely used preservative is 70 per cent alcohol (sometimes 75 or 80 per cent is recommended). The alcohol in question is industrial methylated spirit, which is ethyl alcohol with ten per cent methyl alcohol added and about five per cent water. The would-be purchaser must first obtain an excise licence. Domestic methylated spirit contains a dye and an emetic and is not recommended as a preservative. Iso-propyl alcohol, being non-intoxicating, can be bought without a license and seems to be just as good as 'industrial meths' for most purposes although it is more expensive. Some alcohol-based insect preservatives contain five to ten per cent of acetic acid or glycerol (glycerine) or both, to keep the preserved material supple. The glycerine has the advantage of remaining behind when the alcohol has evaporated after years of neglect, preventing the material from drying out completely.

Individual specimens or groups of specimens with identical data are stored in spirit in glass specimen tubes. It is convenient to use tubes of a standard length, say two inches, but of varying diameters if necessary.

Spirit storage has several advantages over dry storage. The material does not shrivel, it is less easily damaged and not so brittle. Less time is needed to process it and less space is required for storage, particularly of large series with identical data. There are disadvantages too. There is always a loss of colour. Body contents of soft-bodied insects slowly pull away from the thin, transparent cuticle and ultimately disintegrate and chitin itself seems to be slowly dissolved. This last process takes many years and is first seen in the disappearance of the tips of fine setae and in weakening of arthrodial membranes, leading to disarticulation. Maceration of tissues for slide-mounting or genitalia preparations is more difficult and sometimes impossible to achieve completely after long storage in alcohol. Specimens preserved in spirit can be dried out and pinned or otherwise dry-mounted but rarely look as good as those that have never been wetted. Most methods for recovering material from spirit involve intermediate treatment with an organic solvent. Acetone, cellosolve, ethyl acetate and others have been tried with varying success (Sabrosky, 1957, 1966; Vockeroth, 1960). After immersion in these it can be either air-dried or subjected to critical-point drying (Gordh & Hall, 1979). Alternatively, fluid-preserved material can be transferred to water and then freeze-dried in a vacuum chamber.

Preservation on microscope slides

Most Sternorrhyncha are best preserved on microscope slides as many of the diagnostic characters can only be seen in slide preparations viewed at high magnifications. Exceptions are adult Psylloidea, which can be staged, pointed or even carded, and the scales or the hypertrophied females of some Coccoidea, which are so hard and thick that they can only be kept whole. Many adult and immature Coccoidea and the puparia of whiteflies can be preserved dry, on pressed leaves or dry twigs of their host plants, until required for examination but aphids and immature Psylloidea are best kept in spirit if they cannot be mounted straight away. Freshly collected specimens always make better mounts than long-dead ones, particularly if the latter have been stored in alcohol.

The mounting medium recommended depends partly on the particular group of insects studied but mainly on the personal preferences of the specialist. Canada balsam and Euparal have both proved their worth over long periods. Some other media popular with hemipterists have developed unexpected defects after long storage, including drying-out (which should be preventable by careful ringing), excessive darkening and crystallisation. A collection intended for posterity should be mounted in one or other of the two resins named, to be on the safe side.

Various textbooks give accounts of the procedures to be followed in making balsam slide-mounts. They should be followed with an understanding of the reasons for the various stages and an appreciation of the fact that larger specimens need to spend longer in each stage than smaller ones.

First, specimens should be macerated and softened in potassium hydroxide solution, as described for genitalia preparations. The body contents should be completely dissolved away. Then the specimens should be transferred to clean water to wash out the alkali. They may need several changes of water. Waxy specimens, such as whitefly puparia and coccids, need special treatment at this stage to remove the wax. They should be heated (in a water-bath again, and in a well-ventilated place) for a few minutes in chloral-phenol or carbol-xylol. The former is a mixture of equal parts by weight of chloral hydrate and phenol, which needs warming to mix but remains liquid when mixed. The latter is xylol (xylene) with 10 per cent by weight of phenol crystals dissolved in it. Very waxy material may need two or more changes of de-waxing fluid. It may be necessary to bleach dark material or to stain pale material before proceeding. A cold mixture of ammonia (880-volume) and hydrogen peroxide (20-volume) is recommended as a rapid bleaching agent by Martin (1987). The same author recommends staining with acid fuchsin in glacial acetic acid for whitefly puparia.

Macerated and wax-free specimens, bleached or stained if necessary, should be transferred to 70 per cent alcohol, then 90 per cent alcohol and finally to absolute alcohol (not undiluted industrial methylated spirit, which contains some water). The reason for this procedure is to remove the water from the specimens. They will shrivel irremediably if transferred straight from water to strong alcohol. Very delicate material may need to have a spell in 50 per cent alcohol before going on to 70 per cent. Two changes of alcohol, with the specimens remaining in the fluid for five minutes at a time, are recommended at each stage but this may have to be varied in the light of experience. Martin (1987) outlines an alternative dehydration process through glacial acetic acid.

Once dehydrated, the material is transferred to one or two changes of clove oil, cedarwood oil or xylol (xylene), which clears and hardens it and removes the alcohol. It is important to have the specimens spread in their final positions by this stage. Pipetting the absolute alcohol off the specimens and adding the clearing agent to the dish without disturbing them is preferable to lifting them out of one dish and into another. If the specimens become milky at this stage, they have not lost all of their water and should be put back into fresh absolute alcohol again.

Finally, the cleared specimens are transferred from the clearing agent to a drop of balsam in the middle of a slide and covered with a coverslip. The slide should be labelled immediately and laid flat in a drying oven for several weeks until the balsam has hardened. Thick card labels, which can be glued to the glass with polyvinyl alcohol (PVA glue), should be placed one at each end of the slide. If they are thicker than the coverslip and balsam together, they afford a degree of protection to the specimen. Euparal mounts are made in much the same way except that euparal essence is used as the clearing agent.

Labelling

All the data relating to the capture of an insect, or enough information to indicate the observation for which it is a voucher, should be permanently associated with the specimen. The information should be copied in full in indelible ink (never ball-pen ink) onto one or, if necessary, more than one label mounted on the same pin as the specimen or inserted into its tube if it is preserved in spirit or glued to its slide if it is made into a microscope mount.

The data recorded should be unambiguous and as full as possible. Localities should be recorded in descending order of size of geographical area, starting with country (if the collection is not restricted to one country only), then county or Watsonian vice-county (see Dandy, 1969), nearest town, parish (if appropriate) and finally collecting site. If the locality is within a nature reserve, National Park, Site of Special Scientific Interest or other designated area, this should be noted as well. The county should not be omitted from the label. A Kentish collector may be perfectly familiar with Leeds, Stoke, Luton and Eccles, but people from other counties who receive loans and exchanges from him may be seriously misled unless they know that the labels refer to the Kentish sites rather than their better-known namesakes. Latitude and longitude or the British or Irish National Grid reference should also be added.

The date, at least the year and month, should be recorded. Months should be given either as Roman numerals or in words (abbreviated to save space if necessary), since 7.2.1945 can mean either February the seventh or July the second; both 7.ii.1945 and ii.7.1945 unambiguously refer to the former. The time of day will be important in some circumstances (e.g. if the specimen was taken in flight) but not in others (e.g. if it was sieved from litter in the winter).

Some indication of the habitat should be given and the wording should be chosen so as to be both concise and informative. Once the collector has learnt to see the environment from the insect's point of view, succinct descriptive phrases should come easily. 'Grazed chalk turf' and 'lower branches of sycamore in heavy shade', in this context, convey more information than 'grubbing on Downs' and 'beating Sycamore in wood'.

Field notebooks, although they should never be discarded, should not be used as a substitute for informative labels. Cross-reference from specimens to notebooks by means of code-numbers will only work until the notebooks are mislaid or destroyed, which may happen upon moving house and is almost a certainty following the collector's death.

If a specimen has featured in a publication, either as a figure or as the basis of an observation, the name of the relevant book or periodical, volume, page number, figure number (if any) and date of publication should be recorded on another label, together with the name actually used for it in the publication. This label should never be altered to conform with any change in the name currently considered correct, whether as a result of synonymy or a correction of a misidentification. Determination labels should bear the name of the insect, the name of the person who made the identification and the date on which the identification was made. They should never be discarded, even if the identification is later shown to be incorrect. It is not unusual for specimens in reference collections to bear two, three or even more determination labels, acquired at different dates.

Sometimes more than one specimen is preserved on a single mount. The reasons for this practice vary. It may simply be that the collector wants to save space or to save time in writing out labels. If this is so, the label should bear an indication of the number of specimens. This is particularly important where there are two specimens of different sexes on the same mount, because it is often assumed that specimens so mounted are taken in copula. If they really were in copula, this should be stated, to remove any doubt. Such information is valuable in establishing the timing of the breeding season at the locality where the specimens were taken. It is also strong, but not quite conclusive, evidence that they belong to the same species. Sometimes predator-prey or symbiotic relationships are preserved by mounting the participants together. In this case, it should be clearly stated on the label what the relationship was. It is all to easy to forget, after the lapse of a few years, if the mirid was preying on the ant, or the ant was preying on the mirid, or the mirid was living unmolested among the ants.

Storage

Dry collections are best kept in cork-lined store boxes or foamed plastic-lined ones (polystyrene is not suitable because it does not grip the pins well). Display cases open to the light are not suitable for scientific purposes because the colours of the specimens rapidly fade. Reasonably light-tight cabinets with glass-topped drawers are more elegant than boxes and just as good for storage but they are less adaptable in use and much more expensive. Each box (or drawer) should have a 'camphor-cell' filled with flake naphthalene, which should keep dermestid beetles at bay (though it does not usually kill them or prevent their breeding if they do get into the collection). After beetles, damp is the main enemy. It favours the growth of mould and the multiplication of booklice, both of which can devastate a collection if unchecked. Accordingly, the collection should be housed in a well ventilated room that can be heated to reduce the humidity.

Fluid-preserved material should be kept in a cool, dark place. Light bleaches specimens in alcohol even faster than dry ones and warmth speeds up the other processes of deterioration that affect them. The main danger is of drying out. A convenient method of avoiding this is to place the tubes in short, wide jars that are themselves filled with alcohol. Then only a few jars, instead of hundreds of small tubes, need be checked and topped up from time to time. Retrieval of material from storage can be facilitated by storing the tubes bottom-up (any spare space in the jars to be filled with cotton wool) with a uniquely numbered disc in the bottom of each, held in place by a cotton-wool plug. An index on cards or in a notebook can then be kept with the collection. The individual tubes should, of course, contain the relevant data and determination labels.

Slides are most economically stored on edge in shallow cardboard trays (just over three inches wide and an inch high) with each slide in its own transparent plastic envelope. Thick labels on each side of the coverslip, together with the envelopes, give generally adequate protection and make it possible to dispense with the costly racks that are sold for slide storage. It is important to ensure that the mountant has set properly before storing a slide on edge, or the mount will tend to creep downwards, rolling the specimen between slide and coverslip and damaging it. Horizontal storage in flat trays is both expensive and wasteful of space but it does have the advantage of preventing the mounts creeping. Slide collections can be treated like card indexes, with tagged index cards inserted at appropriate points between the slides.

18

GLOSSARY

Note: plurals of many words of classical origin and form are given in parentheses.

1A: First (most anterior) Anal vein
2A: Second Anal vein
3A: Third Anal vein
acetabulum (acetabula): the socket in the thorax in which the basal segment (coxa) of the leg is attached, walled on the outer side by downward extensions of the pleura
aedeagus: the part of the male genitalia that is inserted into the female during copulation
aeromicropyle: a small hole in the shell of the egg through which sperm enters and gaseous exchange occurs
aestivalis (aestivales): a special summer morph (form) that undergoes aestivation
aestivation: a period of inactivity, usually involving arrested development, during the summer
alata (alatae): any of the fully winged morphs of Aphidoidea and Adelgoidea
alate: fully winged
alatiform aptera: an aptera with some of the characteristics of an alata (e.g. larger eyes than normal) but still completely wingless
alleles: alternative states of a gene expressing a character
anal plate: = subanal plate
anal ring: in several families of Coccoidea, a sclerotized ring surrounding the anus, usually bearing setae and pores
anal tube: the cylindrical tenth abdominal segment bearing the anus at its tip
Anal Veins: the veins of the clavus of the fore wing or behind the fold-line of the hind wing
androconia: male secretory structures that produce a pheromone inducing readiness to copulate in females
andropara (androparae): a parthenogenetic morph producing only males
anepimeron (anepimera): the upper part of the epimeron
anepisternum (anepisterna): the upper part of the episternum
anholocyclic: of Aphidoidea and Adelgoidea, omitting the sexual generation from the annual cycle
annulated: ringed (e.g. with alternating bands of colour)
annulus: a small ring, especially one between the main segments of the antennae, as found in some Heteroptera
anteclypeus: the anterior part of the clypeus if it is differentiated
antennal tubercle: = antennifer
antennifer: a projection of the head to which the antenna is attached
Anterior Cubitus: one of the longitudinal wing veins, situated in front of the claval furrow (Posterior Cubitus is not present in Hemiptera)
antevannal vein: = glochis
aphidophagous: feeding on aphids
apodeme: a sclerotized process to which muscles are attached
appendix: the part of the fore wing beyond the peripheric vein

aptera (apterae): any completely wingless morph, especially of Aphidoidea and Adelgoidea
apterous: completely without wings
arboreal: living in trees
areolate: divided up into numerous small depressions separated by ridges
arolium (arolia): outgrowths of the membrane at the tip of the tarsus
arthrodial membrane: thin, membranous cuticle between the sclerites
articulatory plate: = basal apparatus
asexual reproduction: reproduction not involving fertilization of eggs
Auchenorrhyncha: leafhoppers, planthoppers and their relatives (but not Psylloidea, which also hop)
auricle: the raised margins of the opening of the metathoracic gland
autoecious: not alternating between different host plants (see heteroecious)
autotomy: self-amputation

basal plate: = basal apparatus
basal apparatus: a median and often complex structure in the male genitalia to which the aedeagus and often also the parameres are articulated
basipulvillus: the basal, non-adhesive stalk of the pulvillus if it is differentiated
biological control: the use of natural enemies to reduce pest poulations to economically insignificant levels
bisexual: of a species or population, having individuals of both sexes (i.e. not parthenogenetic or hermaphroditic)
bivoltine: having two generations in the course of one year
brachypter: a brachypterous individual
brachyptera (brachypterae): a rarely encountered morph, in aphids, which has partially developed but useless wings
brachypterous: (1) with the fore wings only about half as long as those of individuals capable of flight and the hind wings even shorter; (2) with any degree of wing reduction apart from complete winglessness
brachyptery: the state of being brachypterous
brochosome: a minute, perforated, spherical body, probably an excretory product, voided in large numbers in the faeces of many leafhoppers and often found adhering to their wings and bodies

C: Costa
caecum (caeca): a tubular outgrowth of the intestine
callar region: the part of the pronotum where the calli are situated, behind the collar (if any) and in front of the anterior lobe
callus (calli): one of a pair of depressions in the pronotum, directly above the anterior coxae
carinate: keeled
cauda: a median abdominal projection above the anus in Aphidoidea and Adelgoidea
cephalotheca: the anterior part of the body of mature larval, pupal and female adult Strepsiptera that protrudes from the host's body
chelate: with an opposable claw, like the pincer of a crab
chorion: outer layer (often a thick shell) surrounding egg
cibarial muscles: muscles operating the food-sucking pump
circulative plant pathogen: one that migrates to the salivary gland of the vector
clasper: = paramere or style
claval furrow: impressed line separating clavus from rest of wing

claval commissure: the line along which the posterior margins of the clavi meet (along the midline of the body) when the fore wings are folded over the body at rest
clavate: thickened at one end, like a club
clavus (clavi): posterior basal region of the fore wing, marked off from the rest of the wing by the claval furrow
claw: paired or (in Coccoidea) single, usually hook-like structure articulated to apex of tarsus (rarely fused with it)
cleptoparasite: of Sphecidae, a species whose females usurp the nests and food supplies of others for their own offspring
clone: asexually produced progeny of a single female (may include more than one generation)
clypeus (clypei): the central sclerite at the front of the head
coleopteroid: beetle-like, referring to the fore wings of some Heteroptera in which the clavus is more elongated than in the macropterous form and the membrane is greatly reduced or lacking altogether
collar: the anterior margin of the pronotum if it is demarcated from the rest of it by a furrow
compound eye: the main organ of vision, with several or many facets (lenses). See ommatidium
conjunctiva: membranous and usually eversible distal part of aedeagus
connective: = basal apparatus, a pair of structures at the base of the aedeagus and claspers in Auchenorrhyncha
connexivum: series of small plates along margins of abdomen (see laterotergites, laterosternites)
coriaceous: thickened but flexible
corium: major part of fore wing of Heteroptera, excluding clavus, cuneus and embolium if they are demarcated by furrows, and excluding membrane
cornicle: one of a pair of conical or cylindrical structures with large, terminal apertures, situated on the hinder part of the abdominal dorsum of most Aphidoidea
Costa: vein running close to anterior margin of wing, often absent in Hemiptera
costal: of the anterior margin of the wing
costal fracture: a nick in the anterior margin of the fore wing of some Heteroptera, running towards the centre of the wing as a line of flexion cutting off the cuneus from the apex of the corium
coxa: the basal segment of the leg, articulating with the thorax
coxal cavity: socket on the thorax into which the coxa is at least partially recessed
coxal cleft: suture in outer wall of acetabulum between its episternal and epimeral elements
crypsis: resemblance to background or to an inanimate object, conferring an advantage such as protection from predators
crypt: small out-pocketing from intestine
CuA: Anterior Cubitus
Cubitus: (in Hemiptera) = Anterior Cubitus
cuneal fracture: = costal fracture
cuneus: part of fore wing of some Heteroptera, bounded proximally by costal fracture and distally by membrane

dentate: toothed
diapause: condition of temporarily suspended development, growth or maturation
dicotyledonous plants, dicotyledons or Dicotyledones: the largest group of flowering plants, including all native trees and shrubs (except conifers) and most herbaceous plants apart from grasses, lilies and their relatives

dimorphism: the condition of occurring in either of two forms differing in wing length, colour, the need to diapause etc. - a special case of polymorphism; sexual dimorphism, having obvious differences between the sexes in addition to the structures directly associated with reproduction
dioecious: of aphids, = heteroecious
disc: the middle part of a plate-like stucture (wing, pronotum etc.)
discal: situated in the middle part
discal cells: cells of the fore wing between R and M, closed apically by cross-vein r–m, and between M and CuA, closed basally and apically by two m-cu cross-veins
distal: of or towards the apex
distipulvillus: the distal, adhesive part of the pulvillus if it is differentiated
dorsal: referring to the morphologically upper side
dorsoventrally flattened: wider than high
dorsum (dorsa): the dorsal surface of the body or of a part of it

ectoparasite: a parasite that does not enter the body of its host
emarginate: with a small nick or concavity in the edge
emboliar groove: an impressed line parallel to and close to the costal margin of the hemelytron in some Heteroptera
embolium (embolia): the narrow strip of the hemelytron anterior to the emboliar groove
endoparasite: a parasite that enters the body of its host to feed
entire: not segmented or split or otherwise divided
epimeron (epimera): posterior part of pleuron, separated from sternum by coxa
epipleur: ventral thickening of costal margin of hemelytron that is exposed ventrally when hemelytron is in resting position
episternum (episterna): anterior part of pleuron, meeting sternum and often continuous with it
epistomal suture: suture between clypeus and frons
evaporative area: =evaporatorium
evaporatorium (evaporatoria): area of matt, finely sculptured cuticle surrounding opening of metathoracic gland in Heteroptera
exuviae (always plural): 'cast skin', the moulted exoskeleton of a nymph or larva.

facial cone: in Psylloidea, = genal cone
facultative: able to happen if circumstances allow, but not a predetermined necessity
fatbody: mass of storage tissue, mainly in abdomen
femur (femora): long segment of leg above the knee joint, articulating basally with the trochanter and apically with the tibia
filiform: thread-like
flagellum: the whole of the antenna beyond the pedicel
flocculent: fluffy
fossa spongiosa: = spongy fossa
frons: front of head, behind clypeus and in front of vertex
frontoclypeus: in Cicadellidae, the fused postclypeus and frons
fundatrix (fundatrices): in Aphidoidea and Adelgoidea, the parthenogenetic female morph resulting from sexual reproduction
fuscous: grey-brown
fusiform: spindle-shaped

gallicola (gallicolae): gall-making morph of Adelgoidea
gena (genae): the side of the head below the eye
genal cone: in Psylloidea, one of a pair of conspicuous projections on the front of the head

genital capsule: in males, the sternum of the ninth abdominal segment, modified to accommodate the genitalia
genitalia: structures involved in copulation
genital plate: in Aphidoidea and Adelgoidea, = subgenital plate
genital plates: in male Cicadellidae, paired ventral appendages of the ninth abdominal segment, concealing the genitalia from below
genital scale: in several groups, a small, triangular sclerite, visible in ventral view, covering the extreme base of the ovipositor
genital valve: in male Cicadellidae, sternum of the ninth abdominal segment, bearing the genital plates distally
glochis (glochides): in the hind wing of some Heteroptera, a spur arising from the posterior side of CuA and directed back towards the base of the wing, running roughly parallel to the vannal fold
gonocoxa (gonocoxae): basal segment of paired, two-segmented appendage of eighth or ninth abdominal segment of female; see gonostylus, valvifer
gonoplac: one of a pair of unsegmented appendages of ninth abdominal segment of female forming a sheath for the ovipositor valves; often absent
gonostylus (gonostyli) or **gonostyle**: apical segment of paired, two-segmented appendage of eighth or ninth abdominal segment of female; see gonocoxa, valvula
gravid: full of ripe eggs
gregarious: of bugs, associating in groups; of endoparasites, normally with several individuals developing simultaneously in one host from several eggs laid at the same time (compare superparasitism)
gula: the throat area behind the rostrum in the head of Heteroptera
gynopara (gynoparae): in Aphidoidea and Adelgoidea, a parthenogenetic morph producing only sexually-reproducing females

haemocoel: the main body cavity
haemocyte: blood cell, floating in the haemolymph
haemolymph: fluid that fills the haemocoel, bathing the internal organs
hamulohaltere: rudimentary hind wing of some male Coccoidea
hamus: in the hind wing of Heteroptera, the section of M that traverses the basal cell
hemelytron (hemelytra) or **hemielytron (hemielytra)**: fore wing of Heteroptera if corium is differentiated from membrane
hemimetabolous: with a life-cycle in which sexually mature individuals do not moult and there is no pupal instar. Compare ametabolous or paurometabolous, in which sexually mature individuals continue to moult at intervals, and holometabolous.
heteroecious: of Aphidoidea and Adelgoidea, with some generations on one kind of host plant and others on a different kind, with seasonal migrations between the two
Heteroptera: true bugs, i.e. Hemiptera with a gula
hiemalis (hiemales): specialised overwintering morph in Thelaxidae
holocyclic: of Aphidoidea and Adelgoidea, having a sexual generation in the annual (or biennial) life-cycle
holometabolous: with a life-cycle in which sexually mature individuals do not moult and there is a pupal instar immediately before the last moult. See hemimetabolous
Homoptera: Auchenorrhyncha plus Sternorrhyncha
honeydew: sugar-rich faeces voided by phloem-feeding Hemiptera
host: the animal fed upon by a parasite or whose tissues harbour a micro-organism; a host-plant
host plant: a plant normally fed upon
host range: the spread of different animals or plants normally fed upon

host-specific: = monophagous
hydrofuge: water-repellent
hyperparasite: a parasite whose host is itself a parasite of something else
hypocostal lamina: = epipleur

immunological: descriptive of a method of detecting body proteins of prey in the gut contents of predators by testing the latter against antibodies to the proteins of the prey (prepared by vaccinating vertebrates with extracts of the prey)
incrassate: thickened
instar: a stage in the life-cycle between successive moults
interclaval vein: a vein in the hind wing between two claval furrows or betwen the branches of a forked claval furrow

jugal lobe: the lobe at the posterior basal angle of the hind wing, which folds beneath the rest of the wing in repose
jugum (juga): one of the pair of lobes flanking the tylus in Heteroptera (= paraclypeus)

katepimeron (katepimera): the lower part of the epimeron
katepisternum (katepisterna): the lower part of the episternum

labium: ventral appendage of the head, situated behind the mouth and, in Hemiptera, in the form of a usually jointed tube housing the stylets
labrum: small, median appendage of the head, situated in front of the mouth, ventral to the clypeus (tylus) and in Hemiptera closing stylet channel of labium anteriorly at its base
lamellate: extended as a thin plate
larva (larvae): immature stage, especially of Sternorrhyncha and holometabolous insects
laterosternite: small plate cut off from edge of sternite by a longitudinal suture
laterotergite: small plate cut off from edge of tergite by a longitudinal suture
lingula: in Aleyrodidae, a small sclerite bearing the anus
lora (lorae) or lorum (lora): one of a pair of small plates at the side of the head above the origin of the rostrum

M: Media
macropter: a macropterous individual
macropterous: with fully developed wings
macroptery: the state of being macropterous
mandible: one of the first pair of mouthparts, which are biting jaws in most insects but which, in Hemiptera, are modified into the outer pair of stylets
mandibular plate: = lora (laterally) and jugum or paraclypeus (dorsally or frontally)
maxilla (maxillae): one of the second pair of mouthparts, which are shredding jaws in most insects but which, in Hemiptera, are modified into the inner pair of stylets
maxillary plate: a sclerite sometimes differentiated in front of and below the gena in Heteroptera
m–cu: a cross-vein between M and CuA
Media: a longitudinal vein in the wing, behind the Radius and in front of the Anterior Cubitus
medial fracture: a flexion-line running from the base of the hemelytron or tegmen to about its middle, usually running between R and M and terminating just before the cross-vein that joins them
median frontal prominence: a small projection on the front of the head in some aphids
membrane: in Heteroptera, the apical, thin, flexible and usually transparent part of the hemelytron

meniscus: the surface film
meso-: prefix applied to any of the sclerites of the mesothorax
mesophyll: non-vascular and non-epidermal tissue of vegetative parts of plants
meta-: prefix applied to any of the sclerites of the metathorax
micropter: a micropterous individual
micropterous: with greatly shortened wings, not meeting beyond the scutellum
micropylar: relating to the presence of micropyles or aeromicropyles
micropyle: small aperture in shell of egg through which sperm enters
mimic: an organism that benefits from its resemblance to another, for example an animal that is is shunned by predators because of its resemblance to an animal that is repellent to them
mimicry: a situation in which one organism (the mimic, q.v.) benefits from its resemblance to another (the model, q.v.), by being confused with it by an operator such as a predator
model: an organism that a mimic (q.v.) resembles and with which it is confused by an operator such as a predator
monocotyledonous plants, monocotyledons or **Monocotyledones**: group of flowering plants that includes grasses, lilies and their relatives
monoecious: of Aphidoidea and Adelgoidea, = autoecious
monophagous: feeding on only one kind of host plant or host
morph: one of several physically or physiologically different forms that may occur in a single species. See polymorphism
mummified: converted into a mummy
mummy: in Aphidoidea and some other Sternorrhyncha, the empty exoskeleton, often lined with silk and containing the mature larva or pupa of the parasite that has consumed its body contents
mycetocyte: a fat-body cell containing symbiotic micro-organisms
mycetome: a cluster of mycetocytes
mycorrhizal: involving a symbiotic relationship between plant roots and fungi
myrmecophilous: living in a symbiotic relationship with ants

nodal line: an alignment of cross-veins across the apical half of a wing
nomenclature: the application of names
noncirculative plant pathogen: one that is borne on the stylets of the vector and does not migrate to its salivary glands
notum (nota): the whole dorsal surface of a thoracic segment
nymph: immature instar of hemimetabolous insect

obligate: predetermined, unavoidable
occipital region: the back of the head
ocellus (ocelli): simple eye, with only one lens
ocular tubercle: = triommatidium
ommatidium (ommatidia): single element of a compound eye, with a single lens
operculum (opercula): in Aleyrodidae, flap in the vasiform orifice, overlying base or all of lingula
operculum (opercula): of egg, lid that can be burst open along a predetermined line of weakness in the chorion
ovipara (oviparae): in Aphidoidea, the sexually reproducing female that lays the winter eggs
oviparity: the condition of laying eggs rather than giving birth to nymphs or larvae
oviparous: laying eggs
oviposition: the act of laying eggs

ovipositor: the organ that places eggs in position as they are laid
ovipositor sheath: = gonoplac

pala (palae): the anterior tarsus of Corixidae
Palaearctic Region: large biogeographical region including Europe, North Africa and non-tropical Asia
palar pegs: a row of conical setae on the pala of male Corixidae
palette: a leaf-like expansion of the distal part of the antennal flagellum in some Cicadellidae
palisade mesophyll: the upper layer of photosynthetic cells in a leaf
palp: in mouthparts, a sensory and usually segmented appendage to maxilla or labium in most insects but lacking in Hemiptera
palp: in Psylloidea, = gonoplac
paraclypeus (paraclypei): one of the paired lobes at the front of the head, flanking the clypeus
paramere: one of the almost always paired ventral appendages of the ninth abdominal sternum of the male
parasite: an organism that feeds repeatedly or continuously on another one without immediately killing it
parasitic: of the relationship between a parasite and its host; living as a parasite
parasitism, percentage: the percentage of a host population that is parasitised
parasternite: = laterosternite
paratergite: = laterotergite, particularly of female eighth and ninth abdominal segments
parempodium (parempodia): one of a pair of usually bristle-like structures inserted apically on the unguitractor plate (between the claws)
parthenogenesis: kind of reproduction in which unfertilised females produce viable eggs or young (hence parthenogenetic, parthenogenetically)
paryptera (parypterae): in Psylloidea, a small thoracic sclerite just in front of tegula
pathogen: micro-organism that induces symptoms of disease
pathogenic: causing disease
pedicel: the second segment of the antenna, divided into two segments in adult Pentatomoidea (shieldbugs)
penis: aedeagus
peripheric vein: a marginal or partly submarginal vein linking the apices of all the longitudinal veins and their branches
peritreme: the region of specialized structures surrounding the opening of the metathoracic gland in Heteroptera
phallotheca: sclerotized base of the aedeagus
phallus: aedeagus
pheromone: a chemical substance produced by one animal that influences the behaviour of others of the same or different species
phloem: vascular tissue transporting sugars, amino acids and other metabolites in solution around plant
phoresy: relationship in which a less mobile organism clings to a more mobile one in the interests of its own dispersal
phoretic: relating to phoresy
phragma (phragmata): internal projection from anterior or posterior margin of a segment, usually serving as an apodeme for the attachment of muscles
phytophagous: feeding on plants
piceous: the colour of pitch
placoid sensillum (sensilla): = rhinarium

pleural suture: suture on pleuron at the junction of episternum and epimeron
pleuron (pleura): the whole side wall of a thoracic segment
polymorphism: discontinuous variation within a species, other than the normal differences between instars and sexes
polyphagous: feeding on a variety of different host-plants, hosts or prey
polyvoltine or plurivoltine: having several generations in the course of a year. Sometimes given as multivoltine
postclypeus: the posterior part of the clypeus if it is differentiated
posterior lobe of the pronotum: in most adult Heteroptera and some Auchenorrhyncha, a broad, posterior projection of the pronotum roofing over most of the mesonotum or even more of the body
postnotum: the most posterior of the four regions into which the mesonotum or metanotum is divided by transverse sutures
praescutum: the most anterior of the four regions into which the mesonotum or metanotum is divided by transverse sutures
precipitin test: a technique used in immunological studies
predaceous or **predatory**: of the relationship between a predator and its prey; living as a predator
predator: an animal that kills and eats other animals (compare parasite)
pregenital segments of the abdomen: the segments anterior to those that bear the genital appendages, i.e. the first seven of females and the first eight of males
pretarsus: the claws and associated structures at the tip of the tarsus
primary host: in Aphidoidea and Adelgoidea, the host plant on which sexual reproduction occurs
primary sensillum (sensilla): in Aphidoidea and Adelgoidea, a rhinarium in either of the primary positions: near the apex of the penultimate antennal segment or part-way along the last segment, at the base of the terminal process
pro-: prefix applied to any of the sclerites of the prothorax (compare meso-, meta-)
processus terminalis (processus terminales): in Aphidoidea and Adelgoidea, the slender apical part of the last antennal segment
proctiger: = anal tube; in female Psylloidea, the large, dorsal plate at the apex of the abdomen, bearing the anus
propagative plant pathogen: a circulative one that multiplies in the body of the vector
proximal: at or towards the base
pseudopygidium: in some Coccoidea, coalesced apical abdominal segments still showing segmentation (compare pygidium)
pterostigma: a sclerotized and often pigmented area on the costal margin of a mainly membranous fore wing
pubescence: a covering of hairs
pulvillus (pulvilli): an adhesive pad attached to a claw
pupillarial: of adult female Diaspididae (Coccoidea), remaining within the scale of the last larval instar
puparium (puparia): of Aleyrodidae, the last larval stage in its post-feeding state when its tissues are being reorganised into the adult form
puparium (puparia): of Diptera, the modified and hardened skin of the last larval stage inside which the true pupa is formed
pygidium: in adult female and nymphal Diaspididae (Coccoidea), unsegmented structure resulting from coalescence of the last five abdominal segments
pygofer: in male Cicadellidae, the enlarged ninth abdominal tergite bearing the genital structures and proctiger; in other groups, = genital capsule
pygophore: = genital capsule

R: Radius
R1: first branch of Radius
radicicola (radicicolae): in *Phylloxera*, the subterranean morph
Radius: a longitudinal wing vein behind the Subcosta and in front of the Media
ramus (rami): a grooved, rod-like sclerite on the first or second valvula, the two rami locking the valvulae together in a sliding fit
raptorial: of a predator, one that seizes its prey with its legs; of a limb, one adapted for this purpose
remigium: the whole of the fore wing apart from the clavus, or of the hind wing apart from the jugal lobe
reticulate or **reticulated**: with a net-like pattern
rhinarium (rhinaria): a sensillum on the antenna of Aphidoidea and Adelgoidea consisting of a shallow pit with a thin, membranous floor
ring-segment: a small ring separated from the base or apex of an antennal segment by membrane
r–m: a cross-vein linking R and M
rostrum: the externally visible mouthparts of Hemiptera, comprising labium, labrum and stylets but often used in a sense that refers only to the labium

sacciform: of hymenopteran larva, short and fat like a well-
filled bag
Sc: Subcosta
scale insects: Coccoidea, especially the less mobile ones
scape: the first (basal) segment of the antenna
sclerite: plate of sclerotized cuticle, bounded by sutures or areas of arthrodial membrane
sclerotized: of cuticle, thickened and toughened by impregnation with tanned proteins (hence sclerotization)
scutellum (scutella): the third of the four regions into which the mesonotum or metanotum is divided by transverse sutures; in adult Heteroptera and Auchenorrhyncha, when unqualified by prefix, = mesoscutellum, the triangular or rounded plate visible between the bases of the hemelytra or tegmina in repose and sometimes spreading out to cover them
scutum (scuta): the second of the four regions into which the mesonotum or metanotum is divided by transverse sutures
secondary host: in heteroecious Aphidoidea and Adelgoidea, host plant on which only asexual reproduction occurs
secondary sensillum (sensilla): on antenna of Aphidoidea and Adelgoidea, any rhinarium other than the primary sensilla
sensillum (sensilla): a sense-organ; see also primary sensillum, secondary sensillum
seta (setae): a hair or bristle
setiferous, setose: bearing setae
sexuales (plural): in Aphidoidea and Adelgoidea, males and sexually reproducing females
sexupara (sexuparae): in Aphidoidea and Adelgoidea, parent morph of sexuales
shield bugs: = Pentatomoidea
sinuate: with a broad concavity in the edge
siphunculus (siphunculi): = cornicle
spongy fossa (fossae): in some predaceous Heteroptera, groove in tibia, floored with flexible cuticle bearing numerous short setae, serving to retain prey
staphylinoid or **staphyliniform brachypter**: one with fore wings truncate straight across, without membrane and resembling wing cases of staphylinid beetle
stenopterous: with the wings narrowed in comparison with the fully winged form

Sternorrhyncha: Coccoidea, Aphidoidea, Adelgoidea, Psylloidea and Aleyrodoidea
sternite: the main sclerite of a sternum
sternum (sterna): the whole morphologically ventral surface of a thoracic or abdominal segment
stigma: = pterostigma
stridulate: to produce sound by stridulation
stridulation: the production of sounds by a rasping mechanism
style: in Cicadellidae, one of the pair of appendages articulating with the genital plate; in other Hemiptera = paramere
stylet: one of the inner or outer pair of slender, tough, flexible mouth parts that run in the stylet channel in the labium and whose tips, at least, enter the tissues of the host plant or prey in feeding
stylopization: condition of genital malformation and reduced fertility or sterility caused by the presence of a strepsipteran parasite
subanal plate: in Aphidoidea and Adelgoidea, small plate situated below the anus
subbrachypterous: with the wings reduced to the extent that the insect is unable to fly but still covering most of the abdomen (in Heteroptera) or extending beyond it (in Auchenorrhyncha) (hence subbrachypter)
Subcosta: longitudinal vein of wing running behind Costa (if present) and in front of Radius
subgenital plate: in females of some Sternorrhyncha, the seventh abdominal sternite, projecting below the ovipositor
submacropterous: with the wings slightly reduced in length so that sustained flight is not possible (hence submacropter)
sulcate: grooved
superparasitism: parasitism of an individual host by more than one individual parasite, resulting from more than one episode of oviposition (unlike gregarious parasites, which hatch from a batch of eggs laid at the same time)
suture: an impressed line marking the junction of two sclerites
symbiosis: relationship between two different organisms to the benefit of one or both and to the net detriment of neither (hence symbiotic)
symbiote: a partner in a symbiotic relationship
synonym: one of two or more names used for the same thing or concept (hence synonymy, synonymous)
systematics: the arrangement of species, genera etc into a classification

tarsus (tarsi): the foot, which may be entire or divided into two or three segments (not counting the claws and associated structures)
taxonomy: recogniton of genera, species etc
tegmen (tegmina): the coriaceous fore wing of Auchenorrhyncha
tegula (tegulae): a small, flap-like sclerite on the mesothorax just anterior to the base of the fore wing
tergite: the main sclerite of a tergum
tergum (terga): the whole morphologically dorsal surface of an abdominal segment
terminal process: = processus terminalis
tertiary parasite: a parasite whose host is itself a hyperparasite
test: in Coccoidea, a uniform, glassy, roof-like covering of secreted wax (as in Asterolecaniidae and male puparia of Coccidae) or an envelope of felted, filamentous wax (as in many Eriococcidae)
thyridium (thyridia): one of a pair of spots at or close to the junction of the top of the head and the face, differing in texture and usually in pigmentation from the surrounding area, in some Cicadellidae

tibia (tibiae): long segment of leg below the knee joint, articulating basally with the femur and apically with the tarsus

timid predator: one that makes no attempt to subdue prey, taking only inert or sluggish organisms

transmission: the process whereby a micro-organism is transferred from one host to another

transovarial transmission: the passage of a micro-organism from the mother's body to the eggs or embryos developing inside her

traumatic insemination: copulation in which the body wall of the female is pierced and sperm enters her body cavity

trichobothrium (trichobothria): sense-organ consisting of a fine hair arising from a pit

triommatidium (triommatidia): in Aphidoidea and Adelgoidea, a cluster of 3 ommatidia different from the normal ommatidia of the compound eye

triploid: condition in which the cells of the body all have 3 sets of chromosomes (instead of the usual two sets)

triungulin: the active, first-instar larva of a strepsipteran

trivoltine: having 3 generations in the course of a year

trochanter: a small segment of the leg, articulating basally with the coxa and apically (and usually not so freely) with the femur

trochantin: a small sclerite in the membrane connecting the coxa (the basal segment of the leg) with the thorax

trophamnion: = trophic membrane

trophic membrane: layer of cells just inside chorion (shell) of egg of some parasitic Hymenoptera, absorbing nutrients from the host's haemolymph, causing the egg to grow in size

truncate: ending abruptly, as if cut off

tylus: median lobe at front of head in Heteroptera (= clypeus)

tymbal: taut membrane or plate that can be vibrated by muscles attached to it to produce a sound

tympanal organ: sound-producing organ incorporating a tymbal

unguitractor plate: a plate inserted into the apex of the tarsus between the claws

univoltine: having a single generation in the course of a year

valvifer: sclerite of ovipositor to which muscles are attached and whose movement operates the attached valvula (**first valvifer** = gonocoxa of eighth abdominal segment; **second valvifer** = gonocoxa of ninth abdominal segment)

valvula (valvulae): valve of ovipositor, guiding egg into position as it is laid or (third valvula) protecting other valvulae that do this (**first valvula** = gonostylus of eighth abdominal segment; **second valvula** = gonostylus of ninth abdominal segment; **third valvula** = gonoplac)

vannal fold: line along which jugal lobe of hind wing folds under remigium in repose

vasiform orifice: in Aleyrodidae, depression in dorsum, housing anus and associated structures

vector: insect able to carry disease organisms from one plant to another

vein: tubular or channel-like rib in wing

venation: the arrangement of veins in the wing

venter: the ventral surface of the body or of a part of it

ventral: referring to the morphological under side

vertex: top of head between eyes

vesica: apical, sclerotized part of aedeagus beyond conjunctiva in some Heteroptera

virginopara (virginoparae): in Aphidoidea and Adelgoidea, asexually reproducing morph whose progeny are themselves asexually reproducing females

virgino-sexupara (virgino-sexuparae): in *Phylloxerina*, a morph whose progeny include both asexually and sexually reproducing morphs

vivipara (viviparae): a viviparous morph

viviparity: the condition of giving birth to nymphs or larvae rather than laying eggs

viviparous: giving birth to nymphs or larvae

wing-polymorphism: polymorphism involving morphs differing in the degree of development of the wings (usually a dimorphism involving a macropterous morph and either an apterous one or some other flightless morph)

xylem: vascular tissue conveying mostly water and minerals from roots to aerial parts of plant

Y-vein: in clavus of some Auchenorrhyncha, a vein compounded of the two Anal veins which run separately in the basal part and then unite to run together to its apex

19

INFORMATION SOURCES

The works referred to in the preparation of this book are listed in the bibliography. In order to keep the text easy to read, references have not been cited for every piece of information each time it is presented. Major reference works covering more than one family are cited in the introductory material to the order and to each of the three suborders. To check the source of a piece of information on, say, the feeding habits of a particular family, it may be necessary to consult the references cited in the account of the biology of that family and in the introductory material on its suborder and on the Hemiptera as a whole. In the case of Sternorrhyncha, general works are cited in the introductory material to the superfamilies as well. If it is present in none of these works, then either the information was given to us by one of the many colleagues whose assistance has been so freely given during the compilation of the book, or it is one of our own observations.

The problems of keeping up with the enormous amount of information published on insects are considerably reduced by consulting some or all of the various abstracting journals that are intended to provide a guide to the literature. The best known of these is the *Zoological Record*, which attempts to deal with all aspects of zoology and is particularly good in the fields of taxonomy and biogeography. It is published jointly by BIOSIS and the Zoological Society of London. Hemiptera are covered by Section 13, Part F. Other comprehensive abstracting journals are *Entomology Abstracts* (London), and *Abstracts of Entomology* (Philadelphia). The *Review of Applied Entomology* (London), which is published in two series: (A) Agriculture and (B) Medical and Veterinary, is restricted to coverage of pests and methods of dealing with them. The *Annual Review of Entomology* (Stanford and Palo Alto, California) is composed entirely of well-researched reviews on entomological topics, each with an extensive and up-to-date bibliography.

Three entomological societies based in London provide a means of contact between entomologists and publish entomological journals. These societies are the Royal Entomologica. Society, the British Entomological and Natural History Society and the Amateur Entomologists' Society. The first of these publishes, among others, the quarterly journal *Antenna*, each issue of which has a feature entitled 'The British Insect Fauna' which is intended to keep readers up to date with new additions to the British fauna and with identification guides and papers on systematics and faunistics relevant to Britain and Ireland. The Systematics Association publishes, from time to time, a list of identification guides to the fauna and flora of Britain and, in the most recent volume, adjacent parts of Europe (Simms, Freeman & Hawksworth, 1988).

The Biological Records Centre is currently co-ordinating schemes to map the distribution of Auchenorrhyncha and of Heteroptera, both terrestrial and aquatic. Details of the schemes, addresses of the amateur coordinators and occasional newsletters are available from the BRC, which is based at the Institute for Terrestrial Ecology, Monks Wood Experimental Station, Abbots Ripton, Huntingdon.

A number of papers relevant to British Hemiptera appear in specialist entomological journals, including those of the three societies mentioned above. Examination of the bibliography below will give an idea of the relative frequency with which papers on Hemiptera appear in them. The *Entomologist's Monthly Magazine* carries by far the greatest number of papers on this order. Other British journals are the *Entomologist's Gazette*, the *Entomologist* and the *Entomologist's Record and Journal of Variation*. The publications of the British Ecological Society also carry papers on Hemiptera from time to time. Numerous local natural history societies exist and their addresses can usually be found at local public libraries. Some of them have members interested in Hemiptera and their journals sometimes carry faunistic papers of local interest.

Papers on Irish Hemiptera sometimes appear in journals published in Britain but Ireland is also served by three journals of its own: *Proceedings of the Royal Irish Academy*, *Irish Naturalists' Journal* and *Bulletin of the Irish Biogeographical Society*.

There are several newsletters of world-wide coverage designed to keep specialists in touch with each other and with recent developments in their areas of interest. For those interested in Auchenorrhyncha there is *Tymbal* edited by M. R. Wilson, IIE, c/o The Natural History Museum, London. *The Heteropterists' Newsletter* is edited by C. W. Schaefer, University of Connecticut, Storrs, Connecticut. Aphidologists are served by the *Aphidologists' Newsletter*, edited by C. S. Wood-Baker, 10 Green Lane, Chislehurst, Kent, and coccidologists by *The Scale*, published by the USDA Systematic Entomology Laboratory, Beltsville, Maryland.

REFERENCES

Adenuga, A.O. 1971. Observations on three species of Aphrophorinae (Hem., Cercopidae) reared in the greenhouse. *Entomologist's Monthly Magazine* **107**: 30–33.

Aldrich, J.R. 1988. Chemical ecology of the Heteroptera. *Annual Review of Entomology* **33**: 211–238.

Alexander, K.N.A. 1981. A second British record of *Issus muscaeformis* (Schrk.) (Homoptera, Issidae) from North Lancashire. *Entomologist's Monthly Magazine* **117**: 144.

Allen, A.A. 1964. *Coenosia lineatipes* Zett. (Dipt., Muscidae) preying on *Typhlocyba cruenta* H.-S. (Hem., Cicadellidae). *Entomologist's Monthly Magazine* **99**: 212.

Aoki, S. & Kurosu, U. 1986. Soldiers of a European Gall Aphid, *Pemphigus spyrotecae* (Homoptera: Aphidoidea): Why do they moult? *Journal of Ethology* **4**: 97–104.

Asche, M. & Remane, R. 1982. Beiträge zur Delphaciden-Fauna Greichlands I. (Homoptera Cicadina Delphacidae). *Marburger Entomologische Publikationen* **1** (6): 231–290.

Askew, R.R. 1968. Hymenoptera Chalcidoidea. Elasmidae and Eulophidae (Elachertinae, Eulophinae, Euderinae). *Handbooks for the Identification of British Insects* **8** (2b). 39 pp.

Averill, A.W. 1945. Supplement to food-plant catalogue of Aphids of the World. 51 pp. *Bulletin of the Maine Agricultural Experiment Station* **393** (S).

Avidov, Z. 1956. Bionomics of the Tobacco White Fly (*Bemisia tabaci* Gennad.) in Israel. *Ktavim. Records of the Agricultural Research Station, Rehovot, Israel* **7**: 25–41.

Bährmann, R. 1973. Anatomisch-morphologische und histologische Untersuchungen an den Saisenformen von *Aleurochiton complanatus* (Baerensprung) (Homoptera, Aleyrodina). *Zoologische Jahrbücher* (Abt. Systematik) **100**: 107–169.

Bährmann, R. 1980. Untersuchungen zur saisonalen Entwicklung der Aleyrodiden (Homoptera, Insecta). *Wissenschaftliche Zeitschrift der Friedrich-Schiller Universität, Jena* (Mathemat-Naturwiss.) **29** : 169–177.

Bakkendorf, O. 1934. Biological investigations on some Danish Hymenopterous egg-parasites, especially in Homopterous and Heteropterous eggs, with taxonomic remarks and descriptions of new species. *Entomologiske Meddelelser* **19**: 1–135.

Balachowsky, A. 1950. Les Kermes (Hom. Coccoidea) des chênes en Europe et dans le Bassin Méditerranéen. *Proceedings of the Eighth International Congress of Entomology, Stockholm*: 739–754.

Balachowsky, A. 1954. Les cochenilles paléarctiques de la tribu des Diaspidini. *Mémoires Scientifiques de l'Institut Pasteur.* **1954**: 1–450.

Balfour-Browne, F. 1953. Coleoptera Hydradephaga. *Handbooks for the Identification of British Insects* **4** (3). 33 pp.

Barson, G. & Carter, C.I. 1972. A species of Phylloxeridae, *Moritziella corticalis* (Kalt.) (Homoptera) new to Britain, and a key to the British Oak-feeding phylloxerids. *Entomologist* **105**: 130–134.

Baumert, D. 1958. Mehrjährige Zuchten einheimischer Strepsipteren an Homopteren. 1. Larven und Puppen von Elenchus tenuicornis Kirby. *Zoologische Beiträge* **3**: 365–421.

Baumert, D. 1959. Mehrjährigen Zuchten einheimischer Strepsipteren an Homopteren. 2 Hälfte. Imagines, Lebenszyklus und Artbestimmung von Elenchus tenuicornis Kirby. *Zoologische Beiträge* **4**: 343–409.

Baumert, D. & Behrisch, A. 1960a. Der Einfluss des Strepsipteren-Parasitismus auf die Geschlechtsorgane einer Homoptera. Teil 1. Entwicklungsdauer von Wirt und Parasit sowie Reduktionserscheinungen beim Wirt. *Zoologische Beiträge* **6**: 85–126.

Baumert, D. & Behrisch, A. 1960b. Der Einfluss des Strepsipteren-Parasitismus auf die Geschlechtsorgane einer Homoptera. Teil 2. Abstufung der Reduktionserscheinungen und Diskussion der Wirkungsweise des Parasiten. *Zoologische Beiträge* **6**: 291–332.

Baylac, M. 1986. Observations sur la biologie et l'écologie de *Lestodiplosis* sp. (*Dipt. Cecidomyiidae*), prédateur de la cochenille du hêtre *Cryptococcus fagi* (Hom. Coccoidea). *Annales de la Société Entomologique de France* (N.S.) **22**: 375–386.

Bei-Bienko, G.Y. [Ed.] 1964. Keys to the Insects of the European USSR. Volume 1. Apterygota, Palaeoptera, Hemimetabola. *Opredeliteli po Faune SSSR* **84**: 1–936. [In Russian. English translation, 1967, by J. Salkind, vii + 1214 pp. Jerusalem. Israel Program for Scientific Translations.]

Besseling, A.J. 1964. Die Nederlandse Watermijten (Hydrachnellae: Latreille, 1802). *Monographieën van de Nederlandsche Entomologische Vereeninging* **1**. 199 pp.
Bink-Moenen, R.M. 1989. A new species and new records of European whiteflies (Homoptera: Aleyrodidae) from heathers (*Erica* spp.). *Entomologist's Gazette* **40**: 173–181.
Blackman, R. 1974. *Aphids*. 175 pp, 8 pls. London and Aylesbury.
Blackman, R. 1977. The existence of two species of *Euceraphis* (Homoptera: Aphididae) on birch in Western Europe, and a key to European and North American species of the genus. *Systematic Entomology* **2**: 1–8.
Blackman, R. 1984. Two species of Aphididae (Hem.) new to Britain. *Entomologist's Monthly Magazine* **120**: 185–186.
Blackman, R. 1989. Cytological and morphological differences within Palaearctic *Glyphina* (Homoptera: Aphididae), and their taxonomic significance. *Systematic Entomology* **14**: 7–13.
Blackman, R. & Eastop, V.F. 1984. *Aphids on the World's crops: an identification and information guide*. vii + 466 pp. Chichester.
Booij, C.J.H. 1982. Biosystematics of the *Muellerianella* complex (Homoptera, Delphacidae): host-plants, habitats and phenology. *Ecological Entomology* **7**: 9–18.
Boratynski, K.L. 1952. *Matsucoccus pini* (Green, 1925) (Homoptera, Coccoidea: Margarodidae): bionomics and external anatomy with reference to the variability of some taxonomic characters. *Transactions of the Royal Entomological Society of London* **103**: 285–326; 4 pls.
Boratynski, K.L. 1957. On the two species of the genus *Carulaspis* MacGillivray (Homoptera: Coccoidea, Diaspidini) in Britain. *Entomologist's Monthly Magazine* **93**: 246–251.
Boratynski, K.L. 1961. A note on the species of *Asterolecanium* Targioni-Tozzetti, 1869 (Homoptera, Coccoidea, Asterolecaniidae) on Oak in Britain. *Proceedings of the Royal Entomological Society of London* (B) **30**: 4–14.
Borchsenius, N.S. 1957. Insecta. Homoptera. IX. Coccoidea. Coccidae. *Fauna SSSR* (N.S.) **66**: 1–493. [In Russian]
Börner, C. 1952. Europae centralis Aphides. *Mitteilungen der Thüringischen Botanischen Gesellschaft* (Beiheft) **3**: 1–484. [Addenda, pp 485–488, circulated in 1953 with reprints but not published in the journal.]
Bouček, Z. 1954. Chalcidologicke poznamky I, Pteromalidae, Torymidae, Eurytomidae, Chalcididae (Hymenoptera). *Sborník Entomologického Oddělení Národního Musea v Praze* **29** (426): 49–80.
Brindley, M.D. 1939. Observations on the life-history of *Euphorus pallipes* (Curtis) (Hym.: Braconidae), a parasite of Hemiptera-Heteroptera. *Proceedings of the Royal Entomological Society of London* (A) **14**: 51–56.
Buckley, R. 1987. Ant-plant-Homopteran interactions. *Advances in Ecological Research* **16**: 53–85.
Buchner, P. 1965. *Endosymbiosis of animals with plant microorganisms*. (Revised English edition.) xvii + 909 pp. New York.
Burckhardt, D. 1987. Jumping plant lice (Homoptera: Psylloidea) of the temperate Neotropical region. Part 1: Psyllidae (subfamilies Aphalarinae, Rhinocolinae and Aphalaroidinae). *Zoological Journal of the Linnean Society* **89**: 299–392.
Butler, C.G. 1938. On the ecology of *Aleurodes brassicae* Walk. (Hemiptera). *Transactions of the Royal Entomological Society of London* **87**: 291–311.
Butler, E.A. 1923. *A biology of the British Hemiptera-Heteroptera*. viii + 682 pp; 7 pls. London.
Cantwell, G.E. (Ed.) 1974. *Insect diseases*. 2 Vols. xi pp + pp 1–300 + 21 pp and xi pp + pp 301–595. Marcel Dekker. New York.
Carayon, J. 1984. Les androconies de certains Hemipteres Scutelleridae. *Annales de la Société Entomologique de France* (N.S.) **20** (4): 113–134.
Carter, C.I. 1971. Conifer wooly aphids (Adelgidae) in Britain. *Bulletin of the Forestry Commission* **42**. 51 pp.
Carter, C.I. 1976. A gall forming Adelgid (*Pineus similis* (Gill.)) new to Britain, with a key to the Adelgid galls on Sitka Spruce. *Entomologist's Monthly Magazine* **111**: 29–32.
Carter, C.I. & Eastop, V.F.E. 1973. *Mindarus obliquus* (Chol.) (Homoptera, Aphidoidea) new to Britain and records of two other aphids recently found feeding on conifers. *Entomologist's Monthly Magazine* **108**: 202–204.
Carter, C.I. & Fourt, D.F. 1984. The Lupin Aphid's arrival and consequences. *Antenna* **8**: 129–132.
Carter, C.I. & Maslen, N.R. 1982. Conifer Lachnids. *Forestry Commission Bulletin* **58**. 76 pp.

REFERENCES

Chandler, A.E.F. 1968. A preliminary key to the eggs of the commoner aphidophagous Syrphidae (Diptera) occurring in Britain. *Transactions of the Royal Entomological Society of London* **120**: 199–218.

Cheng, L. & Birch, M.C. 1978. Insect flotsam: an unstudied marine resource. *Ecological Entomology* **3**: 87–97.

China, W.E. 1925. Notes on the life-history of *Tricephora vulnerata* Illiger (Hom., Cercopidae). *Entomologist's Monthly Magazine* **61**: 133–134.

Claridge, M.F. 1959. A new species of trichogrammatid (Hymenoptera, Chalcidoidea) parasitic in mirid eggs (Hemiptera-Heteroptera). *Procceedings of the Royal Entomological Society of London* (B) **28**: 128–131.

Claridge, M.F. & Nixon, G. A. 1981. *Oncopsis* leafhoppers on British trees: polymorphism in adult *O. flavicollis* (L.). *Acta Entomologica Fennica* **38**: 15–19.

Claridge, M.F. & Nixon, G.A. 1986. *Oncopsis flavicollis* (L.) associated with tree birches (*Betula*): a complex of biological species or a host plant utilization polymorphism? *Biological Journal of the Linnean Society* **27**: 381–397.

Claridge, M.F. & Reynolds, W.J. 1972. Host plant specificity, oviposition behaviour and egg parasitism in some woodland leafhoppers of the genus *Oncopsis* (Hemiptera Homoptera: Cicadellidae). *Transactions of the Royal Entomological Society of London* **124**: 149–166.

Claridge, M.F. & Reynolds, W.J. 1973. Male courtship songs and sibling species in the *Oncopsis flavicollis* species group (Hemiptera: Cicadellidae) *Journal of Entomology* (B) **42**: 29–39.

Claridge, M.F., Reynolds, W.J. & Wilson, M.R. 1977. Oviposition behaviour and food plant discrimination in leafhoppers of the genus *Oncopsis*. *Ecological Entomology* **2**: 19–25.

Claridge, M.F. & Wilson, M.R. 1976. Diversity and distribution patterns of some mesophyll-feeding leafhoppers of temperate woodland canopy. *Ecological Entomology* **1**: 231–250.

Claridge, M.F. & Wilson, M.R. 1978a. Oviposition behaviour as an ecological factor in woodland canopy leafhoppers. *Entomologia Experimentalis et Applicata* **24**: 101–109.

Claridge, M.F. & Wilson, M.R. 1978b. Seasonal changes and alternation of food plant preferences in oligophagous mesophyll-feeding leafhoppers. *Oecologia* **37**: 247–255.

Claridge, M.F. & Wilson, M.R. 1981. Host plant associations, diversity and species-area relationships of mesophyll-feeding leafhoppers of trees and shrubs in Britain. *Ecological Entomology* **6**: 217–238.

Clausen, C.P. 1940. *Entomophagous insects*. x + 688 pp. McGraw-Hill. New York and London.

Cloudsley-Thompson, J.L. & Sankey, J.H.P. 1958. Some aspects of the fauna of the district around Juniper Hall, Mickleham, Surrey – IV. *Entomologist's Monthly Magazine* **94**: 43–47.

Cobben, R.H. 1960. The larvae of Corixidae and an attempt to key the last larval instar of the Dutch species (Hem., Heteroptera). *Hydrobiologia* **16**: 323–356.

Cobben, R.H. 1968. Evolutionary trends in Heteroptera. Part I. Eggs, architecture of the shell, gross embryology and eclosion. *Mededelingen van de Landbouwhogeschool te Wageningen* **151**. (8) + 475 pp.

Cobben, R.H. 1978. Evolutionary trends in Heteroptera. Part II. Mouthpart-structures and feeding strategies. *Mededelingen van de Landbouwhogeschool te Wageningen* **78–5**. 407 pp.

Coe, R.L. 1966. Diptera Pipunculidae. *Handbooks for the Identification of British Insects* **10** (2e). 83 pp; 1 pl.

Collin, J.E. 1966. The British species of *Chamaemyia* Mg. (*Ochthiphila* Fln.) (Diptera). *Transactions of the Society for British Entomology* **17**: 121–128.

Conci, C., Tamanini, L. & Burckhardt, D. 1985. *Aphorma lichenoides*, new for Italy, and revision of the genus (Homoptera, Psylloidea). *Bollettino del Museo Civico di Storia Naturale, Verona* **10**: 445–458.

Cook, D.R. 1974. Water mite genera and subgenera. *Memoirs of the American Entomological Institute* **21**. 860 pp.

Couch, J.N. 1938. *The genus Septobasidium*. ix + 480 pp. Chapel Hill, North Carolina.

Crisp, D.T. 1961. A study of egg mortality in *Corixa germari* Fieb. (Hem., Corixidae). *Entomologist's Monthly Magazine* **96**: 131–132.

Crowson, R.A. 1956. Coleoptera: introduction and keys to families. *Handbooks for the Identification of British Insects* **4** (1). 159 pp.

Crowson, R.A. 1976. On a possible female of *Halictophagus curtisi* Dale (Col., Stylopidae). *Entomologist's Monthly Magazine* **111**: 62.

Dandy, J.E. 1969. *Watsonian vice-counties of Great Britain*. Ray Society, London.

Danks, H.V. 1971. Biology of some stem-nesting aculeate Hymenoptera. *Transactions of the Royal Entomological Society of London* **122**: 323–399.

Danzig, E.M. 1964. Suborder Coccinea – Coccids or mealy bugs and scale insects. Pp 616–654. [In Russian. English translation, 1967, pp 800–850.] *In* Bei-Bienko, 1964, q.v.

Davids, C. & Schoots, C.J. 1975. The influence of the water mite species *Hydrachna conjecta* and *H. cruenta* (Acari, Hydrachnellae) on the egg production of the Corixidae *Sigara striata* and *Cymatia coleoptrata* (Hemiptera). *Verhandlungen der Internationalen Vereinigung für Theoretische und Angewandte Limnologie* **19**: 3079–3082.

Debauche, H.R. 1948. Étude sur les Mymarommidae et les Mymaridae de la Belgique (Hymenoptera Chalcidoidea). *Mémoires du Musée Royal d' Histoire Naturelle de Belgique* **108**. 248 pp; 24 pls.

De Long, D.M. 1971. The bionomics of leafhoppers. *Annual Review of Entomology* **16**: 179–210.

Disney, R.H.L. 1983. Scuttle Flies Diptera, Phoridae (except *Megaselia*). *Handbooks for the Identification of British Insects* **10** (6). 81 pp.

Dixon, A.F.G. 1973. *Biology of Aphids*. 58 pp. London.

Dixon, A.F.G. 1985. *Aphid Ecology*. ix + 157 pp. Glasgow.

Dixon, J.J. 1960. Key to and descriptions of the third instar larvae of some species of Syrphidae (Diptera) occurring in Britain. *Transactions of the Royal Entomological Society of London* **112**: 345–379.

Dolling, W.R. 1973. Photoperiodically determined phase production and diapause termination in *Notostira elongata* (Geoffroy) (Hemiptera: Miridae). *Entomologist's Gazette* **24**: 75–79.

Dolling, W.R. & Martin, J.H. 1985. *Aleurochiton acerinus* Haupt, a whitefly (Hom., Aleyrodidae) new to Britain. *Entomologist's Monthly Magazine* **121**: 143–144.

Domenichini, G. 1966. Hymenoptera Eulophidae: Palaearctic Tetrastichinae. *In*: Delucchi, V. & Remaudière, G. [Eds] *Index of Entomophagous Insects* **1**. 101 pp. Paris.

Domenichini, G. 1968. Contributo alla conoscenza biologica e tassonomica dei Tetrastichinae paleartici (Hymenoptera Eulophidae) con particolare riguardo ai materiali dell' Istituto di Entomologia dell' Universita di Torino. *Bollettino di Zoologia Agraria e Bachicoltura* (2) **8**: 75–110.

Douglas, J.W. & Scott, J. 1865. *The British Hemiptera. Vol. I. Hemiptera-Heteroptera*. xii + 628 pp; 21 pls.

Drosopoulos, S. 1982. Hemipterological studies in Greece. Part II. Homoptera-Auchenorrhyncha. On the family Delphacidae. *Marburger Entomologische Publikationen* **1** (6): 35–88.

Dunn, J. 1960. The natural enemies of the lettuce root aphid, *Pemphigus bursarius* (L.). *Bulletin of Entomological Research* **51**: 271–278.

Dupuis, C. 1963. Essai monographique sur les *Phasiinae* (Diptères Tachinaires parasites d'Hétéroptères). *Mémoires du Muséum National d'Histoire Naturelle. Paris* (n.s.) Ser. A. Zoologie **26**. 461 pp.

Eastop, V.F. 1953. A study of the Tramini (Homoptera – Aphididae). *Transactions of the Royal Entomological Society* **104**: 385–413; 1 pl.

Eastop, V.F. 1972. A taxonomic review of the species of *Cinara* Curtis occurring in Britain (Hemiptera: Aphididae). *Bulletin of the British Museum (Natural History)* (Ent.) **27**: 104–186.

Eastop, V.F. & Hille Ris Lambers, D. 1976. *Survey of the World's aphids*. 573 pp. The Hague.

Edgar, W.D. 1970. Prey of the wolf spider *Lycosa lugubris* (Walck.). *Entomologist's Monthly Magazine* **106**: 71–73.

Edwards, J. 1894–1896. *The Hemiptera-Homoptera of the British Isles*. 271 pp; 30 pls. London.

Evans, H.F. 1976. the role of predator-prey size ratio in determining the efficiency of capture by *Anthocoris nemorum* and the escape reactions of its prey, *Acyrthosiphon pisum*. *Ecological Entomology* **1**: 85–90.

Evlakhova, A.A. 1974. *Entomogeneous Fungi: classification, biology, practical significance*. 260 pp. Leningrad. [In Russian]

Eyre, S.R. 1963. *Vegetation and soils: a world picture*. xvi + 324 pp; 32 pls. London.

Fergusson, N.D.M. 1980. A revision of the British species of the genus *Dendrocerus* Ratzeburg (Hymenoptera: Ceraphronoidea) with a review of their biology as aphid hyperparasites. *Bulletin of the British Museum (Natural History)* (Entomology) **41**: 255–314.

Fergusson, N.D.M. 1986. Charipidae, Ibaliidae & Figitidae (Hymenoptera: Cynipoidea). *Handbooks for the Identification of British Insects* **8** (1c). 55 pp.

Ferrière, C. 1965. Hymenoptera Aphelinidae d'Europe et du Bassin Méditerranéen. *Faune de l'Europe et du Bassin Méditerranéen* **1**. 206 pp.

Ferrière, C. & Kerrich, G.J. 1958. Hymenoptera Chalcidoidea: Agaontidae, Leucospidae, Chalcididae, Eucharitidae, Perilampidae, Cleonymidae and Thysanidae. *Handbooks for the Identification of British Insects* **8** (2a). 40 pp.

Fitton, M. G., Graham, M.W.R. DeV., Bouček, Z.R.J., Fergusson, N.D.M., Huddleston, T., Quinlan, J. & Richards, O W. 1978. See Kloet, G.S. & Hincks, C. 1978.

Foster, W.A. 1975. The life history and population biology of an intertidal aphid, *Pemphigus trehernei* Foster. *Transactions of the Royal Entomological Society of London* **127**: 193–207.

Fraser, F.C. 1959. Mecoptera, Megaloptera, Neuroptera. *Handbooks for the Identification of British Insects* **1** (12–13). 40 pp.

Fulmek, L. 1943. Wirtsindex der Aleyrodiden- und Cocciden-Parasiten. *Entomologische Beihefte aus Berlin-Dahlem* **10**: 1–100.

Furk, C. & Prior, R.N.B. 1976. On the life cycle of *Pemphigus* (*Pemphigus*) *populi* Courchet with a key to British species of *Pemphigus* Hartig (Homoptera: Aphidoidea). *Journal of Entomology* (B) **44**: 265–280.

Gauld, I.D. & Bolton, B. (Eds) 1988. *The Hymenoptera*. xii + 332 pp. Oxford, London.

Gibson, D.O. 1976. A new form of Cercopis vulnerata Ill. (Hemiptera-Homoptera: Cicadoidea). *Entomologist's Record and Journal of Variation* **88**: 261.

Giller, P.S. 1986. The natural diet of the Notonectidae: field trials using electrophoresis. *Ecological Entomology* **11**: 163–172.

Giustina, W. Della. 1989. Homoptères Cicadellidae. Vol. 3. (Compléments). *Faune de France* **73**. xi + 350 pp.

Glen, D.M. 1973. The food requirements of *Blepharidopterus angulatus* (Heteroptera: Miridae) as a predator of the lime aphid, *Eucallipterus tiliae*. *Entomlogia Experimentalis et Applicata* **16**: 255–267.

Glen, D.M. 1977. Ecology of the parasites of a predatory bug, *Blepharidopterus angulatus* (Fall.). *Ecological Entomology* **2**: 47–55.

Goeldlin de Tiefenau, P. 1974. Contribution à l'étude systématique et écologique des Syrphidae (Dipt.) de la Suisse occidentale. *Mitteilungen der Schweizerischen Entomlogischen Gesellschaft* **47**: 151–252.

Gordh, G. & Hall, J.C. 1979. A critical point drier used as a method of mounting insects from alcohol. *Entomological News* **90**: 57–59.

Graham, M.W.R.DeV. 1969. The Pteromalidae of North Western Europe (Hymenoptera: Chalcidoidea). *Bulletin of the British Museum (Natural History)* (Entomology) Suppl. **16**. 908 pp.

Graham, M.W.R.DeV. 1976. The British species of *Aphelinus* with notes and descriptions of other European Aphelinidae (Hymenoptera). *Systematic Entomology* **1**: 123–146.

Graham, M.W.R.DeV. 1982. The Haliday collection of Mymaridae (Insecta, Hymenoptera, Chalcidoidea) with taxonomic notes on some material in other collections. *Proceedings of the Royal Irish Academy* (B) **82**: 189–243.

Graham, M.W.R.DeV. 1987. A reclassification of the European Tetrastichinae (Hymenoptera: Eulophidae), with a revision of certain genera. *Bulletin of the British Museum (Natural History)* (Entomology) **55**: 1–392.

Granger, C. 1944. Notes biologiques sur les Hyménoptères de France. *Bulletin de la Société Entomologique de France* **49**: 91–92.

Greathead, D.J. 1989. Biological control as an introduction phenomenon: a preliminary examination of programmes against Homoptera. *Entomologist* **108**: 28–37.

Green, E.E. 1927a. A brief review of the indigenous Coccidae of the British Islands, with emendations and additions. *Entomologist's Record and Journal of Variation* **39**: sep. pag. 1–4; pl. 4.

Green, E.E. 1927b. A brief review of the indigenous Coccidae of the British Islands, with emendations and additions. (Concluded) *Entomologist's Record and Journal of Variation* **40**: sep. pag. 5–14; pls 1, 3, 4.

Günthardt, M.S. & Wanner, H. 1981. The feeding behaviour of two leafhoppers on *Vicia faba*. *Ecological Entomology* **6**: 17–22.

Günthart, H. 1987. Oekologische Untersuchungen in Unterengadin. 12. Lieferung. D8. Zikaden (Auchenorrhyncha). *Ergebnisse der wissenschaftlichen Untersuchungen im Schweizerischen Nationalpark* **12**: 203–299.

Halbert, N. 1935. A list of Irish Hemiptera (Heteroptera and Cicadina). *Proceedings of the Royal Irish Academy* **42** (B): 211–318.

Halkka, O., Halkka, L., Hovinen, R., Raatikainen, M. & Vasarainen, A. 1975. Genetics of *Philaenus* colour polymorphism: the 28 genotypes. *Hereditas* **79**: 308–310.

Halkka, O., Halkka, L., Raatikainen, M. & Hovinen, R. 1973. The genetic basis of balanced polymorphism in *Philaenus* (Homoptera). *Hereditas* **74**: 69–80.

Halstead, A.J. 1981. A whitefly pest of sweet bay. *Plant Pathology* **30**: 123.

Hammond, P.M. & Barham, C.S. 1982. *Laricobius erichsoni* Rosenhauer (Coleoptera: Derodontidae), a species and superfamily new to Britain. *Entomologist's Gazette* **33**: 35–40.

Hardy, A.C. & Cheng, L. 1986. Studies in the distribution of insects by aerial currents. III. Insect drift over the sea. *Ecological Entomology* **11**: 283–290.

Harris, K.F. & Maramorosch, K. (Eds). 1977. *Aphids as virus vectors.* xvi + 559 pp. New York.

Harris, K.F. & Maramorosch, K. (Eds). 1980. *Vectors of plant pathogens.* xiv + 467 pp. New York.

Harris, K.M. 1968. A systematic revision and biological review of the cecidomyiid predators (Diptera: Cecidomyiidae) on world Coccoidea (Hemiptera: Homoptera). *Transactions of the Royal Entomological Society of London* **119**: 401–494.

Harris, K.M. 1973. Aphidophagous Cecidomyiidae (Diptera): taxonomy, biology and assessments of field populations. *Bulletin of Entomological Research* **63**: 305–325.

Harris, K.M. 1982. The Aphid Midge: a brief history. *Antenna* **6**: 286–289.

Harrison, J.W.H. 1916a. Notes and records. Homoptera. Scale insects. *Vasculum* **2**: 28.

Harrison, J.W.H. 1916b. Notes and records. Coccidae. Scale insects. *Vasculum* **2**: 93.

Hassan, A.L. 1939. The biology of some British Delphacidae (Homopt.) and their parasites with special reference to the Strepsiptera. *Transactions of the Royal Entomological Society of London* **89**: 345–384.

Heads, P.A. 1986. Bracken, ants and extrafloral nectaries. IV. Do Wood Ants (*Formica lugubris*) protect the plant against insect herbivores? *Journal of Animal Ecology* **55**: 795–809.

Heie, O.E. 1980. The Aphidoidea (Hemiptera) of Fennoscandia and Denmark. I. General part. The families Mindaridae, Hormaphididae, Thelaxidae, Anoeciidae, and Pemphigidae. *Fauna Entomologica Scandinavica* **9**. 236 pp.

Heie, O.E. 1982. The Aphidoidea (Hemiptera) of Fennoscandia and Denmark. II. The family Drepanosiphidae. *Fauna Entomologica Scandinavica* **11**. 176 pp, 4 pls.

Heie, O.E. 1986. The Aphidoidea (Hemiptera) of Fennoscandia and Denmark. III. Family Aphididae: subfamily Pterocommatinae and tribe Aphidini of subfamily Aphidinae. *Fauna Entomologica Scandinavica* **17**. 314 pp.

Heinze, K. 1960. Systematik der mitteleuropäischen Myzinae mit besonderer Berücksichtigung der im Deutschen Entomologischen Institut befindlichen Sammlung Carl Börner (Homoptera: Aphidoidea – Aphididae). *Beiträge zur Entomologie* **10**: 744–842.

Heinze, K. 1961. Ibid (concl.) *Beiträge zur Entomologie* **11**: 24–96.

Heinze, K. 1962. Pflanzenschädliche Blattlausarten der Familien Lachnidae, Adelgidae und Phylloxeridae, eine systematisch-faunistische Studie. *Deutsche Entomologische Zeitschrift* (N.F.) **9**: 143–227.

Heiss, E. & Péricart, J. 1975. Introduction à une revision des *Piesma* palearctiques. Étude du matériel-type; établissement de diverses synonymies et de nouveaux regroupements (Hemiptera Piesmatidae). *Annales de la Société Entomologique de France* (N.S.) **11**: 517–540.

Herting, B. 1971. *A catalogue of parasites and predators of terrestrial arthropods.* (Section A) Host or prey/enemy. 1. Arachnida to Heteroptera. 129 pp. CIBC. Farnham Royal.

Herting, B. 1972a. Ibid. (A) 2. Homoptera. 210 pp.

Herting, B. 1972b. Ibid. (C) Bibliography. 1. A–L. Pp 1–265.

Herting, B. 1972c. Ibid. (C) Bibliography. 2. M–Z. Pp 266–497.

Herting, B. 1973. Ibid. (A) 3. Coleoptera to Strepsiptera. 185 pp.

Herting, B. 1977. Ibid. (A) 4. Hymenoptera. 206 pp.

Herting, B. 1978. Ibid. (A) 5. Neuroptera, Diptera, Siphonaptera. 156 pp.

Herting, B. 1980. Ibid. (B) Enemy/host or prey. 1. All except Hymenoptera Terebrantia. 178 pp.

Hille Ris Lambers, D. 1933. Notes on Theobald's 'The Plantlice or Aphididae of Great Britain'. *Stylops* **2**: 169–176.

Hille Ris Lambers, D. 1934. Notes on Theobald's 'The Plantlice or Aphididae of Great Britain'. *Stylops* **2**: 25–33.

Hille Ris Lambers, D. 1938. Contributions to a monograph of the Aphididae of Europe. I. *Temminckia* **3**: 1–44; 4 pls.

Hille Ris Lambers, D. 1939. Contributions to a monograph of the Aphididae of Europe. II. *Temminckia* **4**: 1–134; 6 pls.
Hille Ris Lambers, D. 1947. Contributions to a monograph of the Aphididae of Europe. III. *Temminckia* **7**: 179–319; 7 pls.
Hille Ris Lambers, D. 1949. Contributions to a monograph of the Aphididae of Europe. IV. *Temminckia* **8**: 182–329; 6 pls.
Hille Ris Lambers, D. 1953. Contributions to a monograph of the Aphididae of Europe. V. *Temminckia* **9**: 1–176; 6 pls.
Hillyard, P.D. & Sankey, J.L.P. 1990. Harvestmen. *Synopsies of the British Fauna* (New Series) **5** (second edition). 120pp.
Hille Ris Lambers, D. 1966. Polymorphism in Aphididae. *Annual Review of Entomology* **11**: 47–78.
Hincks, W.D. 1950. Notes on some British Mymaridae (Hym.). *Transactions of the Society for British Entomology* **10**: 167–207.
Hincks, W.D. 1952. The British species of the genus *Ooctonus* Haliday, with a note on some recent work on the fairy flies (Hym., Mymaridae). *Transactions of the Society for British Entomology* **11**: 153–163.
Hincks, W.D. 1959. The British species of the genus *Alaptus* Haliday *in* Walker (Hym., Chalc., Mymaridae). *Transactions of the Society for British Entomology* **13**: 137–148.
Hodek, I. 1967. Bionomics and ecology of predaceous Coccinellidae. *Annual Review of Entomlogy* **12**: 79–104.
Hodkinson, I.D. 1973. The biology of *Strophingia ericae* (Curtis) (Homoptera, Psylloidea) with notes on its primary parasite *Tetrastichus actis* (Walker) (Hym., Eulophidae). *Norsk Entomologisk Tidsskrift* **20**: 237–243.
Hodkinson, I.D. 1974. The biology of the Psylloidea (Homoptera): a review. *Bulletin of Entomological Research* **64**: 325–339.
Hodkinson, I.D. & Hollis, D. 1980. *Floria variegata* Löw (Homoptera: Psylloidea) in Britain. *Entomologist's Gazette* **31**: 171–172.
Hodkinson, I.D. & White, I. M. 1979. Homoptera Psylloidea. *Handbooks for the Identification of British Insects* **2** (5a). iv + 94 pp.
Houk, E.J. & Griffiths, G.W. 1980. Intracellular symbiotes of the Homoptera. *Annual Review of Entomology* **25**: 161–187.
Hueber, J.J. 1986. Systematics, biology, and hosts of the Mymaridae and Mymarommatidae (Insecta: Hymenoptera): 1758–1984. *Entomography* **4**: 185–243.
Iheagwam, E.U. 1977. Comparative flight performance of the seasonal morphs of the cabbage whitefly, *Aleyrodes brassicae* (Wlk.), in the laboratory. *Ecological Entomology* **2**: 267–271.
Itô, Y. 1989. The evolutionary biology of sterile soldiers in aphids. *Trends in Ecology and Evolution* **4**: 69–73.
Jansson, A. 1986. The Corixidae (Heteroptera) of Europe and some adjacent regions. *Acta Entomologica Fennica* **47**. 94 pp.
Janvier, H. 1960. Recherches sur les Hymenopteres nidifiants aphidivores. I. *Annales des Sciences Naturelles* (Zool.) **2**: 281–321.
Janvier, H. 1961a. Ibid. II. *Annales des Sciences Naturelles* (Zool.) **3**: 1–51.
Janvier, H. 1961b. Ibid. III. *Annales des Sciences Naturelles* (Zool.) **3**: 847–883.
Janvier, H. 1962. Ibid. IV-VII. *Annales des Sciences Naturelles* (Zool.) **4**: 489–514.
Jensen, D.D. 1957. Parasites of the Psyllidae. *Hilgardia* **27**: 71–99.
Jervis, M.A. 1977. A new key for the identification of the British species of *Aphelopus* (Hym; Dryinidae). *Systematic Entomology* **2**: 301–303.
Jervis, M.A. 1979. Courtship, mating and "swarming" in *Aphelopus melaleucus* (Dalman) (Hymenoptera: Dryinidae). *Entomologist's Gazette* **30**: 191–193.
Jervis, M.A. 1980a. Studies on oviposition behaviour and larval development in species of *Chalarus* (Diptera, Pipunculidae), parasites of typhlocybine leafhoppers. *Journal of Natural History* **14**: 759–768.
Jervis, M.A. 1980b. Life history studies on *Aphelopus* species (Hymenoptera, Dryinidae) and *Chalarus* species (Diptera, Pipunculidae), primary parasites of typhlocybine leafhoppers (Homoptera, Cicadellidae). *Journal of Natural History* **14**: 769–780.
Jervis, M.A. 1980c. Ecological studies on the parasite complex associated with typhlocybine leafhoppers (Homoptera, Cicadellidae). *Ecological Entomology* **5**: 123–136.

John, B. & Claridge, M. 1974. Chromosome variations in British populations of *Oncopsis* (Homoptera: Cicadellidae). *Chromosoma* **46**: 77–89.
Joy, N.H. 1932. *A practical handbook of British beetles*. Vol. **1**, xxvii + 622 pp; Vol. **2**, 194 pp. London.
Kerzhner, I.M. 1981. Hemiptera of the family Nabidae. *Fauna of the USSR* (New series) **124** (=Insecta **13** (2)). 326 pp. [In Russian]
Kerzhner, I.M. & Jaczewski, T.L. 1964. Order Hemiptera (Heteroptera). Pp 656–845. [In Russian. English translation, 1967, pp 851–1118.] *In*: Bei-Bienko, 1964, q.v.
Killington, F.J. 1936. *A monograph of the British Neuroptera*. 2 vols. 575 pp; 30 pls. Ray Society. London.
Kloet, G.S. & Hincks, W. D. 1964. A check list of British insects. Second edition (completely revised). Part 1. *Handbooks for the Identification of British Insects* **11** (1). xv + 119 pp.
Kloet, G.S. & Hincks, C. 1978. A check list of British insects. Second ed. (completely revised). Part 4: Hymenoptera. *Handbooks for the Identification of British Insects* **11** (4). ix + 159 pp.
Knight, W.J. 1965. Techniques for use in the identification of leafhoppers (Homoptera: Cicadellidae). *Entomologist's Gazette* **16**: 129–136.
Knight, W.J. 1966. A preliminary list of leaf-hopper species (Homoptera: Cicadellidae) occurring on plants of economic importance in Britain. *Entomologist's Monthly Magazine* **101**: 94–109.
Kosztarab, M. & Kosztarab, M.P. 1988. Studies on the morphology and systematics of scale insects. No. 14. A selected bibliography of the Coccoidea (Homoptera). Third supplement (1970–1985). *Virginia Polytechnic State University Agricultural Experiment Station Bulletin* **88** (1). ii + 252 pp.
Kosztarab, M. & Kozár, F. 1988. Scale insects of Central Europe. *Series Entomologica* **40**. 456 pp.
Kozlov, M.A. & Kononova, S.V. 1983. Telenomine fauna of the USSR (Hymenoptera, Scelionidae, Telemoninae). *Opredeliteli po Faune SSSR* **136**. 335 pp. [In Russian]
Kryger, J.P. 1950. The European Mymaridae comprising the genera known up to c. 1930. *Entomologiske Meddeleser* **26**. 97 pp.
Kukashev, D.Sh. 1983. The water bug Sigara concinna – intermediate host of the cestode Tatria biremis (Cestoda, Amabiliidae). *Parazitologiya* **17**: 165–167. [In Russian]
Lal, K.B. 1934. Insect parasites of Psyllidae. *Parasitology*, Cambridge **26**: 325–334.
Lansbury, I. 1961. Comments on the genus *Cimex* (Hem. Het. Cimicidae) in the British Isles. *Entomologist* **94**: 133–134.
Lansbury, I. 1983. Notes on the Australian species of *Cymatia* Flor s.l. (Insecta: Heteroptera: Corixidae). *Transactions of the Royal Society of South Australia* **107**: 51–57.
Laurence, B.R. 1952. The prey of some Empididae and Dolichopodidae (Dipt.). *Entomologist's Monthly Magazine* **88**: 156–157.
Lauterer, P. & Baudys, E. 1968. Description of a new gall on Chamaenerion angustifolium (L.) Scop. produced by the larva of Craspedolepta subpunctata (Forst.), with notes on the bionomics of this psyllid. *Casopis moravského Musea* **53**: 243–248, 2 pls.
Leatherdale, D. 1970. The arthropod hosts of entomogeneous fungi in Britain. *Entomophaga* **15**: 419–435.
Lees, A.D. 1966. The control of polymorphism in aphids. Pp 207–277. *In*: Beament, J., Treherne, J.E. & Wigglesworth, V.B. (Eds) *Advances in Insect Physiology* **3**. 382 pp.
Lees, D.R. & Dent, C.S. 1983. Industrial melanism in the spittlebug *Philaenus spumarius* (L.) (Homoptera: Aphrophoridae). *Biological Journal of the Linnean Society* **19**: 115–129.
Le Quesne, W.J. 1960. Hemiptera (Fulgoromorpha). *Handbooks for the Identification of British Insects* **2** (3). 68 pp.
Le Quesne, W.J. 1965a. Trigonocranus emmeae (Hem., Cixiidae) new to Britain. *Entomologist's Monthly Magazine* **100**: 117.
Le Quesne, W.J. 1965b. Hemiptera (Cicadomorpha) (excluding Deltocephalinae and Typhlocybinae). *Handbooks for the Identification of British Insects* **2** (2a). 64 pp.
Le Quesne, W.J. 1969. Hemiptera (Cicadomorpha – Deltocephalinae). *Handbooks for the Identification of British Insects* **2** (2b). 148 pp.
Le Quesne, W.J. 1983. *Cicadula flori* (Sahlberg), new to Britain (Hem., Cicadellidae). *Entomologist's Monthly Magazine* **119**: 177.
Le Quesne, W.J. 1987. *Cicadella lasiocarpae* Ossiannilsson (Hemiptera: Cicadellidae) new to Britain. *Entomologist's Gazette* **38**: 87–89.

Le Quesne, W.J. & Payne, W.R. 1981. Cicadellidae (Typhlocybinae) with a check list of the British Auchenorrhyncha (Hemiptera, Heteroptera) *Handbooks for the Identification of British Insects* **2** (2c). 95 pp.
Le Quesne, W.J. & Woodroffe, G.E. 1976. Geographical variation in the genitalia of three species of Cicadellidae (Hemiptera). *Systematic Entomology* **1**: 169–172.
Lesne, P. 1905. Les relations des fourmis avec les Hémiptères Homoptères de la famille des Fulgorides; domestication des Tettigometra. *Bulletin de la Société Entomologique de France* **1905**: 161–164
Leston, D. 1961. Observations on the Mirid (Hem.) hosts of Braconidae (Hym.) in Britain. *Entomologist's Monthly Magazine* **97**: 65–71.
Lindroth, C.H. 1974. Coleoptera Carabidae. *Handbooks for the Identification of British Insects* **4** (2). 148 pp.
Lipa, J.J. 1966. Miscellaneous observations on Protozoan infections of *Nepa cinerea* Linnaeus including descriptions of two previously unknown species of Microsporidia, *Nosema bialoviesianae* sp. n. and *Thelohania nepae* sp. n. *Journal of Invertebrate Pathology* **8**: 158–166.
Loan, C.C. 1965. Life cycle and development of *Leiophron pallipes* Curtis (Hymenoptera: Braconidae, Euphorinae) in five mirid hosts in the Belleville district. *Proceedings of the Entomological Society of Ontario* **95**: 115–121.
Loan, C.C. 1974. The European species of *Leiophron* Nees and *Peristenus* Foerster (Hymenoptera: Braconidae, Euphorinae). *Transactions of the Royal Entomological Society of London* **126**: 207–238.
Loan, C.C. 1980. Plant bug hosts (Heteroptera: Miridae) of some euphorine parasites (Hymenoptera: Braconidae) near Belleville, Ontario, Canada. *Naturaliste Canadien* **107**: 87–93.
Loan, C.C. 1983. Host and generic relations of the Euphorini (Hymenoptera: Braconidae). *Contributions of the American Entomological Institute* **20**: 388–397.
Loan, C.C. & Bilewicz-Pawinska, T. 1973. Systematics and biology of four Polish species of *Peristenus* Foerster (Hymenoptera: Braconidae, Euphorinae). *Environmental Entomology* **2**: 271–278.
Locket, G.H. & Millidge, A.F. 1951. *British spiders* Vol. **1**. ix + 310 pp; frontisp. Ray Society. London.
Locket, G.H. & Millidge, A.F. 1953. *British spiders*. Vol. **2**. vii + 449 pp.
Locket, G.H., Millidge, A.F. & Merrett, P. 1974. *British spiders*. Vol. **3**. ix. + 314 pp.
Loginova, M.M. 1979. New species of jumping plant lice (Homoptera, Psylloidea) in the arid zone of the USSR *Trudy Zoologicheskogo Instituta Akademiya Nauk SSSR* **38**: 15–25. [In Russian]
Lundblad, O. 1927. Die Hydracarinen Schwedens. 1. Beitrag zur Systematik, Embryologie, Okologie und Verbreitungsgeschichte der Schwedischen Arten. *Zoologiska Bidrag fran Uppsala* **11**: 181–540.
McAlpine, J.F 1960. A new species of *Leucopis* (*Leucopella*) from Chile and a key to the World genera and subgenera of Chamaemyiidae (Diptera). *Canadian Entomologist* **92**: 51–58.
Macan, T.T. 1956. A revised key to the British water bugs (Hemiptera-Heteroptera). *Freshwater Biological Association Scientific Publication* **16**. 74 pp. 2nd ed., 1965. 78 pp. See also Savage, 1989.
Macgillivray, A.D. 1921. *The Coccidae*. vii + 502 pp. Illinois.
Mackauer, M. & Starý, P. 1967. Hym. Ichneumonoidea: World Aphidiidae. In: Delucchi, V. & Remaudière, G. (Eds) *Index of Entomophagous Insects*. 195 pp. Paris.
Madelin, M.F. 1966. Fungal parasites of insects. *Annual Review of Entomology* **11**: 423–448.
Manawadu, D. 1986. A new species of *Eriopeltis* Signoret (Homoptera: Coccidae) from Britain. *Systematic Entomology* **11**: 317–326.
Maramorosch, K. & Harris, K.F. (Eds) 1979. *Leafhopper vectors and plant pathogens*. xvi + 654 pp. London
Marshall, J., Jessop, L. & Hammond, P.M. (in prep.) Beetle larvae – a guide to families. *Handbooks for the Identification of British Insects* **5** (18).
Martin, J.H. 1978. Aleurochiton complanatus (Baerensprung) (Homoptera, Aleyrodidae) – confirmation of occurrence in Britain. *Entomologist's Monthly Magazine* **113**: 7.
Martin, J.H. 1981. A new species of *Acyrthosiphon* (Homoptera, Aphididae) from *Primula* in Britain. *Systematic Entomology* **6**: 97–101.
Martin, J.H. 1987. An identification guide to common whitefly pest species of the world (Homoptera, Aleyrodidae). *Tropical Pest Management* **33**: 298–322.
Massee, A.M. 1955. The county distribution of the British Hemiptera-Heteroptera. Second ed. *Entomologist's Monthly Magazine* **91**: 7–27.
Massee, A.M. 1960. Massed flights of Aradus depressus (F.) (Hem., Aradidae) noted at East Malling, Kent. *Entomologist's Monthly Magazine* **96**: 7.

Matthews, M.J. 1986. The British species of *Gonatocerus* Nees (Hymenoptera: Mymaridae), egg parasitoids of Homoptera. *Systematic Entomology* **11**: 213–229.

Mauri, G. 1982. Note sulla biologia della cicaletta nerorossa (Cercopis sanguinea Geoffr.) (Auchen., Cercopidae) e possibilità di lotta. *Mitteilungen der Schweizerischen Entomologischen Gesellschaft* **55**: 87–92.

Mayakovsky, V. 1929. *The Bedbug.* Moscow. [In Russian]

Measday, A.V. 1979. A rare form of Cercopis vulnerata Ill. (Hemiptera – Homoptera: Cicadoidea). *Entomologist's Record and Journal of Variation* **91**: 285.

Mercet, R.G. 1921. *Fauna Iberica. Himenopteros Fam. Encirtidos.* xi + 732 pp. Madrid.

Miller, D.R. & Kosztarab, M. 1979. Recent advances in the study of scale insects. *Annual Review of Entomology* **24**: 1–27.

Minks, A.K. & Harrewijn, P. (Eds) 1987. Aphids. Their biology, natural enemies and control. Vol. A. xx + 450 pp. *World Crop Pests* **2** (A). Amsterdam.

Minks, A.K. & Harrewijn, P. (Eds) 1988. *Ibid.* Vol. B. xix + 364 pp. *World Crop Pests* **2** (B). Amsterdam.

Minks, A.K. & Harrewijn, P. (Eds) 1989. *Ibid.* Vol. C. xvi + 312 pp. *World Crop Pests* **2** (C). Amsterdam.

Mochida, O. 1973. The characters of two wing-forms of *Javesella pellucida* (F.) (Homoptera: Delphacidae) with special reference to reproduction. *Transactions of the Royal Entomological Society of London* **125**: 177–225.

Morley, C. 1905. *The Hemiptera of Suffolk.* x + 34 pp. Plymouth.

Morris, M.G. 1972. Distributional and ecological notes on *Ulopa trivia* Germar (Hem., Cicadellidae). *Entomologist's Monthly Magazine* **107**: 174–181.

Morris, M.G. 1990. Orthocerous weevils. Coleoptera: Curculionoidea (Nemonychidae, Anthribidae, Urodontidae, Attelabidae and Apionidae). *Handbooks for the Identification of British Insects* **5** (14). 108 pp.

Morrison, H. & Morrison, E.R. 1965. A selected bibliography of the Coccoidea. First Supplement. *Miscellaneous Publications. United States Department of Agriculture* **987**. 44 pp.

Morrison, H. & Renk, A.V. 1957. A selected bibliography of the Coccoidea. *Miscellaneous Publications. United States Department of Agriculture* **734**. 222 pp.

Motas, C. 1928. Contribution à la connaissance des Hydracariens français particulièrement du Sud-Est de la France. *Travaux du Laboratoire d'Hydrobiologie et Pisciculture de l'Université de Grenoble* **20**. 373 pp.

Mound, L.A. 1962. *Aleurotrachelus jelinekii* (Frauen.) (Homoptera, Aleyrodidae) in southern England. *Entomologist's Monthly Magazine* **97**: 196–197.

Mound, L.A. 1966. A revision of the British Aleyrodidae (Hemiptera – Homoptera). *Bulletin of the British Museum (Natural History) (Entomology)* **17**: 397–428.

Mound, L.A. & Halsey, S.H. 1978. *A systematic catalogue of the Aleyrodidae (Homoptera) with host plant and natural enemy data.* 340 pp. Chichester.

Müller, F.P. 1973. Aphiden an Moosen (Homoptera, Aphididae). *Entomologische Abhandlungen und Berichte aus dem Staatlichen Museum für Tierkunde in Dresden* **39**: 205–242.

Müller, F.P. 1975. Bestimmungsschlüssel für geflügelte Blattläuse in Gelbschalen. *Archiv für Phytopathologie und Pflanzenschutz* **11**: 49–77.

Müller, H.J. 1942. Ueber Bau und Funktion des Legeapparatus der Zikaden (Homoptera Cicadina). *Zeitschrift für Morphologie und Ökologie der Tiere* **38**: 534–629.

Müller, H.J. 1954. Der Saisondimorphismus bei Zikaden der Gattung *Euscelis* Brullé (Homoptera Auchenorrhyncha). *Beiträge zur Entomologie* **4**: 1–56.

Müller H.J. 1974. Farb-Polymorphismus bei Larven der Jasside *Mocydia crocea* H.S. (Homoptera, Auchenorrhyncha). *Zoologischer Anzeiger* **192**: 303–315.

Müller, H.J. 1979. Zur weiteren Analyse des larvalen Polymorphismus der Jasside *Mocydia crocea* H.S. (Homoptera Auchenorrhyncha). *Zoologische Jahrbücher* (Syst.) **106**: 311–343.

Müller, H.J. 1982. On the larval polymorphism in *Mocydia crocea* H.S. *Acta Entomologica Fennica* **38**: 28–29.

Müller, H.J. 1984a. Ueber den Voltinismus der Dornzikade *Centrotus cornutus* (L.) (Homoptera Auchenorrhyncha: Membracidae) und die Einnischung mehrjähriger Insekten. *Zoologische Jahrbücher* (Syst.) **111**: 321–337.

Müller, H.J. 1984b. Zur Entwicklung und Lebensweise der Larven der Dornzikade *Centrotus cornutus* (L.) (Homoptera Auchenorrhyncha: Membracidae) unter besonderer Berücksichtigung der Kotschleuder. *Zoologische Jahrbücher* (Anat.) **111**: 385–399.

Müller, H.J. 1987. Ueber die Vitalität der Larvenformen der Jasside *Mocydia crocea* (H.-S.) (Homoptera Auchenorrhyncha) und ihre ökologische Bedeutung. *Zoologischer Jahresbericht (Systematik)* **114**: 105–129; 1 col. pl.

Müller-Kögler, E. 1965. *Pilzkrankheiten bei Insekten.* 444 pp. Berlin and Hamburg.

Nast, J. 1972. *Palaearctic Auchenorrhyncha (Homoptera) an annotated check list.* 550 pp. Warsaw.

Nault, L.R. 1987. Origin and evolution of Auchenorrhyncha-transmitted plant infecting viruses. Pp 131–149. *In*: Wilson, M.R. & Nault, L.R. (Eds) *Proceedings of the Second International Workshop on Leafhoppers and Planthoppers of Economic Importance* Provo, Utah

Nault, L.R. & Ammar, El D. 1989. Leafhopper and planthopper transmission of plant viruses. *Annual Review of Entomology* **34**: 503–530.

Nault, L.R. & Rodriguez, J.G. (Eds) 1985. *The leafhoppers and planthoppers.* xvi + 500 pp. New York.

New, T.R. 1975. The biology of Chrysopidae and Hemerobiidae (Neuroptera), with reference to their usage as biocontrol agents: a review. *Transactions of the Royal Entomological Society of London* **127**: 115–140.

Newstead, R. 1901. *Monograph of the Coccidae of the British Isles.* Vol. **1**. xii + 220 pp; pls A-E + I-XXXIV. London.

Newstead, R. 1903. *Monograph of the Coccidae of the British Isles.* Vol. **2**. viii + 270 pp; pls F + XXXV-LXXV. London.

Nikol'skaya, M.N. 1952. The chalcids in the fauna of the USSR (Chalcidoidea). *Opredeliteli po Faune SSSR* **44**: 1–575. [In Russian. English translation: Israel program for scientific translations, Jerusalem, 1963: 1–593.]

Nixon, G.E.J. 1957. Hymenoptera Proctotrupoidea: Diapriidae subfamily Belytinae. *Handbooks for the Identification of British Insects* **8** (3) (dii). 107 pp.

Noyes, J.S. 1978. New Chalcids recorded from Great Britain and notes on some Walker specimens (Hym., Encyrtidae, Eulophidae). *Entomologist's Monthly Magazine* **113**: 9–13.

Olmi, M. 1984. A revision of the Dryinidae (Hymenoptera). *Memoirs of the American Entomological Institute* **37**. 2 Vols. 1913 pp.

Onillon, J.C. 1969. Étude du complexe parasitaire *Trioza urticae* L. (Homoptère Psyllidae), *Tetrastichus upis* Walk. (Hymenopt. Tetrastichidae). *Annales de Zoologie – Ecologie Animale* **1**: 55–65.

Ossiannilsson, F. 1978. The Auchenorrhyncha (Homoptera) of Fennoscandia and Denmark. Part 1: Introduction, infraorder Fulgoromorpha. *Fauna Entomologica Scandinavica* **7** (1): 1–222; 4 pls.

Ossiannilsson, F. 1981. The Auchenorrhyncha (Homoptera) of Fennoscandia and Denmark. Part 2: The Families Cicadidae, Cercopidae, Membracidae, and Cicadellidae (excl. Deltocephalinae). *Fauna Entomologica Scandinavica* **7** (2): 223–593.

Ossiannilsson, F. 1983. The Auchenorrhyncha (Homoptera) of Fennoscandia and Denmark. Part 3: The Family Cicadellidae: Deltocephalinae, Catalogue, Literature and Index. *Fauna Entomologica Scandinavica* **7** (3): 594–979.

Oudemans, A.C. 1912. Die bis jetzt bekannten Larven von Trombidiidae und Erythraeidae. *Zoologische Jahrbücher* Suppl. **14**. 230 pp.

Palmer, J.M. 1986. Thrips in English Oak trees. *Entomologist's Gazette* **37**: 245–252.

Parker, N.J.B. 1975. An investigation of reproductive diapause in two British populations of *Anthocoris nemorum* (Hemiptera: Anthocoridae). *Journal of Entomology* (A) **49**: 173–178.

Payne, K. 1973. A survey of the *Spartina*-feeding insects in Poole Harbour, Dorset. *Entomologist's Monthly Magazine* **108**: 66–79.

Payne, K. 1979. Auchenorrhyncha (Homoptera) of Gait Barrows National Nature Reserve. *Entomologist's Monthly Magazine* **114**: 210.

Patch, E.M. 1938. Food plant catalogue of the aphids of the World. *Maine Agricultural Experiment Station Bulletin* **393**: 1–431. [For index, see Averill, 1945]

Peck, O., Bouček, Z. & Hoffer, A. 1964. Keys to the Chalcidoidea of Czechoslovakia (Insecta: Hymenoptera). *Memoirs of the Entomological Society of Canada* **34**. 120 pp.

Péricart, J. 1972. Hémiptères Anthocoridae, Cimicidae et Microphysidae de l'Ouest-Paléarctique. *Faune de l'Europe et du Bassin Méditerranéen* **7**. 402 pp.

Péricart, J. 1983. Hémiptères Tingidae euro-méditerranéens. *Faune de France* **69**. x + 622 pp; 6 pls.

Péricart, J. 1984. Hémiptères Berytidae euro-méditerranéens. *Faune de France* **70**. viii + 172 pp.

Péricart, J. 1987. Hémiptères Nabidae d'Europe occidentale et du Maghreb. *Faune de France* **71**. xi + 188 pp; 3 pls.

Péricart, J. 1990. Hémiptères Saldidae et Leptopodidae d'Europe occidentale et du Maghreb. *Faune de France* **71**. 238 pp.
Perkins, J.F. 1976. Hymenoptera Bethyloidea (except Chrysididae). *Handbooks for the Identification of British Insects* **6** (3a). 38 pp.
Poinar, G.O. 1975. *Entomogeneous nematodes: a manual and host-list of insect-nematode associations* vi + 317 pp. Leiden.
Poinar, G.O. & Thomas, G.M. 1984. *Laboratory guide to insect pathogens and parasites* xvi + 392 pp. New York and London.
Poisson, R. 1933. Trois nouvelles espèces de nématodes de la cavité générale d'hémiptères aquatiques. *Annales de Parasitologie Humaine et Comparée* **11**: 463–466.
Poisson, R. 1957. Hétéroptères aquatiques. *Faune de France* **61**. 263 pp.
Poisson, R. & Pesson, P. 1951. Super-ordre des Hémiptéroïdes. Pp 1385–1803 (Pesson, Ordre des Homoptères, pp 1390–1647; Poisson, Ordre des Hétéroptères, pp 1657–1803). *In*: Grassé, P. P. (ed.) *Traité de Zoologie. Anatomie, Systématique, Biologie* **10** (2): 976–1948. Paris.
Polaszek, A. 1986. Aestivating sexual morphs in the aphid genus *Thelaxes* (Insecta: Homoptera). *Journal of Natural History* **20**: 1333–1338.
Polaszek, A. & Cotman, H.E. 1983. An aphid new to Europe from *Robinia pseudacacia* L. in London. *Entomologist's Monthly Magazine* **119**: 251–252.
Pontin, A.J. 1959. Some records of predators and parasites adapted to attack aphids attended by ants. *Entomologist's Monthly Magazine* **95**: 154–155.
Pontin, A.J. 1961. The prey of *Lasius niger* (L.) and *L. flavus* (F.) (Hym., Formicidae). *Entomologist's Monthly Magazine* **97**: 135–137.
Pontin, A.J. 1978. The numbers and distribution of subterranean aphids and their exploitation by the ant *Lasius flavus* (Fabr.). *Ecological Entomology* **3**: 203–207.
Pope, R.D. 1953. Coleoptera Coccinellidae and Sphindidae. *Handbooks for the Identification of British Insects* **5** (7). 12 pp.
Pope, R.D. 1973. The species of *Scymnus* (s. str.), *Scymnus* (*Pullus*) and *Nephus* (Col., Coccinellidae) occurring in the British Isles. *Entomologist's Monthly Magazine* **109**: 3–39.
Poulton, E.B. 1906. Predaceous insects and their prey. *Transactions of the Entomological Society of London* **1906**: 323–409.
Prasad, V. & Cook, D.R. 1972. Water mite larvae. *Memoirs of the American Entomological Institute* **18**. 326 pp.
Prestidge, R.A. 1982. Instar duration, adult consumption, oviposition and nitrogen utilization efficiencies of leafhoppers feeding on different quality food (Auchenorrhyncha: Homoptera). *Ecological Entomology* **7**: 91–101.
Prior, R.N.B. 1965. Two new techniques used in leafhopper taxonomy which may also be applicable to other orders of small insects requiring maceration and partial dissection. *Entomologist's Monthly Magazine* **100**: 246–249.
Prior, R.N.B. 1972. A note on the capturing and feeding behaviour of *Platypalpus notata* Meigen (Dipt., Empididae) on the leafhopper prey *Macrosteles sexnotatus* (Fall.) (Hem., Cicadelllidae). *Entomologist's Monthly Magazine* **107**: 183–184.
Prior, R.N.B. 1976. *Metopolophium festucae* (Theob.) (Homoptera, Aphidoidea) accidentally parasitised by larvae of *Elenchus tenuicornis* (Kirby) (Strepsiptera, Stylopoidea). *Entomologist's Monthly Magazine* **111**: 91; pl. 3.
Pungerl, N.B. 1986. Morphometric and electrophoretic study of *Aphidius* species (Hymenoptera: Aphidiidae) reared from a variety of aphid hosts. *Systematic Entomology* **11**: 327–354; 1 fold-out table.
Ribaut, H. 1936. Homoptères Auchenorrhynches. I. (Typhlocybinae). *Faune de France* **31**. 228 pp.
Ribaut, H. 1952. Homoptères Auchenorrhynches. II. (Jassidae). *Faune de France* **57**. 474 pp.
Richards, O.W. 1956. Hymenoptera: introduction and keys to families. *Handbooks for the Identification of British Insects* **6** (1). 94 pp. 2nd ed, 1977. 100 pp.
Richards, O.W. 1980. Scolioidea, Vespoidea and Sphecoidea; Hymenoptera, Aculeata. *Handbooks for the Identification of British Insects* **6** (3b). 118 pp.
Richards, W.R. 1976. A host index for species of Aphidoidea described during 1935 to 1969. *Canadian Entomologist* **108**: 499–550.

Robinson, D.M. 1961a. The parasites of Psyllidae – 2. Parapsyllaephagus adulticolus gen. et sp. nov., the first hymenopterous parasite of an adult psyllid (Homoptera). *Annals and Magazine of Natural History* (13) **4**: 117–121.

Robinson, D.M. 1961b. The parasites of Psyllidae – 3. Some notes on the biology and host relationships of *Parapsyllaephagus adulticolus* Robinson (Hymenoptera). *Annals and Magazine of Natural History* (13) **4**: 155–159.

Rosen, D. (Ed.) 1990. Armored scale insects. Their biology, natural enemies and control. Vol. A. xvi + 384 pp. *World Crop Pests* **4** (A). Amsterdam.

Rotheray, G.E. 1987. The larvae and puparia of five species of aphidophagous Syrphidae (Diptera). *Entomologist's Monthly Magazine* **123**: 121–125.

Rothschild, G.H.L. 1964. The biology of *Conomelus anceps* Germar (Homoptera: Delphacidae). *Transactions of the Society for British Entomology* **16**: 135–148.

Rothschild, G.H.L. 1966. A study of a natural population of Conomelus anceps (Germar) (Homoptera: Delphacidae) including observations on predation using the precipitin test. *Journal of Animal Ecology* **35**: 413–434.

Russell, L.M., Kosztarab, M. & Kosztarab, M.P. 1974. A selected bibliography of the Coccoidea. Second supplement. *Miscellaneous Publications. United States Department of Agriculture* **1281**. 122 pp.

Russell, R.J. 1970. The effectiveness of *Anthocoris nemorum* and *A. confusus* (Hemiptera: Anthocoridae) as predators of the sycamore aphid, *Drepanosiphum platanoides* I. The number of aphids consumed during development. *Entomologia Experimentalis et Applicata* **13**: 194–207.

Sabrosky, C.S. 1957. On mounting Diptera from fluid. *Bulletin of the Entomological Society of America* **3** (1): 38.

Sabrosky, C.S. 1966. Mounting insects from alcohol. *Bulletin of the Entomological Society of America* **12**: 349.

Samson, R.A., Evans, H.C. & Latgé, J.P. 1988. Atlas of entomopathogenic Fungi. xi + 187 pp. Berlin.

Saunders, E. 1892. *The Hemiptera Heteroptera of the British Islands*. vi + 350 pp; 32 pls. London.

Savage, A.A. 1989. Adults of the British aquatic Hemiptera Heteroptera. A key with ecological notes. *Freshwater Biological Association Scientific Publication* **50**. 173 pp.

Schaefer, M. 1982. Chelicerata. Pp 99–135. *In*: Tischler, W. 1982, q.v.

Schauff, M.E. 1984. The Holarctic genera of Mymaridae (Hymenoptera: Chalcidoidea). *Memoirs of the Entomological Society of Washington* **12**: 1–67.

Schmuck, R. 1987. Aggregation behavior of the Fire Bug, Pyrrhocoris apterus, with special reference to its assembling scent (Heteroptera: Pyrrhocoridae). *Entomologia Generalis* **12**: 155–169.

Schmutterer, H. 1952a. Die Ökologie der Cocciden (Homoptera, Coccoidea) Frankens. 1. Abschnitt. *Zeitschrift für Angewandte Entomologie* **33**: 369–420.

Schmutterer, H. 1952b. Ibid. 2. Abschnitt. *Zeitschrift für Angewandte Entomologie* **33**: 544–584.

Schmutterer, H. 1952c. Ibid. 3. Abschnitt. *Zeitschrift für Angewandte Entomologie* **34**: 65–100.

Schmutterer, H. 1959. Schildläuse oder Coccoidea. I. Deckschildläuse oder Diaspididae. *Tierwelt Deutschlands und der Angrenzenden Meeresteile* **45**: 1–260.

Séguy, E. 1934. Muscidae Acalypterae et Scatophagidae. *Faune de France* **28**. 832 pp; 27 pls.

Shaposhnikov, G.K. 1964. Suborder Aphidinea. Pp 489–616. [In Russian. English translation 1967, pp 616–799.] *In*: Bei-Bienko, 1964, q.v.

Shaw, M.W. 1964. A basic list of the Scottish Aphididae. *Transactions of the Society for British Entomology* **16**: 49–92.

Shaw, S.R. 1985. A phylogenetic study of the subfamilies Meteorinae and Euphorinae (Hymenoptera: Braconidae). *Entomography* **3**: 277–370.

Shirt, D.B. 1987. *British red data books* **2**. Insecta. Nature Conservancy Council. xliv + 402 pp.

Sikes, E.K. 1928. The external morphology and life-history of the coccid bug, *Orthezia urticae* Linn. *Proceedings of the Zoological Society of London* **1928**: 269–305; 2 pls.

Silvestri, F. 1919. Contribuzioni alla conoscenza degli insetti dannosi e dei loro simbionti. V. La cocciniglia del nocciuolo (*Eulecanium coryli* L.). *Bollettino del Laboratorio di Zoologia Generale e Agraria della R. Scuola Superiore d'Agricoltura. Portici* **13**: 127–192; 6 additional tables.

Simms, R.W., Freeman, P. & Hawksworth, D.L. 1988. Key works to the fauna and flora of the British Isles and North-western Europe. Fifth ed. *Systematics Association Special Vol.* **33**. xii + 312 pp.

Smith, K.G.V. 1963. A short synopsis of the British Chamaemyiidae (Dipt.). *Transactions of the Society for British Entomology* **15**: 103–115.
Southwood, T.R.E. 1957. The zoogeography of the British Hemiptera Heteroptera. *Proceedings of the South London Entomological and Natural History Society* **1956**: 111–136.
Southwood, T.R.E. 1966. *Ecological methods, with particular reference to the study of insect populations.* xviii + 931 pp. London.
Southwood, T.R.E. & Leston, D. 1959. *Land and water bugs of the British Isles.* 436 pp; 63 pls. London.
Soyka, W. 1956. Monographie der Polynemagruppe. *Abhandlungen der Zoologisch-Botanischen Gesellschaft in Wien* **19**: 1–115.
Sparing, I. 1959. Die Larven der Hydrachnellae, ihre parasitische Entwicklung und ihre Systematik. *Parasitologische Schriftenreihe* **10**: 1–168.
Staddon, B.W. & Griffiths, D. 1968. Some observations on the food of *Aeshna juncea* (L.) nymphs (Odonata) with particular reference to Corixidae (Hemiptera). *Entomologist's Monthly Magazine* **103**: 226–230.
Starý, P. 1966. *Aphid parasites of Czechoslovakia.* 242 pp; 21 pls. The Hague.
Starý, P. 1970. *Biology of aphid parasites (Hymenoptera: Aphidiidae) with respect to integrated control.* 643 pp. The Hague.
Starý, P. 1987. Subject bibliography of aphid parasitoids (Hymenoptera: Aphidiidae) of the World, 1758–1982. *Monographs to Applied Entomology* **25**. 101 pp. Hamburg.
Starý, P., Remaudière, G. & Leclant, F. 1971. Les Aphidiidae (Hym.) de France et leurs hôtes (Hom. Aphididae). *Entomophaga* (hors ser.) **5**. 73 pp.
Steinhaus, E.A. 1946. *Insect microbiology.* xi + 256 pp. Comstock. Ithaca, N. Y.
Steinhaus, E.A. 1949. *Principles of insect pathology.* xi + 219 pp. New York, Toronto and London.
Steinhaus, E.A. 1963. *Insect pathology: an advanced treatise.* 2 vols, xvii + 661 and xiv + 689 pp. New York and London.
Sternlicht, M. 1972. A new species of *Kermes* Boitard (Coccoidea, Kermesidae) from England on oak. *Entomologist's Gazette* **23**: 259–266.
Stewart, A.J.A. 1986a. The inheritance of nymphal colour/pattern polymorphism in the leafhoppers *Eupteryx urticae* (F.) and *E. cyclops* Mats. (Hemiptera: Auchenorrhyncha). *Biological Journal of the Linnean Society* **27**: 57–77.
Stewart, A.J.A. 1986b. Nymphal colour/pattern polymorphism in the leafhoppers *Eupteryx urticae* (F.) and *E. cyclops* (Mats.) (Hemiptera: Auchenorrhyncha): spatial and temporal variation in morph frequencies. *Biological Journal of the Linnean Society* **27**: 79–101.
Stewart, A.J.A. 1986c. Descriptions and key to nymphs of *Eupteryx* (Curtis) leafhoppers (Homoptera: Cicadellidae) occurring in Britain. *Systematic Entomology* **11**: 365–376.
Stewart, A.J.A. 1988. Patterns of host-plant utilization by leafhoppers in the genus *Eupteryx* (Hemiptera: Cicadellidae) in Britain. *Journal of Natural History* **22**: 357–379.
Stewart, R.K. 1969. The biology of *Lygus rugulipennis* Poppius (Hemiptera: Miridae) in Scotland. *Transactions of the Royal Entomological Society of London* **120**: 473–457.
Stichel, W. 1955–1962. *Illustrierte Bestimmungstabellen der Wanzen. II. Europa (Hemiptera-Heteroptera Europae).* 4 vols. Berlin-Hermsdorf. (Vol. 1. 1955–1956, pp 1–168. Vol. 2. 1956–1958, pp 169–907. Vol. 3. 1958–1960, 428 pp. Vol. 4. 1957–1962, 838 pp. General-Index. 1962–1962, 112 pp.)
Stiling, P. 1980. Host plant specificity, oviposition behaviour and egg parasitism in some leafhoppers of the genus *Eupteryx* (Hemiptera: Cicadellidae). *Ecological Entomology* **5**: 79–85.
Stoner, A. 1973. Incidence of *Wesmalia pendula* (Hymenoptera Braconidae), a parasite of male *Nabis* species in Arizona. *Annals of the Entomological Society of America* **66**: 471–473.
Stork, N.E. & Hammond, P.M. (in prep.) The arthropod fauna of oak trees in Richmond Park (U. K.) as revealed by knockdown insecticide sampling: I. Materials and methods. *Ecological Entomology.*
Strand, M.R., Meola, S.M. & Vinson, S.B. 1986. Correlating pathological symptoms in *Heliothis virescens* eggs with development of the parasitoid *Telenomus heliothidis. Journal of Insect Physiology* **32**: 389–402.
Stroyan, H.L.G. 1950. Recent additions to the British aphid fauna. Part I: *Dactynotus* Rafinesque to *Rhopalosiphum* Koch, C. L. *Transactions of the Royal Entomlogical Society of London* **101**: 89–124.
Stroyan, H.L.G. 1952. The identification of aphids of economic importance. *Plant Pathology* **1**: 9–14, 42–48, 92–99, 123–129.

Stroyan, H.L.G. 1955. Recent additions to the British aphid fauna. Part II. *Transactions of the Royal Entomological Society of London* **106**: 283–340.
Stroyan, H.L.G. 1957. Further additions to the British aphid fauna. *Transactions of the Royal Entomological Society of London* **109**: 311–359.
Stroyan, H.L.G. 1964. Notes on hitherto unrecorded or overlooked British aphid species. *Transactions of the Royal Entomological Society of London* **116**: 29–72.
Stroyan, H.L.G. 1972. Additions and amendments to the check list of British aphids (Homoptera: Aphidoidea). *Transactions of the Royal Entomological Society of London* **124**: 37–79.
Stroyan, H.L.G. 1977. Aphidoidea – Chaitophoridae and Callaphididae. *Handbooks for the Identification of British Insects* **2** (4a). 130 pp.
Stroyan, H.L.G. 1979. Additions to the British aphid fauna. *Zoological Journal of the Linnean Society* **65**: 1–54.
Stroyan, H.L.G. 1981. A North American lupin aphid found in Britain. *Plant Pathology* **30**: 253.
Stroyan, H.L.G. 1984. Aphids – Pterocommatinae and Aphidinae (Aphidini). *Handbooks for the Identification of British Insects* **2** (6). 232 pp.
Stubbs, A.E. & Falk, S.J. 1983. *British hoverflies: an illustrated identification guide.* 253 pp; frontisp. + 13 pls. London.
Sunderland, K.D., Fraser, A.M. & Dixon, A.F.G. 1986. Field and laboratory studies of money spiders as predators of cereal aphids. *Journal of Applied Ecology* **23**: 433–447.
Sunderland, K.D. & Vickerman, G.P. 1980. Aphid feeding by some polyphagous predators in relation to aphid density in cereal fields. *Journal of Applied Ecology* **12**: 755–766.
Sweet, M.H. 1960. The seed bugs: a contribution to the feeding habits of the Lygaeidae (Hemiptera: Heteroptera). *Annals of the Entomological Society of America* **53**: 317–321.
Tanasiychuk, V.N. 1970. Palaearctic species of the genus *Chamaemyia* Panzer. *Entomological Review, Washington* **49**: 128–140. [Translation; original, in Russian, in *Entomologicheskoe Obozrenie*]
Tanasiychuk, V.N. 1986. Mikhi-Serebryanki (Chamaemyiidae). *Fauna SSSR* **14** (7). 335 pp. [In Russian]
Theobald, F.V. 1926. *The Plant Lice or Aphididae of Great Britain.* **1**. ix + 372 pp. Ashford and London.
Theobald, F.V. 1927. *The Plant Lice or Aphididae of Great Britain.* **2**. vi + 411 pp. Ashford and London.
Theobald, F.V. 1929. *The Plant Lice or Aphididae of Great Britain.* **3**. vi + 364 pp. Ashford and London.
Thompson, R.T. 1958. Coleoptera Phalacridae. *Handbooks for the Identification of British Insects* **5** (5b). 17 pp.
Thompson, W.R. 1950. *A catalogue of the parasites and predators of insect pests.* Section 1. Parasite host catalogue. Part 3. Hemiptera. iv. + 149 pp. CIBC. Ottawa.
Thompson, W.R. 1958. *Ibid.* Section 2. Host parasite catalogue. 5 parts, 698 pp.
Thompson, W.R. 1964. *Ibid.* Section 3. Predator host catalogue. v + 204 pp. CIBC. Farningham Royal.
Thompson, W.R. 1965. *Ibid.* Section 4. Host predator catalogue. iv + 198 pp.
Thomson, H.M. 1960. A list and brief descriptions of the Microsporidia infecting insects. *Journal of Insect Pathology* **2**: 346–385.
Thorpe, W.H. 1968. *Orthezia cataphracta* (Shaw) (Hem., Coccidae) feeding on a Basidiomycete fungus *Collybia* sp. *Entomologist's Monthly Magazine* **103**: 155.
Tischler, W. 1982. *Fauna von Deutschland* 15th ed. [1st ed. by P. Brohmer]. x + 582 pp. Heidelberg.
Todd, V. 1950. Prey of harvestmen (Arachnida, Opiliones). *Entomologist's Monthly Magazine* **86**: 252–254.
Trehan, K.A. 1940. Studies on the British White-flies (Homoptera, Aleyrodidae). *Transactions of the Royal Entomological Society of London* **90**: 575–625.
Trjapitzin, V.A. 1979. A new species of *Ooencyrtus* Ashmead 1900 (Hymenoptera, Encyrtidae) – parasite of *Coreus marginatus* L. (Hemiptera, Coreidae) eggs in Moldavia. *Trudy Vsesoyuznogo Entomologicheskogo Obshchestva* **61**: 160–161. [In Russian]
Trjapitzin, V.A. et al. 1978. Hymenoptera Part II. *In*: Medvedev, G. S. [Ed.] Key to the insects of the European USSR. Vol. III. *Opredeliteli po Faune SSSR* **120**: 1–756 + [ii]. [In Russian]
Turner, B.D. 1984. Predation pressure on the arboreal epiphytic herbivores of larch trees in southern England. *Ecological Entomology* **9**: 91–100.
Tuxen, S.L. (Ed.) 1970. *Taxonomist's glossary of genitalia in insects.* Second ed. 359 pp. Copenhagen.
Usinger, R.L. 1966. Monograph of Cimicidae (Hemiptera-Heteroptera). *Thomas Say Foundation Monograph* **7**. xi + 585 pp.

Van Doesburg, P.H. 1980. Genitalia vials from PVC tube. *Entomologische Berichten, Amsterdam* **40**: 177–178.
Van Emden, F.I. 1942. A key to the genera of larval Carabidae. *Transactions of the Royal Entomological Society of London* **92**: 1–99.
Van Emden, F.I. 1949. The larvae of British beetles: VII (Coccinellidae). *Entomologist's Monthly Magazine* **85**: 265.
Van Emden, F.I. 1954. Diptera Cyclorrhapha (I) Section (a). Tachinidae and Calliphoridae. *Handbooks for the Identification of British Insects* **10** (4a). 133 pp.
Van Emden, H.F. 1972. *Aphid Technology*. xiv + 344 pp. London and New York.
Vepsäläinen, K. & Krajewski, S. 1986. Identification of the waterstrider (Gerridae) nymphs of Northern Europe. *Annales Entomologici Fennici* **52**: 63–77.
Vidano, C. 1965. A contribution to the chorological and oecological knowledge of the European Dikraneurini (Homoptera Auchenorrhyncha). *Zoologische Beitrage* (N.S.) **11**: 343–367.
Viets, K. 1936. Wassermilben oder Hydracarina (Hydrachnellae und Halacaridae). *Tierwelt Deutschlands* **31–32**. 574 pp.
Viggiani, G. 1971. Ricerche sugli Hymenoptera Chalcidoidea XXIX. Descrizione del *Tetrastichus ledrae* n. sp. (Eulophidae), parassita oofago di *Ledra aurita* (L.) (Hom. Cicadellidae). *Bollettino del Laboratorio di Entomologia Agraria 'Filippo Silvestri'* **29**: 260–269.
Vilbaste, J. 1968. Preliminary key for the identification of the nymphs of North European Homoptera Cicadinea. I. Delphacidae. *Annales Entomologici Fennici* **32**: 65–74.
Vilbaste, J. 1982. Preliminary key for the identification of the nymphs of North European Homoptera Cicadinea. II. Cicadelloidea. *Annales Zoologici Fennici* **19**: 1–20.
Villiers, A. 1945. *Atlas des Hémiptères de France. I. Hétéroptères Gymnocérates*. 83 pp, 12 pls. Paris.
Villiers, A. 1947. *Atlas des Hémiptères de France. II. Hétéroptères Cryptocérates, Homoptères, Thysanoptères*. 113 pp, 12 pls. Paris.
Vockeroth, J.R. 1960. A method of mounting insects from alcohol. *Canadian Entomologist* **98**: 67–70.
Wagner, E. 1952. Blindwanzen oder Miriden. *Tierwelt Deutschlands und der Angrenzenden Meeresteile* **41**. iv + 218 pp.
Wagner, E. 1966. Wanzen oder Heteropteren. I. Pentatomorpha. *Tierwelt Deutschlands und der Angrenzenden Meeresteile* **54**. vi + 235 pp.
Wagner, E. 1967. Wanzen oder Heteropteren. II. Cimicomorpha. *Tierwelt Deutschlands und der Angrenzenden Meeresteile* **55**. iv + 179 pp.
Wagner, E. & Weber, H.H. 1964. Hétéroptères Miridae. *Faune de France* **67**. 591 pp.
Wainhouse, D. & Howell, R.S. 1983. Intraspecific variation in beech scale populations and in susceptibility of their host *Fagus sylvatica*. *Ecological Entomology* **8**: 351–359.
Wallace, F.G. 1966. The trypanosomatid parasites of insects and arachnids. *Experimental Parasitology* **18**: 124–193.
Waloff, N. 1967. Biology of three species of *Leiophron* (Hymenoptera: Braconidae, Euphorinae) parasitic on Miridae on Broom. *Transactions of the Royal Entomological Society of London* **119**: 187–213.
Waloff, N. 1968. Studies on the insect fauna on Scotch broom *Sarothamnus scoparius* (L.) Wimmer. *Advances in Ecological Research* **5**: 87–208.
Waloff, N. 1974. Biology and behaviour of some species of Dryinidae (Hymenoptera). *Journal of Entomology* (A) **49**: 97–109.
Waloff, N. 1980. Studies on grassland leafhoppers (Auchenorrhyncha, Homoptera) and their natural enemies. *Advances in Ecological Research* **11**: 81–215.
Waloff, N. 1981. The life history and descriptions of *Halictophagus silwoodensis* sp. n. (Strepsiptera) and its host *Ulopa reticulata* (Cicadellidae) in Britain. *Systematic Entomology* **6**: 103–113.
Waloff, N. & Jervis, M.A. 1987. Communities of parasitoids associated with leafhoppers and planthoppers in Europe. *Advances in Ecological Research* **17**: 281–402.
Waloff, N. & Solomon, M.G. 1973. Leafhoppers (Auchenorrhyncha: Homoptera) of acidic grassland. *Journal of Applied Ecology* **10**: 189–212.
Walter, S. 1975. Larval forms of some Central European Euscelini (Homoptera, Auchenorrhyncha). *Zoologische Jahrbücher* (Syst.) **102**: 241–302.
Walter, S. 1978. Larval forms of some Central European Euscelinae (Homoptera, Auchenorrhyncha). Part II. *Zoologische Jahrbücher* (Syst.) **105**: 102–130.
Waterhouse, D.F. (Ed.) 1970. *The Insects of Australia*. xii + 1029 pp. Melbourne.

Weber, H. 1929. Hemiptera I. Part 31: 1–70. *In*: Schulze, P. (Ed.) *Biologie der Tiere Deutschlands* **29**.
Weber, H. 1930. *Biologie der Hemipteren*. vii + 543 pp. Berlin.
Weber, H. 1931. Hemiptera II. Part 31: 71–208. *In*: Schulze, P. (Ed.) *Biologie der Tiere Deutschlands* **34**.
Weber, H. 1935. Hemiptera III. Part 31: 209–355. *In*: Schulze, P. (Ed.) *Biologie der Tiere Deutschlands* **38**.
Weiser, J. 1961. *Mikrosporidia als Parasiten der Insekten*. 149 pp. Berlin.
Weiser, J. 1969. *An atlas of insect diseases*. 43 pp; 292 pls. Shannon and Prague.
White, I.M. & Hodkinson, I.D. 1982. Psylloidea (nymphal stages) Hemiptera, Homoptera. *Handbooks for the Identification of British Insects* **2** (5b). 50 pp.
Whittaker, J.B. 1965. The biology of *Neophilaenus lineatus* (L.) and *N. exclamationis* (Thunberg) (Homoptera: Cercopidae) on Pennine moorland. *Proceedings of the Royal Entomological Society of London* (A) **40**: 51–60.
Whittaker, J.B. 1981. An experimental investigation of interrelationships between the Wood Ant (*Formica rufa*) and some tree-canopy herbivores. *Journal of Animal Ecology* **50**: 313–326.
Williams, D.J. 1962. The British Pseudococcidae (Homoptera: Coccoidea). *Bulletin of the British Museum (Natural History) (Entomology)* **12**: 1–79.
Williams, D.J. 1978. The anomalous ant-attended mealybugs (Homoptera: Pseudococcidae) of south-east Asia. *Bulletin of the British Museum (Natural History) (Entomology)* **37**: 1–72.
Williams, D.J. 1985. The British and some other European Eriococcidae (Homoptera: Coccoidea). *Bulletin of the British Museum (Natural History) (Entomology)* **51**: 347–393.
Wilson, G. Fox- 1935. Contributions from the Wisley Laboratory. LXXIII. The Rhododendron White Fly. *Journal of the Royal Horticultural Society* **60**: 264–271; plate-figs 75–78.
Wilson, M.R. 1978. Descriptions and key to the genera of the nymphs of British woodland Typhlocybinae (Homoptera). *Systematic Entomology* **3**: 75–90.
Wilson, M.R. 1981. Identification of European *Iassus* species (Homoptera: Cicadellidae) with one species new to Britain. *Systematic Entomology* **6**: 115–118.
Wood-Baker, C.S. 1980. Aphids of Kent. *Transactions of the Kent Field Club* **8**: 3–49.
Woodroffe, G.E. 1958. Biological notes on some Hemiptera-Heteroptera from Devon and Cornwall. *Entomologist's Monthly Magazine* **94**: 24.
Woodroffe, G.E. 1959. Two forms of Coranus subapterus Degeer (Hem., Reduviidae) associated with distinct habitats. *Entomologist* **92**: 125–128.
Woodroffe, G.E. 1961. Some arboreal Mirids (Hem.) on an unusual host tree. *Entomologist's Monthly Magazine* **96**: 128.
Woodroffe, G.E. 1962. Further notes on the Hemiptera of the Isle of Rhum. *Entomologist's Monthly Magazine* **97**: 199–200.
Woodroffe, G.E. 1966a. *Piesma spergulariae* sp. n. (Hem., Piesmatidae) from the Isles of Scilly. *Entomologist* **99**: 107–110.
Woodroffe, G.E. 1966b. A taxonomic note on *Saldula pallipes* (Fab.) and *Saldula palustris* (Doug.) (Hem., Saldidae). *Entomologist* **99**: 190–192.
Wootton, R.J. & Betts, C.R. 1986. Homology and function in the wings of Heteroptera. *Systematic Entomology* **11**: 389–400.
Wratten, S.D. 1973. The effectiveness of the coccinellid beetle, *Adalia bipunctata* (L.), as a predator of the lime aphid, *Eucallipterus tiliae* (L.). *Journal of Animal Ecology* **42**: 785–802.
Wratten, S.D. 1976. Searching by *Adalia bipunctata* (L.) (Coleoptera: Coccinellidae) and escape behaviour of its aphid and cicadellid prey on lime (*Tilia* x *vulgaris* Hayne). *Ecological Entomology* **1**: 139–142.
Yin Yin May, E. 1975. Study of two forms of the adult *Stenocranus minutus*. *Transactions of the Royal Entomological Society of London* **127**: 241–254.
Young, E.C. 1966. Observations on migration in Corixidae (Hemiptera: Heteroptera) in southern England. *Entomologist's Monthly Magazine* **101**: 217–229.

INDEX

A

abdomen 74–79
Acanthosomatidae 93–94, plates 1, 4
Achilidae 62
Aclerdidae 223
Adelgidae 176–179, plate 3
Adelgoidea 176
Aepophilidae 60, 128–129
aestivation 50–52
alarm pheromones 47–48
Aleyrodidae 173–176, plate 7
Aleyrodoidea 173–176
algae 2
alien host plants 58–59
alternation of host plants 10
Alydidae 102–104
annual cycles 50–51
Anoeciidae 197–198
antennae 66
Anthocoridae 23, 118–120, plate 5
Anthribidae 42
ants 15, 24, 25–26, 49, plate 1
Aphelinidae 31–32
Aphelocheiridae 137–138
Aphididae 192–196, plates 1, 2, 7
Aphidiidae 29–30, plate 1
Aphidoidea 182–185
aphids – see Aphidoidea, Adelgoidea
arachnids 21–23
Aradidae 92–93
artificial habitats 59
Asilidae 41
assassin bugs – see Reduviidae
Asterolecaniidae 220, 221

B

bacteria 12–13, 13–14, 16
barkbugs – see Aradidae
batbugs – see Polyctenidae
bats 8–9, 61, 121
bedbugs – see Cimicidae
beetles – see Coleoptera
Belostomatidae 61
Berytidae 110–111
birds 8–9, 118, 120–121
biting in defence 47
blood 8–9

Braconidae 28–29
brochosomes 71, 143
bush cricket 46

C

Callaphididae 190–192
Calophyidae 170
Cantharidae 43
capsid bugs – see Miridae
Carabidae 43
carrion 9
Cecidomyiidae 38–39
Ceratocombidae 127–128
Cercopidae 153–156, plate 6
Chaitophoridae 187–189
Chamaemyiidae 40
Charipidae 30–31
chemical protection from enemies 47–48
Chloropidae 40
Cicadellidae 156–162, plates 1, 2, 6
Cicadidae 151–153
Cimicidae 120–122, plate 1
Cixiidae 148–149
climate 55–56
Coccidae 217–220, plate 8
Coccinellidae 41–42
Coccoidea 205–207
cochineal 206, 223
Coleoptera 41
Coleorrhyncha 60
collecting 225–227
Collembola 46
coloration, cryptic 48, plate 5 (fig. 5)
coloration, warning 48
colour polymorphism 48
Conchaspididae 223
conifers 3–4
Coreidae 100–102, plate 4
Corixidae 140–141
cryptic coloration 48
cuckoo-spit – see Cercopidae
Cydnidae 95–96, plate 4

D

Dactylopiidae 223
daily cycles 50
damselbugs – see Nabidae
defences of plants 10
Delphacidae 149–151

Derodontidae 42
diapause 50–52
Diapriidae 31
Diaspididae 221–223
Dicotyledones 6–8
Dictyopharidae 62
Dinidoridae 61
Dipsocoridae 127
Diptera 36
disease transmission 13–14
displacement 56–57
distribution patterns 55–57
Dolichopodidae 41
dragonfly 46
Drosophilidae 40
Dryinidae 26–27
Dytiscidae 43

E

earwigs 46
Embolemidae 27
Empididae 40–41
Encyrtidae 25, 32–33
endangered species 58
Enicocephalidae 60–61
Erioccidae 214–215
escape from enemies 47
Eulophidae 34
Eupelmidae 34
Eurybrachidae 62
excrement 9
extinctions 58
eyes 64–66

F

faunal changes 57–59
feeding site 8, choice of 9–10, effects plate 2, see also galls
ferns 3
fish 46
flatbugs – see Aradidae
Flatidae 61
flowerbugs – see Anthocoridae
food, choice of 10–11
Formicidae – see ants
froghoppers – see Cercopidae
Fulgoridae 62
fungi 2, 12–13, 13–14, 16–18

G

galls 8, plate 3
genitalia of female 78–79
genitalia of male 74–78
genitalia preparations 230–231
Gerridae 134–136
giant water bugs – see Belostomatidae
grasses 4–6
Greenideidae 62
ground pearls 212
groundbugs – see Lygaeidae

H

habitat, choice of 9–11
Halimicoccidae 223
handling 227–228
harvestmen 22–23
head 63–66
Hebridae 131–132
hibernation 50–52
Homotomidae 172
Hormaphididae 199–200
horsetails 3
hoverflies plate 5, see also Syrphidae
Hydrometridae 132–133
Hymenoptera 24–25

I

Ichneumonidae 30
introductions 58
Issidae 146–148

J

journals 248–249

K

Kermesidae 216

L

labelling 233–234
lac (shellac) 206
lacebugs – see Tingidae
lacewings – see Neuroptera
Lachnidae 187, plate 7
ladybirds plate 5, see also Coccinellidae

Largidae 61
Lathridiidae 43
leafhoppers – see Cicadellidae
legs 67–69
lerps 165
lichens 2
liverworts 2–3
Liviidae 166
Lophopidae 62
Lygaeidae 107–109, plate 5

M

mammals 9, 118, 120–121
man 8–9, 118, 120–121
mapping 57, 248
Margarodidae 210–212
marine bug – see Aepophilidae
marine dispersal 54
maternal care 94, 96
mechanical protection from enemies 47
Meenoplidae 62
Megaspilidae 31
Melyridae 43
Membracidae 153, 154, plate 6
mesophyll 8
Mesoveliidae 130–131
metamorphosis 1
Microphysidae 122–123
midgut caeca 12
mimicry 48–49, plate 5 (fig. 3)
Mindaridae 200–201
Miridae 23, 123–127, plates 2, 5
mites 22–23
Monocotyledones 6
mosses 2–3
mouthparts 1, 63
Muscidae 40
mycetocytes 12–13
mycetomes 12–13
mycoplasmas 13–14
Mymaridae 25, 35

N

Nabidae 23, 116–117, plate 1
Naucoridae 138–139
nematodes 21
Nepidae 136–137
nervous system 1
Neuroptera 45

newsletters 249
Nitidulidae 43
Nogodinidae 62
non-British families 60–62, 223
Notonectidae 139–140

O

Opiliones 22–23
Ortheziidae 208–210
oviposition site, choice of 9–10

P

Palaearctic Region 55
parental care 94, 96
Peloridiidae – see Coleorrhyncha
Pemphigidae 202–205, plate 3
Pentatomidae 97–100, plate 4
Phalacridae 43
phloem 8
Phloeomyzidae 201–202
phoresy 12
Phoridae 40
Phylloxeridae 179–182, plates 2, 7
Piesmidae 111–112
Pipunculidae 37–38
planthoppers – see Delphacidae
Plataspidae 61
Platygasteridae 31
Pleidae 140
pollination 12
Polyctenidae 61
pondskaters – see Gerridae
predaceous habit 8–9
predation by Hemiptera on Hemiptera 23–24
preservation, dry 228–230
preservation in fluids 231
preservation on microscope slides 232–233
prey, detection of 11
Protozoa 18–19
Pseudococcidae 212–214, plate 8
Psyllidae 166–170, plate 7
Psylloidea 164–166
Pteromalidae 33
Pyrrhocoridae 106–107

R

rain trees 8, 154
Raphidiidae 46

Reduviidae 24, 114–116
reptiles 9
Rhopalidae 104–105
Ricaniidae 62
rushes 4–6

S

Saldidae 129–130
saprophagy 9
saucer bug – see Naucoridae
sawflies 24
Scelionidae 25, 36
Schizopteridae 61
Scutelleridae 96
sedges 4–6
Septobasidium 13
shellac 206
shieldbugs 24, 61, see also Acanthosomatidae, Cydnidae, Scutelleridae, Pentatomidae
shorebugs – see Saldidae
Signiphoridae 34
size of predator and prey 47
snakeflies – see Raphidiidae
social wasps 24
societies 248
soldier aphids 203, 205
solitary wasps – see Sphecidae
sphagnum bugs – see Hebridae
Sphecidae 26
spiders 22–23, plate 4
Spondyliaspididae 172–173
Staphylinidae 43
Stenocephalidae 105–106
stiltbugs – see Berytidae
storage 234
Strepsiptera 43–45
symbiosis 12–15
symbiotic microorganisms 12–13
Syrphidae 39

T

Tachinidae 37
tapeworms 21
Tessaratomidae 61
Tettigometridae 145–146
Thelaxidae 198–199
thorax 66–73
Thysanoptera 46
tidal dispersal 54
Tingidae 112–114

Trichogrammatidae 25, 35–36
Triozidae 170–172, plate 3
Tropiduchidae 62

U

Urostylidae 61

V

vegetation 55–56
Veliidae 133–134
vertebrates 8–9, 46, 114
viruses 13–14, 16
vulnerable species 58

W

warning coloration 48
wasps 24
water boatmen – see Notonectidae, Pleidae, Corixidae
water scorpions – see Nepidae
whiteflies – see Aleyrodidae
wings 69–73
world fauna 60–62
worms 21

X

xylem 8

Y

yearly cycles 50–51
yeasts 12–13